THE SKY HARBOR PROJECT

EARLY DESERT FARMING AND IRRIGATION SETTLEMENTS: ARCHAEOLOGICAL INVESTIGATIONS IN THE PHOENIX SKY HARBOR CENTER VOLUME 4: SPECIAL STUDIES, SYNTHESIS, AND CONCLUSIONS

Edited and compiled by

DAVID H. GREENWALD
JEAN H. BALLAGH

With contributions by

RICHARD V. N. AHLSTROM
KIRK C. ANDERSON
MARK L. CHENAULT
LINDA SCOTT CUMMINGS
T. MICHAEL FINK
DAVID H. GREENWALD
DAWN M. GREENWALD
SCOTT KWIATKOWSKI
DAVID SCHALLER
PAT H. STEIN
ARTHUR W. VOKES
MARY-ELLEN WALSH-ANDUZE
M. ZYNIECKI

SWCA, Inc.
Environmental Consultants
Flagstaff and Tucson

SWCA Anthropological Research Paper Number 4

1996

ISBN 1 – 931901 – 08 – 2

This volume is dedicated to the memory of
Julie Hutchinson

General series editor - Robert C. Euler
Assistant editor - Richard V. N. Ahlstrom
Copy editor - Sally P. Bennett
Cover design and production - Christina Watkins

1996
SWCA, Inc. Environmental Consultants
Flagstaff and Tucson, Arizona

TABLE OF CONTENTS

APPENDIXES

List of Figures

List of Tables

PREFACE

Volume 4 of *The Sky Harbor Project: Early Desert Farming and Irrigation Settlements, Archaeological Investigations in the Phoenix Sky Harbor Center* presents the synthesis and interpretations of data recovery investigations conducted during the summer and fall of 1990 and January 1991 at Pueblo Salado (AZ T:12:47, ASM) and Dutch Canal Ruin (AZ T:12:62, ASM) for the City of Phoenix Community and Economic Development Department. This volume expands on the discussions in Volumes 2 and 3, the site reports, and contains special topics, regional comparative studies, and a discussion of project conclusions. Volume 4 also provides an overview of the results of initial testing investigations in the Phoenix Sky Harbor Center begun in February 1989 (see Volume 1 for details).

The four-volume set was originally printed in 1993 as SWCA Archaeological Report No. 93-17 in fulfillment of SWCA's contractual obligations to the City of Phoenix Community and Economic Development Department. The city had undertaken these investigations as part of the preparation for a planned light industrial development adjacent to Phoenix Sky Harbor International Airport. Although the project was driven by the Section 106 process, SWCA's project members have expended extra effort to create a quality product that far exceeds the standards of compliance documents. This is the final volume of the Phoenix Sky Harbor Center Project report, produced as part of the SWCA Anthropological Research Papers series. This volume, along with the previous three, illustrates SWCA's commitment to southwestern archaeology and our support for furthering the growing understanding of the past of this region. As the project evolved from the testing phase, SWCA's investigators found that the archaeological resources in the Phoenix Sky Harbor Center offered some challenging and unique opportunities for study. The importance of the resources derived from both the location and setting of the project area and the prehistoric, historic, and present-day relationships among the natural and cultural resources.

The project setting affected the cultural resources in two ways. First, both Dutch Canal Ruin and Pueblo Salado are located on the geologic floodplain of the Salt River. Years of accumulated alluvial deposits had buried the cultural deposits and disguised their surface indications. Although flood events can result in severe damage to the archaeological record, at Pueblo Salado and Dutch Canal Ruin flooding appears to have deposited deep layers of fine alluvium that helped protect the archaeological remains. The second impact comes from historic and modern use of the area. The effects of urbanization can also disguise surface indications of cultural resources, as well as cause serious damage to them. Considerable archaeological work has been completed within the urban setting of the greater Phoenix metropolitan area, and archaeologists have learned how to glean important information from the preserved remnants of the affected resources. The impacts of urbanization on the archaeological record in the Phoenix Sky Harbor Center varied according to the types of facilities that had been built and according to the depth of the alluvial deposits that covered the archaeological remains. Earlier remains were, in most cases, afforded greater protection than later remains by virtue of having been more deeply buried; as a result, the pre-Classic resources exhibited considerably fewer impacts from historic and modern activities in the project area than did the Classic and post-Classic resources.

Assessing the effects of urbanization is not a simple matter. To evaluate the potential preservation of archaeological resources in a given area of metropolitan Phoenix, four factors should be weighed: (1) location of the area relative to the natural topography; (2) potential depth of accumulated deposits (both cultural and natural); (3) age of the cultural deposits relative to the potential accumulation of later natural deposits; and (4) types of historic and modern activities conducted within the area under consideration. The last factor should be evaluated carefully, as different types of activities will have different levels of impact on the archaeological remains. For example, excavation of a trench for a 6-inch water pipe is likely to have considerably less impact on archaeological deposits than a trench for a 60-inch main line. Depending on the degree of post-occupational sediment accumulation, a relatively less invasive type of disturbance may have little effect on buried resources. This often was the case for the Phoenix Sky Harbor Center archaeological remains.

Prior to the Phoenix Sky Harbor Center Project, few investigations had been completed on the geologic floodplain of the Salt River. The general attitude among archaeologists and others that the floodplain was an unacceptable location for habitation because of high susceptibility to flooding excluded the floodplain from consideration as an important source of archaeological information. To a certain degree the assumption was correct. Large village sites have not been discovered on the floodplain; however, archaeologists have found permanently occupied habitation sites associated with canal irrigation systems. Although flooding was a real threat to residents of the floodplain, they appear to have coped with the ever-present potential for loss of life, homes, and crops and to have persisted in extensive use of the floodplain in the Phoenix Sky Harbor Center during the Classic and post-Classic periods. It is no surprise that the floodplain was used for agricultural pursuits, as it provided deep, fertile alluvium for both prehistoric and historic farmers. However, prior to the testing phase of this project, SWCA's archaeologists did not anticipate the occurrence of compound architecture and a lengthy post-Classic period occupation. Recent investigations for Phoenix Sky Harbor International Airport have demonstrated that occupation and use of the floodplain was even more intensive than the Sky Harbor Center Project revealed.

We wish to thank the City of Phoenix Community and Economic Development Department for the opportunity to have made this contribution to the continuing study of the Hohokam culture.

Steven W. Carothers
President, SWCA, Inc., Environmental Consultants

ACKNOWLEDGMENTS

It is appropriate to acknowledge each individual involved in the preparation of this volume, whether his or her contribution was in the form of analysis, report preparation, editing, drafting, or review. Dr. Richard V. N. Ahlstrom served as the Principal Investigator, contributing also to the writing and editing of some of the chapters. David H. Greenwald, Project Manager, oversaw the general production, editing, and writing of the manuscript and contributed to a number of the chapters. Jean Ballagh also participated in the production and editing of the manuscript, which at times required her to restate the obvious to clarify the intent of the author.

Several analysts contributed to this volume. Pat Stein prepared the section concerning historic irrigation in the Phoenix Basin. Mark L. Chenault, Field Supervisor, contributed a section on the effects of urbanization on Hohokam sites and prepared a chapter on attributes that may be used to define the Polvorón phase. Mark Zyniecki, Field Supervisor, likewise wrote various sections, including a discussion of the place of Pueblo Salado in the Hohokam cultural system and a review of the chronological evidence for the Polvorón phase. Kirk Anderson, Scott Kwiatkowski, Linda Scott Cummings, and David Greenwald jointly prepared a discussion of the agricultural fields discovered at Pueblo Salado. Kirk Anderson also contributed to the chapter on canals. Mary-Ellen Walsh-Anduze analyzed and discussed recent advances in temper studies and ceramic production. Dawn Greenwald examined the lithic assemblages from field house, farmstead, hamlet, and village sites and contributed to the discussion concerning utilization of various resource zones in the project area. T. Michael Fink prepared the human remains discussion, and Arthur W. Vokes contributed the shell analysis.

Behind the scenes, the efforts of Marianne Marek, Laboratory Director, Julie Hutchinson, Charles Sternberg, and Jill Caouette, Draftspersons, and Erica Berk and Margie Smith, Word Processors, provided essential support for the project and the production of this volume. David Abbott, Steven Shackley, Lee Fratt, Douglas Mitchell, Kelley Schroeder, and David Schaller helped to clarify questions that arose during report preparation. The peer review panel consisted of Drs. Judy Brunson and William Doelle, who provided useful comments, direction, and helpful suggestions. Dr. Robert Euler provided additional comments and suggestions. Todd Bostwick, City of Phoenix Archaeologist, also reviewed the volume and provided many helpful suggestions as well as access to reports, papers, and other publications that we generally could not locate. His insights were always useful, and he often intrigued us with his questions and comments. David Greenwald and Jean Ballagh reviewed all comments and editorial suggestions, restructured portions of the volume, and thoroughly edited the report. Dr. Steven Carothers, President of SWCA, provided, as always, support and encouragement, and Mr. Robert Wojtan of the City of Phoenix Community and Economic Development Department provided encouragement and advice and has demonstrated his patience throughout this project.

If anyone who contributed to the production of this volume has been overlooked, we sincerely apologize. We gratefully acknowledge the contributions of all who have been a part of the City of Phoenix Sky Harbor Center Project.

David H. Greenwald
Project Manager

ABSTRACT

This volume, which presents the results of special studies, a project synthesis, and overall conclusions for the Phoenix Sky Harbor Center archaeological investigations, is the last of four. The City of Phoenix Community and Economic Development Department sponsored the project, which was coordinated by Mr. Robert J. Wojtan of the City of Phoenix and David H. Greenwald of SWCA, Inc., Environmental Consultants. The project area, approximately 800 acres located immediately west of Phoenix Sky Harbor International Airport, included portions of two sites, Dutch Canal Ruin (AZ T:12:62, ASM) and Pueblo Salado (AZ T:12:47, ASM).

During the pre-Classic periods, Dutch Canal Ruin consisted of field house settlements scattered along various main canals. The canals fell into disuse, and the Hohokam abandoned the area as an agricultural zone by the Sedentary period. During the Classic and post-Classic periods, they reoccupied the area and constructed new canals. However, the later farmers replaced the field house settlements with sparsely settled farmsteads and hamlets. By contrast, the Hohokam first occupied Pueblo Salado during the Soho phase, residing in small settlements scattered along Canal Salado. The Civano phase witnessed the construction of a compound and population aggregation. Accretionary growth resulted in one of the largest compounds on record in the Phoenix Basin. Occupation of the compound appears to have extended into the Polvorón phase of the post-Classic period, possibly until A.D. 1500 or 1550, with family units occupying pit houses and perhaps continuing to use pre-existing adobe surface structures.

Volume 1 of the series focuses on the testing phase of this project and presents a research design and plan of work for the data recovery phase. Volume 2 presents the results of the field investigations and analyses associated with Dutch Canal Ruin, and Volume 3 those associated with Pueblo Salado. In addition to summarizing the findings of the Phoenix Sky Harbor Center investigations, Volume 4 presents in-depth explorations of selected aspects of the Hohokam occupation at Dutch Canal Ruin and Pueblo Salado in relationship to regional Hohokam settlement and cultural patterns. Issues discussed in Chapters 1-4 are historic and prehistoric irrigation and settlement in the Phoenix Basin, the effects of urbanization and alluviation on Hohokam sites, site autonomy within the Phoenix Basin, and the social implications of field houses. Chapters 5, 6, and 7 consider Hohokam use of the environment as understood through investigation of the buried agricultural fields at Pueblo Salado, project area canals and their hydraulic capacities, and the use of project area resource zones. The discussions in Chapters 8 and 9 explore the attributes and chronology of the post-Classic period Polvorón phase. Chapters 10-13 assess implications of Hohokam material culture within the project area and in regional perspective through analysis of the project burials, ceramic temper studies and whole vessel analysis, intraregional comparison of lithic assemblages, and a diachronic analysis of the project shell assemblage. Chapter 14 summarizes the project findings and discusses the place of Dutch Canal Ruin and Pueblo Salado in the Hohokam cultural continuum. Although the editors prepared each volume in this series to stand alone, the first three volumes should be referred to for more detailed information on the topics discussed in Volume 4.

CHAPTER 1

COMPARISON OF PREHISTORIC AND HISTORIC SETTLEMENT IN THE PHOENIX BASIN

David H. Greenwald
Pat H. Stein

The development and growth of the settlements and irrigation systems of the Hohokam resemble those of the early Euroamerican settlers in several ways, since the early historic farmers and the Hohokam faced similar environmental constraints. The written record from the historic period provides a basis for comparison, and archaeologists can use as analogues those developmental patterns that occurred historically (Ackerly 1991) when exploring problems encountered by the Hohokam. This chapter focuses on the historic and prehistoric development of irrigation, the establishment of settlements, and diachronic change within the Phoenix Basin, with a brief discussion of the effects of historic land use on the prehistoric remains. Further discussion of the historic and modern use of the Phoenix Basin appears in Chapter 2.

A REVIEW OF HISTORIC IRRIGATION IN THE PHOENIX BASIN

The first Euroamerican farmers of the Phoenix Basin faced an environmental challenge well known to the Hohokam centuries earlier: how to control a river that could, at times, be obstinately defiant of authority or restraint. Indeed, the Hohokam had water-control technology similar to that of the nineteenth-century Euroamerican agriculturists. Both relied on fragile brush and rock dams to divert the recalcitrant Salt River into miles of hand-dug canals. Both sought to make the water supply more predictable so that crops could grow and settlements could prosper, creating a human oasis in the Sonoran Desert. But if the environmental challenges, technologies, and hopes for prosperity of these two groups were points in common, their universes were not. Increasingly, historic farmers came to rely on solutions provided by a world much larger than that of the Hohokam to succeed at farming in the Salt River Valley. Social structure, political order, and material culture significantly affected the efforts of historic irrigators to make the waters of the Salt flow smoothly through the Valley of the Sun.

Early Development of Historic Irrigation

The founding of Fort McDowell in 1865 had a catalytic effect on the development of historic irrigation in the Phoenix Basin. Establishment of this military post along the lower Verde River created a need for hay for horses and food for soldiers, a need that intensified over the next decade as conflict with the Apache and Yavapai increased (Reed 1977:19). At first, the fort tried to supply much of its own food by means of a post farm and garden. By June 1866, military personnel had planted sorghum and corn on 120 acres, and were readying an additional quarter-section for a fall planting of wheat and barley. Soldiers tended the fields and dug an irrigation canal.

The post farm and garden drew harsh criticism from opponents, who argued that soldiers who tended fields would lack the energy to fight Indians. Yet viable alternatives were few. The fort could import crops from California, but this practice was both expensive and time-consuming, and supply trains were vulnerable to attack. Alternatively, the fort could purchase crops from the Pima and Maricopa along the Gila River, an option already exercised to some extent; however, critics felt that buying from Indians would impede economic opportunities for pioneers who might want to settle in the area (Mariella 1982). Or the fort could assign the tending of its farm and garden to a civilian (a plan not implemented until April 1868), but even this strategy might not meet the projected needs of the rapidly growing post.

In the mid 1860s, entrepreneurs in Arizona Territory were exploring new ways to make money. The old standard, mining, was then experiencing an economic slump. The situation at Fort McDowell created an opportunity for farming that settlers found irresistible.

Two entrepreneurs quick to realize the opportunity at hand were John Y. T. Smith and Jack Swilling. In 1866, Smith was awarded a hay contract for Fort McDowell, and in the spring of 1867 he established a hay camp (Figure 1.1) along the Salt River about 18 miles southwest of the fort (Luckingham 1989:13; Zarbin 1979:8). Popular lore states that while hauling hay for Smith in the spring or summer of 1867, Swilling developed the notion of cleaning and rebuilding the ancient canals that laced the valley between the hay camp and fort (Figure 1.2) (Zarbin 1979:8). However, several researchers have pointed to chinks in this folktale. Mawn (1977) meticulously traced the activities of Smith and Swilling and doubts the existence of a business relationship between the two in 1867. Cable and Doyel (1986:5–6) have noted that much of Swilling's inspiration might have come from the Pima and Maricopa, whose productive canal systems he would have observed in 1859–1860 while recruiting Indian scouts to protect the Overland Mail. Also, a few settlers were practicing irrigation agriculture successfully in central Arizona outside of the Phoenix Basin before Swilling began his irrigation company. Among these were Charles Adams, who established a farm on the Gila in 1866 (present-day Adamsville), and James Chase, who had a productive operation farther upstream on the Gila (Zarbin 1979:10).

Although Swilling was not "the father of irrigation" (Trimble 1986:149), he was nonetheless the person responsible for irrigation farming taking root in the Salt River Valley. On November 16, 1867, he formed the Swilling Irrigating and Canal Company with a group of miners and merchants from Wickenburg. The company was capitalized with $10,000 worth of stock, divided into 50 shares of $200 each. Each share represented a quarter-mile of canal to be dug. Swilling's irrigating company offered those who could not afford to buy shares a sweat-equity option: they could work at the rate of $66.66 per month until they had accumulated $200 (Zarbin 1979:11). The purpose of the company was to take water from the Salt at a point claimed by Swilling and to use the water to grow crops for Arizona's military forts and civilian settlements. From the outset, then, agriculture in the Phoenix Basin was a commercial rather than merely a subsistence venture.

Swilling's force of canal-builder/colonizers was small, but it accomplished much. Although as many as 31 men attended the organizing meeting of the company, apparently only 16 to 20 accompanied Swilling to the Salt River. After one abortive effort to develop a canal head in rocky strata, the party made swift progress on Swilling's Ditch (Figure 1.1). By March 12, 1868, the company had completed the first part of the canal (Zarbin 1979:12). By July of that year, the settlers had harvested wheat and barley that commanded a healthy price of eight cents a pound at Wickenburg; the corn crop was also reported to be "growing finely" (*Arizona Weekly Miner*, July 8, 1868).

Agricultural Intensification and the Formation of Factions

Early agricultural success spawned population growth and a period in which local residents appeared to go "canal crazy" (Salt River Project 1979:3). By the fall of 1868, approximately 100 people had settled near Swilling's Ditch. Many of them had taken up claims pursuant to the

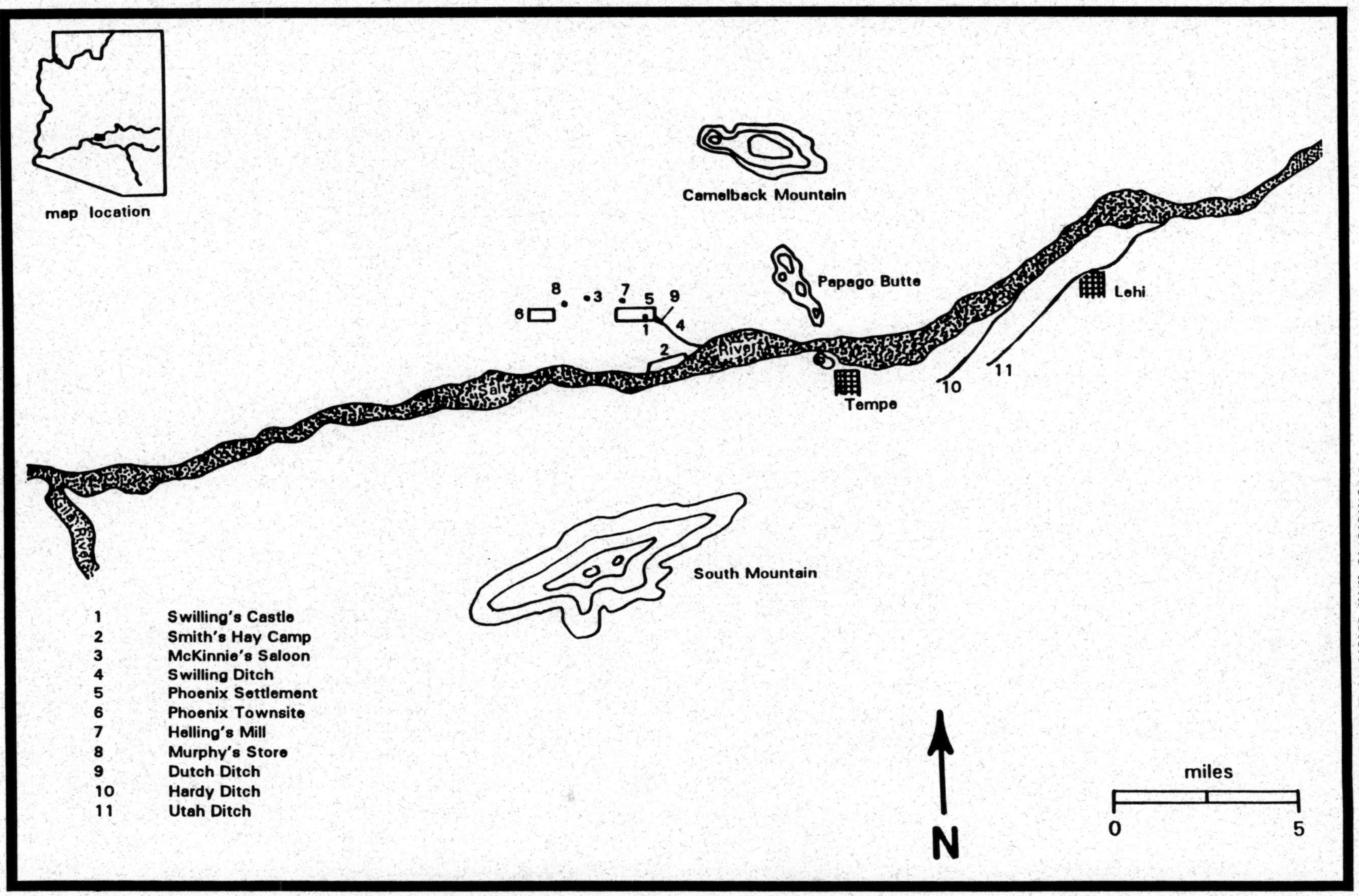

Figure 1.1. Early historic settlement in the Salt River Valley.

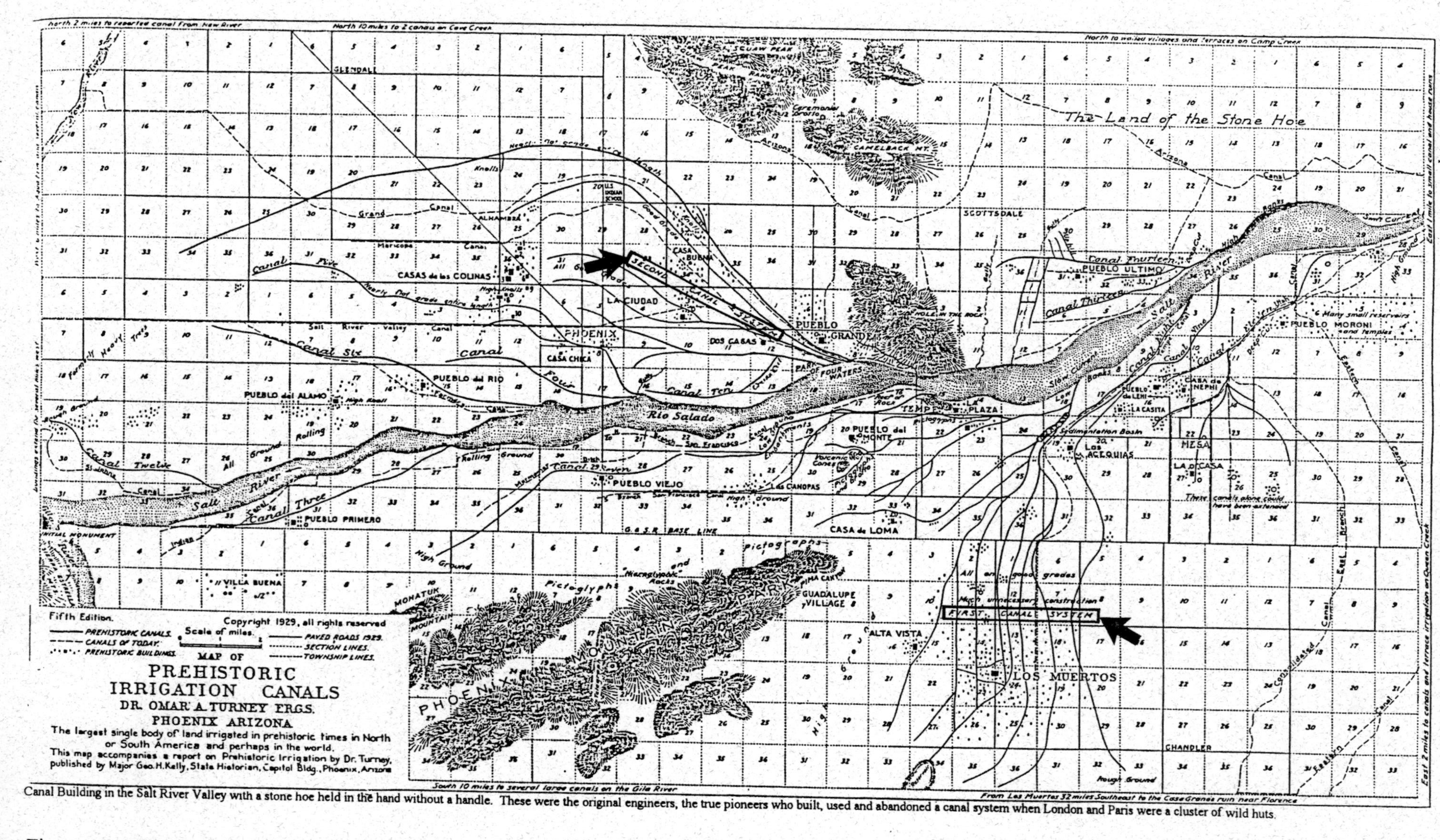

Figure 1.2. Reproduction of Turney's 1929 map of prehistoric canals in the Phoenix Valley. (Boxes identify Canal Sytem 1, south of the Salt River, and Canal System 2, north of the river.)

Homestead Act of 1862, a statute allowing settlers to acquire property in exchange for modest improvements to the land and low government fees. More affluent settlers acquired land through cash entries paid to the United States government.

The construction of additional canals began as early as 1868. In February of that year, the *Arizona Weekly Miner* reported that Joseph Davis was building a ditch below Swilling's. Mr. S. Sawyer, a farmer initially serviced by Davis's ditch, moved farther downstream and started excavating his own ditch in 1869. Completed in 1871, the canal was incorporated by M. P. Griffin and Aaron Barnett (Cable and Doyel 1986:8). The Swilling works, meanwhile, continued to expand. By the fall of 1868, the system included laterals both north and south of the main ditch (Fryman, Woodward, and Garrison 1977:6–7). In late 1868 or 1869 the Dutch Ditch (Figure 1.1) branched off from the Swilling Ditch. In mid 1869, Swilling and Thomas Barnum started a new ditch less than a mile upstream of the one owned and used by the Swilling Irrigating and Canal Company (Zarbin 1979:13). And in 1870–1871, settlers excavated a canal approximately three to four miles upstream of Swilling's operations. Known originally as the McKinny-Kirkland or Hardy Ditch, later the Hayden or Tempe Canal (Figure 1.2), this feature triggered a land boom on the south side of the Salt River in what became Tempe (Fryman, Woodward, and Garrison 1977:6–7; Salt River Project 1979:10–11).

Canal companies that owned the ditches were joint-stock enterprises in which a farmer could acquire a share with cash or by working on construction. In general, a share entitled the owner to sufficient water for a 160-acre field. In addition, the owner paid a fee that financed repairs and paid the salary of an overseer (*zanjero*) who distributed water. Settlers considered shares to be wise investments, for they could be rented to nonowning farmers at a tidy profit (Faulk 1970:169).

The settlement pattern in the Salt River Valley in the late 1860s became one of dispersed farmsteads extending along canals. The federal census of 1870 recorded a population of 164 men and 61 women, ranging from 21 to 30 years old. Ninety-six of the men listed themselves as farmers. The community did not have a single doctor, lawyer, banker, or teacher (Trimble 1986:149). An estimated 1500 to 1800 acres were under irrigation (Luckingham 1989:15; Salt River Project 1979:22), over 6 acres per person.

Occasionally in this landscape of farmsteads settlers built clusters of facilities resembling incipient hamlets or villages (Figure 1.1). One such hamlet developed near Swilling's property and was known as the Phoenix Settlement (Cable and Doyel 1986:8–9) or Punkinsville (Wagoner 1983:193). Centered on present-day 36th Street near Washington, the village featured Swilling's Castle (a large adobe house that doubled as a polling place and post office), a stage station, and William Hancock's general merchandise store. A second hamlet developed near what is now 7th Street and Van Buren and featured the general merchandise store of James Murphy. A third prominent feature of the cultural landscape was James McKinnie's saloon, erected on the southern slope of a platform mound at the prehistoric site of La Ciudad (Cable and Doyel 1986:9), near what is now 20th Street and Van Buren.

The settlement pattern of dispersed farmsteads and fledgling villages might have persisted longer had the agrarian community not realized the advantages of designating a commercial center for the exchange of goods and services with the outside world (Luckingham 1989:15). Many voiced concern that a new economic order was needed to stimulate the export of produce and to promote outside investment in the Valley of the Sun. As a result, settlers formed the Salt River Valley Town Association in 1870 and charged the association with selecting a townsite. The association favored two quarter-sections (Figure 1.1) where downtown Phoenix is located today (Wagoner 1983:193).

The area was on slightly elevated ground, which would make it less susceptible to flooding and thus easier to clear of the mesquite that flourished in low-lying areas (Luckingham 1989:16).

The association's recommendation sparked the emergence of three factions that favored different locations. Faction members shared similar interests in terms of the locations of land they owned and water rights they controlled. The Swilling faction (centered on Swilling's Castle) stressed the economic benefits that would accompany the development of a flour mill proposed for the settlement. In subsequent years, this area became known as Mill City after William Helling's mill (Wagoner 1983:193). A second faction, reputed to be Republicans, favored a location in the vicinity of present-day 16th Street and Van Buren near McKinnie's saloon, an area known as Mowry's Ranch. The third faction, composed mainly of Democrats, supported the recommendation of the Town Association (Cable and Doyel 1986:10). To decide the issue, the community in 1870 held a democratic vote, an election in which the Town Association faction prevailed. Disagreement among the same factions flared again during an 1871 election to determine the seat of Maricopa County. The Phoenix townsite again emerged the winner, with 212 votes over Mill City's 150 and Mowry Ranch's 64.

Emerging Irrigation Problems

In 1872, nine major irrigation ditches in the Salt River Valley supplied water to about 8100 acres. Approximately 4000 acres were planted in barley, 2500 in wheat, and the rest in alfalfa, hay, corn, and vegetables (Salt River Project 1979:22). Census data from 1870 indicate approximately 34 irrigated acres per resident. Canals and laterals continued to be built at a furious pace.

The name "Phoenix," after the bird of Egyptian mythology that rose from its own ashes, was apt, because the new irrigation civilization had arisen from the remains of ancient canals. In the course of developing the new agrarian order, however, farmers damaged scores of prehistoric ruins and canals. The ancient heritage of the valley did not go unnoticed and unappreciated by intellectual leaders such as Judge Joseph T. Alsap:

> That this whole Valley has at some time been densely populated cannot be doubted although neither history, tradition nor legend give any account of who the inhabitants were, from whence they came or whither they have gone. The whole Valley is dotted with the ruins of ancient towns and buildings. The great canal, commonly called the "Montezuma Acequia," intersects the river near the upper end of the Valley and runs thence in a northwesterly direction for several miles. . . . Smaller ditches leading out from it at convenient distances show that it was used for purposes of irrigation and the whole Valley has been under cultivation [Alsap 1872, quoted in Salt River Project 1979:27].

A series of agricultural setbacks beginning in the 1870s and intensifying through the end of the century made many residents wonder if the new civilization would perish as the old one had. Some of these obstacles were sociopolitical in origin, while others were more directly related to the environment. So severe were some that they might have set minds like Alsap's musing that "Daedalus," after the figure in Cretan mythology who flew too close to the sun, might have been a more appropriate name for the new civilization in the scorching desert.

First, when the national depression of 1873–1876 caused mining operations, military posts, and other markets to reduce expenses by cutting orders, farmers in the Salt River Valley felt the pinch (Luckingham 1989:20). Then in 1874, in the middle of the depression, a severe flood hit the valley, wiping out diversion dams and breaching canal banks. The natural catastrophe underscored both the difficulty of controlling the Salt and the necessity of doing so if the new civilization was to survive.

Irrigation canals were sometimes the scene of danger stemming from conflicts over water shortages or other sources of contention between the users of the canals. The unscrupulous would attempt to steal water or to disrupt the flow in rivals' canals, and violence could ensue if they were caught in the act. To protect their water rights from such incursions, farmers often toted and did not hesitate to use rifles, shotguns, and sidearms as they tended canals and crops (Salt River Project n.d.a:3).

Mormon Colonization

In January 1877, the Church of Jesus Christ of Latter-day Saints dispatched 83 of its members from St. George, Utah, to settle previously scouted land in the Salt River Valley upstream from Phoenix and Tempe. Led by Daniel Webster Jones, the colonizers arrived at their destination on March 6, 1877. Although exhausted by their trip through the wintry expanses of southern Utah and northern Arizona, the Mormons started canal construction the day after their arrival. They chose for their canal head a spot recommended by Winchester Miller in 1865, upstream of his ranch. They named their canal the Utah Ditch and their settlement Fort Utah, later renamed Utahville, Jonesville, and, finally, Lehi (Figures 1.1 and 1.2) (City of Mesa 1965:1–32; McClintock 1921:201–209).

According to McClintock (1921), the Lehi community had a distinctive town plan from its inception, although the settlers never fully executed that plan. They built their city along the cross lines of four quarter-sections. Archaeologists have had difficulty confirming reports that the settlement was enclosed within an adobe wall (McClintock 1921:204). A well 25 feet deep provided drinking water.

Jones's colony split into two factions almost immediately, and for reasons completely different from those that had attended the Phoenix townsite controversy seven years before. The Lehi dispute centered on the role that the Pima and Maricopa should play in the religious community. Jones believed that his Indian neighbors should be accepted on an equal footing with other members of his congregation. Most of his followers disagreed so heatedly that in August they departed southward to start a new colony on the San Pedro River, leaving the fort unfinished and Lehi destitute. So poor were the survivors that they could not afford a plow at planting time and were forced to use digging sticks. Without the assistance of Charles Hayden, the Tempe entrepreneur who employed several Mormons and advanced them credit to buy seed, the community of Lehi might not have survived (McClintock 1921:206–207).

A mere month after the departure of Lehi's dissidents, the church dispatched a second party of 83 colonizers from Bear Lake County, Idaho, and Salt Lake County, Utah. Upon their arrival in February 1878 the newcomers settled a tableland south of Lehi and dug a ditch along the ruins of a promising ancient canal shown to them by Jones (McClintock 1921:212–213). This canal could be traced for about 20 miles and maintained an ideal gradient throughout, with laterals spreading widely from the main ditch (Figure 1.2). The settlers wasted little time in making the ancient ditch functional. By utilizing a prehistoric canal to lay their irrigation system, the Mormons saved an estimated $20,000 in labor and expenses (McClintock 1921:214). The entire canal cost approximately $43,000–48,000 but was well worth the expense: the finished channel (Figure 1.2; see also Figure 3.2) carried an impressive 7000 cubic inches of water per second. Hard labor and the great ditch made the new community of Mesa prosper. This effort was the forerunner of the building of Canals 3, 4, 5, and 6 within the modern Salt River Project main canal system.

Worsening Problems and Temporary Solutions

By the late 1870s, settlers frequently voiced concern that the Salt River could not possibly supply all the valley's needs. Construction of the Utah Ditch in 1877 and the Grand Canal and Mesa Ditch in 1878 greatly heightened this concern. In 1878, the year of Swilling's death, water claims to the Salt exceeded 5.5 times the river's average annual flow (Zarbin cited in Myers and Rogge 1986:2-5). Eleven canal companies operated on the north side and 11 on the south side, and additional small, independently owned ditches headed on the Salt as well. Farmers had placed approximately 100,000 acres under cultivation and were clearing additional land daily (Salt River Project 1979:58). New settlers arrived following passage of the national Desert Land Act of 1877, which enabled farmers to acquire large parcels in exchange for irrigation works and small government fees. By the early 1880s, farmers all along the Verde and Salt rivers were coming to an alarming realization: their rivers carried adequate flow only one in three years.

Canal companies and independent operators alike were stressed by flooding, erratic flows, and undercapitalization. Increasingly, both looked to larger solutions to deal with escalating problems.

In 1882 the Arizona Canal Company formulated a bold plan for expanding irrigation. A new canal would head about one mile below the junction of the Verde with the Salt, and it would bring water to approximately 100,000 acres of the northern Salt River Valley that had never been tilled, by either prehistoric or historic cultivators (Rodgers and Greenwald 1988a:27; Salt River Project n.d.a:5). Planned to be 40.75 miles in length, the Arizona Canal (Figure 1.2) would be the longest in the valley's history. Furthermore, unlike its predecessors, it would be capitalized largely by outside investors. The Arizona Canal Company hired William J. Murphy, a former contractor for the Atlantic and Pacific Railroad, to supervise the construction, which involved 20 crews and a cost of about $650,000 (65 times what the Swilling Ditch had cost just 14 years earlier). When the canal company had difficulty paying Murphy, he took his pay in bonds and eventually assumed control of the enterprise.

Murphy started construction in 1883 and finished the canal in 1885 (Rodgers and Greenwald 1988a:27-28). Its head washed out in a spring flood of 1886 but was rebuilt by January 1887. To reconstruct the head, Murphy used an innovative technique consisting of wooden cribs filled with rock, and it was the only dam to survive the flood of 1891 (Salt River Project n.d.a:5). The irrigators eventually extended the canal 5.4 miles westward, bringing its total length to just over 45 miles (Rodgers and Greenwald 1988a:27-28; Salt River Project n.d.a:5).

Murphy meanwhile set about taking charge of irrigation on the north side of the Salt. He organized and became president of the Arizona Improvement Company, which soon gained controlling interest over four northside canals. In 1888 he built the Crosscut Canal to link the four and thus improve delivery (Trimble 1986:152).

But even a "big mover" such as Murphy found himself troubled by events that did not bode well for the future of agriculture in the valley. Irrigation was clearly a feast or famine proposition, usually the latter. Perhaps worse than the frequent years of inadequate flow were the periodic floods, which could destroy heads and dams, choke canals with sediment, and even change the channel of the river, leaving some canal systems high and dry. With the exception of the Arizona Canal after 1886, most ditches of the day had diversion dams of rock and brush; as Ackerly's

(1991:43–48) research has shown, floods exceeding a magnitude of approximately 8,000–12,000 cubic feet per second were sufficient to knock out most of these features. Unfortunately for the historic farmers, "freshets" of this magnitude were quite common.

Recent archaeological research by Hackbarth and Henderson (1992) along the Lower Verde has documented the labor-intensive nature of maintenance associated with early historic canal systems. Floods caused the heads of the Government and Jones ditches (near Fort McDowell) to wash out four to five times a year. Similarly, during its 30 years of operation (circa 1875 to 1905), the Velasco Ditch lost its headgates no fewer than four times and was rerouted at least five times (Hackbarth 1992:499-500). Freshets impacted not only canal heads and diversion dams but also segments that crossed natural drainages. In warmer months, algae choked ditch channels and had to be cleaned out. During one episode to repair damage to the Jones Ditch from a minor flood, a work force of 20 persons took nine days to clean 5688 linear feet of a canal that was 3.5 feet deep and 4.0 feet wide (Hackbarth and Lancaster 1992:431).

Concern over the uncertain water supply deepened with a drought that began in the 1890s and continued into the early twentieth century. The drought was punctuated by severe floods resulting from spring runoff and late-summer thunderstorms. It seemed that water was never available when the farmers most needed it.

Many realized that the situation could be remedied if water could be impounded. The Maricopa County Board of Trade appointed a committee to study the feasibility of a water storage system. The committee reported that an ideal dam site existed 80 miles northeast of Phoenix where Tonto Creek joined the Salt River. Dam building was not new to valley residents; they had, after all, built diversion dams for their irrigation canals. However, a water storage dam capable of withstanding the region's floods would have to be far more massive than anything the valley had ever seen. Massive construction meant enormous capitalization: by the committee's estimate, two to five million dollars. Many doubted that the valley contained sufficient wealth to finance such a project, an assessment that proved accurate.

Some entrepreneurs attempted to build water storage dams in central Arizona, with little success. An earthen dam on Walnut Grove near Wickenburg gained the dubious distinction of causing Arizona's most celebrated catastrophe when it washed out in the flood of February 22, 1890 (Giclas 1985:1). In 1893, the Hudson Reservoir and Canal Company announced plans for dams and reservoirs on the Salt but ran out of funds before construction even began. Backed by Ohio investors, W. H. Beardsley in 1895 started a diversion dam and canal on the Agua Fria but did not complete the project until the 1920s (Myers and Rogge 1986:2–8).

A decade into the drought, a disturbing statistic emerged to suggest that irrigation had passed its zenith in the valley and was now on the decline. Whereas farmers had irrigated 127,512 acres in the valley in 1896, by 1905 they watered only 96,863 acres (Luckingham 1989:47–48). As the drought-stricken land passed its carrying capacity, many farmers gave up and moved away (Salt River Project n.d.b:3). As in the time of the Hohokam, the Phoenix Basin faced the possibility of a postclassic period.

The Birth of Modern Irrigation

New hope for a water storage dam appeared when Congress passed the Hansbrough-Newlands Bill in 1902. The Newlands Act gave western irrigation a shot in the arm by providing federal loans for public and private water control projects. The construction costs incurred by the Reclamation

Service, the government agency responsible for engineering and building the projects, would be repaid out of water project revenues; in 1906 Congress amended the act to permit repayment with revenues derived from hydroelectric power generation (Salt River Project n.d.b:4).

Before the United States government would lend money, however, it had to receive assurance of repayment, and to provide this, landowners had to settle local differences over water rights. The federal government was unwilling to deal with water users individually (Salt River Project n.d.b:4). Also, while western states could assume this type of loan, Arizona, as a territory, lacked authority to do so. Clearly, the settlers needed a new socioeconomic structure to enter into an agreement with the United States government. Such an organization emerged in the form of the Salt River Valley Water Users' Association (SRVWUA), incorporated in 1903 to settle local land and water rights problems and acquire the loan for a storage dam.

The population of the Phoenix Basin quickly rallied behind the SRVWUA. By midsummer of 1903, the association's membership exceeded 4000. Members pledged approximately 200,000 acres as collateral for the government loan. Each acre represented a share of stock in the association (Salt River Project n.d.b:4; Wagoner 1970:425). The association assured that water stored in the reservoir behind the dam would be equally available to all members. The SRVWUA would take responsibility for repaying the government by assessing members on a per-acre basis and enforcing the collection of payments. The group also provided a central body that could assume operation of the dam and related facilities at a later date.

The United States government found this arrangement satisfactory and entered into an agreement with the SRVWUA on June 25, 1904. Both railroads and local boosters heralded this first irrigation project under the 1902 Reclamation Act. Construction of Roosevelt Dam began in 1905, and the Reclamation Service completed the dam in 1911. In the meantime, the agency established other aspects of the new irrigation system such as the completion of Granite Reef diversion dam in May 1908. In addition, the federal government purchased north- and southside canals from private interests (Luckingham 1989:47). Also, the Reclamation Service installed and upgraded facilities for the generation of hydroelectric power.

Certain legal matters remained to be settled, however. With the competition for water so keen, lawsuits had arisen that sometimes took years to resolve. Early laws under which historic irrigators operated were vague (Faulk 1970:169). For example, the Howell Code of 1864, which established the legal code for the territory, stated, "All rivers, creeks, and streams of running water in the Territory of Arizona are hereby declared public," which irrigators interpreted to mean that water was public only until diverted. At that point, they argued, the water became the property of the appropriator and did not belong to the land itself (Faulk 1970:169). They further argued that the first appropriator in time was also the first in right; that is, in years of shortage, water would first go to the irrigator with the earlier appropriation. These issues lay at the heart of the case of *M. Wormser et al. vs. Salt River Valley Canal Company,* filed in 1887 (Salt River Project 1979:85). The plaintiff charged that the canal company treated the public water supply as corporate property, allocating water to its own members and denying use to prior appropriators. The suit was not decided until 1892, when the Federal District Court ruled in favor of the plaintiff, asserting that water belongs to the land and is not a commodity that can be bought and sold apart from the land. The ruling also stated that early water users had priority over later ones (Wagoner 1970:420–421), so long as the earlier users were making productive use of it (Faulk 1970:169–170). Written by Judge Joseph H. Kibbey, this became known as the Kibbey Decision, which established the doctrine of prior appropriation (Wagoner 1970:421; Salt River Project 1979:85).

A second landmark case was that of *Hurley vs. Abbott*, filed in 1905. Hurley sought to establish "the water rights pertaining to each parcel of land and the date each landowner first used water for irrigation" (Salt River Project n.d.b:7). Early landowners believed they should receive more water than newcomers based on their prior use of the water. It took five years for all evidence to be gathered, argued, and deliberated. In his decision of March 1, 1910, Judge Edward Kent established the relative rights of valley landowners to water from the Salt River based on chronology and continuous beneficial use. The Kent Decree also formalized the principal of normal-flow rights (rights to water that naturally flows down a river) and reaffirmed the Kibbey Decision regarding the doctrine of prior appropriation (Salt River Project n.d.b:7).

As Roosevelt Dam neared completion, development of the valley appeared to be back on track. Data collected in 1909 indicated an increase in irrigated land to 126,717 acres, while 1910 census figures recorded a doubling of the population in one decade (Luckingham 1989:48). SRVWUA constructed additional storage reservoirs and diversion dams in central Arizona during successive decades, again using government assistance: Horse Mesa, Mormon Flat, Stewart, Bartlett, Horseshoe, and Waddell dams. By building the modern canal and water storage system, Euroamerican settlers narrowly averted a postclassic period of the historic settlement of the Phoenix Basin. The Hohokam lacked the capacity to build either diversion-type dams or storage reservoirs on a large scale and the ability to construct canals through rock (Gregory 1991:187) and thus had less technological control over their environment and their ultimate destiny in this area.

DEVELOPMENT OF PREHISTORIC IRRIGATION

Archaeologists have yet to determine precisely when the Hohokam began irrigation agriculture. Existing evidence, however, supports the premise that irrigation systems developed during the Pioneer period (A.D. 300–750). By the Snaketown phase (A.D. 650–750), the last phase of the Pioneer period, Hohokam farmers had constructed large-scale canals on the first and second terraces of the Salt River (Cable and Doyel 1985a, 1985b; Greenwald and Ciolek-Torrello 1988a; Howard 1990, 1991a; Midvale n.d.; Schroeder 1991; Wilcox 1987). Few available data confirm earlier canal development (Ackerly and Henderson 1989) in the Phoenix Basin, partly because of the reuse and enlargement of earlier canals during canal expansion. The establishment of large-scale canals does show some correlation with the establishment of village sites (Howard 1991a). Furthermore, indirect evidence derived from the occurrence of ceramics predating the Snaketown phase at habitation sites in Canal System 2 may suggest the presence of a canal on the second terrace before the Snaketown phase (Howard 1991a:5.10). Prior to this time, canals may have been ditches having little capacity, similar to the one discovered in the Tempe Outer Loop (Ackerly and Henderson 1989), that serviced small areas restricted to the first terrace or the geologic floodplain.

Sedentism and the Establishment of Hohokam Irrigation Settlements

Archaeologists have few data currently available from which to make evaluations of the origins of the Hohokam, although they generally accept the assumption that the Hohokam emerged from indigenous Archaic populations. Based on Cable and Doyel's (1987) investigations at the site of Pueblo Patricio and their re-evaluation of Morris's (1969) investigations at the site of Red Mountain, east of Phoenix, Hohokam scholars have equated the initial post-Archaic colonization of the lowland deserts, namely along the major streams and waterways, with the Red Mountain phase (A.D. 1–200) (Cable 1991). Sedentism, the establishment of seasonal or permanent habitation loci, appears to have developed with the adoption of agricultural strategies. Agriculture provides a settled population access to resources through production, as opposed to the foraging strategies and mobility required to access wild resources on a regular basis. The adoption of agriculture, in essence, makes resource management possible and provides resource availability through concentrated group effort. These efforts require planning and can result in greater production than the efforts associated with foraging, although agricultural groups generally use mixed strategies that include hunting and collection of wild plants. Permanent or semipermanent settlements result from increased labor investment and dependence on agriculture. In the Salt and Gila basins, mobile groups were unwilling to make capital investments such as the construction and maintenance of canals. These features were designed for multiyear use and required substantial labor investment during the growing season.

Irrigation settlements are individual habitation loci associated with irrigation facilities or with sites related to canal irrigation. Sites may consist of field houses, farmsteads, hamlets, or villages (see Volume 1:Chapter 1), which denote variations in function, permanency of occupation, and population size and composition. The Hohokam established irrigation settlements along the Salt and Gila rivers prior to the Snaketown phase of the Pioneer period, pre–A.D. 650 (Dean 1991:Figure 3.6). Cable (1991:122) indicates that the Hohokam were practicing irrigation agriculture by at least the Estrella phase, A.D. 500, and perhaps as early as the Vahki phase, A.D. 300 (Dean 1991:Figure 3.6). Clearly, the Hohokam legacy of major village sites, extensive irrigation systems, and complex social organization grew from a simple attempt to provide a reliable subsistence base in an unpredictable environment, just as the historic settlement did. It is as difficult to reconstruct the specific events that led to the establishment of canal irrigation by the Hohokam as it is to determine what sociopolitical mechanisms the Hohokam used to manage aspects of their irrigation system such as canal construction, maintenance, and operation and water allocation. The simple ditch systems used by Native American groups today are not comparable to the Hohokam canals of the Phoenix Basin because of the differences in size (Doolittle 1990); therefore, the development of historic irrigation in the Phoenix Basin provides a better analogue for interpreting Hohokam irrigation organization and structure.

The Historic Analogue as a Means of Exploring Hohokam Settlement and Irrigation Requirements

Reconstruction of Hohokam sociopolitical organization as it related to the construction and operation of canals and the process of water allocation for irrigation and domestic use is a difficult task. Many of the cultural attributes needed for such a reconstruction are intangible, although several researchers have presented models that address the development and use of canals (Howard and Wilcox 1988; Nicholas 1981; Patrick 1903; Turney 1924; Upham and Rice 1980); canal system growth, organization of labor, and sociopolitical organization (Doyel 1980; Grady 1976; Grebinger 1976; Haury 1962, 1976; Howard 1990, 1991a; McAllister 1980; Nicholas 1981; Nicholas and Neitzel 1984; Woodbury 1961a; Woodbury and Ressler 1962); and reaction to environmental stress and agricultural intensification (Cable and Mitchell 1988; Grady 1976; Mitchell 1989b; Neitzel 1987; Weaver 1972). Several researchers have argued that the size of the irrigation systems, especially Canal Systems 1 and 2 (Figure 1.2 [First Canal System and Second Canal System]), required substantial labor investment for their construction and maintenance and a sophisticated sociopolitical system designed for canal operation, distribution and allocation of water, and resolution of disputes (Doyel 1981; Grebinger 1976; Nicholas 1981).

Howard (1990, 1991a) has examined labor requirements for the construction of canals through time in Canal System 2, concluding that labor requirements were greatest during the late Pioneer and early Colonial periods, resulting from rapid expansion and use of canal irrigation methods. After this period of optimal conditions (Graybill 1989; Nials, Gregory, and Graybill 1989), unstable discharge patterns of the Salt River led the Hohokam to rebuild and expand on the lower bajada during the late Colonial period (Graybill 1989). They abandoned the geologic floodplain due to the threat of flooding (Volume 2:Chapter 19; Greenwald and Ciolek-Torrello 1988a) but retained their stable irrigation system on the bajada during the Sedentary period. System expansion occurred again during the Classic period, with a general trend toward locating new canals higher on the terraces and abandoning lower elevations (Howard 1991b:5.33). However, the Hohokam also built new canals, such as Canal Salado and Canal Nuevo (Alignment 8537), on the geologic floodplain during this time (Volume 1:Chapter 8), suggesting renewed interest in an area that had been abandoned for nearly 400 years. The Hohokam may have shifted to previously unused or rejuvenated arable soils when they abandoned areas of Canal System 2 utilized during the Colonial and Sedentary periods.

Through analogy, researchers can examine aspects of models such as those mentioned previously (Ackerly 1982, 1991; Howard 1990). As historic records demonstrate, from the first irrigation ditches until the construction and use of modern water storage and diversionary devices, historic farmers confronted problems similar to those faced by the Hohokam. The following discussion utilizes this historic record in an attempt to understand the socioeconomic structure of the Phoenix Basin Hohokam.

The Initial Requirement

Historic settlement of south-central Arizona was part of the general expansionary movement of the mid 1800s. Importing food to the newly established military posts and communities in the Phoenix Basin was expensive, and immigrants almost immediately turned their attention to developing irrigated farms. As with the Hohokam, early historic residents exploited the Salt River and its adjacent floodplains. The river offered a perennial water supply, and the floodplains provided deep alluvial soils suitable for a variety of crops. Both prehistoric and historic farmers had to construct canals and diversion dams and had to clear fields, although historic farmers were able to take advantage of prehistoric canals in constructing their own.

The Hohokam obviously would have constructed canals and their associated features differently than did the historic farmers because of differences in technology and culture. Howard (1991c) has calculated that each Hohokam canal builder could excavate 3 m^3 of canal per day using stone tools and digging sticks. Historic canal builders had metal tools, explosives, and horse-drawn equipment, enabling them to excavate far greater volumes in shorter periods of time (Hackbarth and Lancaster 1992). Whereas Howard (1991b:Table 5.2) indicates that many of the prehistoric canals may have taken several years to construct, historic canals were often completed in a few months (Zarbin 1979). A basic economic difference probably existed between these two groups as well; the prehistoric farmers were engaged in subsistence agriculture, while historic canal builders formed canal companies for commercial ventures to supply civilian and military needs.

Canal Expansion and Competition

With the success of their initial canal construction, the Hohokam enlarged and lengthened canals and built lateral ditches to provide even larger tracts with water. They started new canals adjacent to existing canals as well as in other areas along the river. They also established new settlements and sometimes located settlements along the same canal alignment or within the same system. Competition for water resulted—not only along the same canal, but also from the construction of other canals that diverted water from the river farther upstream. Recognizing the importance of water distribution, especially in years of low discharge, archaeologists have posited a Hohokam sociopolitical structure that provided water allocation within canal systems and across systems as well as along individual canals (Gregory 1991; Gregory and Nials 1985). The size of Canal Systems 1 and 2, with their numerous irrigation settlements, would have required far greater organizational measures than smaller systems to ensure the maintenance of canals and canal heads and the equitable distribution of water. As Howard (1991a, 1991b) has argued, growth within Canal System 2 continued during the Classic period, albeit at a slower pace than during the preceding Sedentary period. Hohokam settlement generally expanded away from the river, although farmers again used the floodplain for agriculture during the Civano and Polvorón phases.

With what mechanisms or processes did the Hohokam manage water allocation, canal construction and maintenance, and access to farmland? As Gregory (1991:170–174) points out, canal systems in the Salt and Gila valleys contain from one to five platform mound villages, as well as smaller settlements, that relied on the distribution of water for irrigation and domestic use through the canal systems. The complexities associated with water use, especially in systems that contained multiple villages, would have required a strategy of allocation organized through some form of social or political structure. This sociopolitical structure may have included religious aspects, with religious leaders participating in the decision-making process that governed day-to-day activities within canal systems (Bostwick 1992). Perhaps religious connotations attached to cyclic events enabled the sociopolitical system to evolve and operate, resulting in the structuring and organizing of aspects of Hohokam subsistence strategies. Political ties between irrigation communities and canal systems may have been strengthened through the establishment of affinal connections or marriage alliances and trade relationships. Such coalitions would have had particular importance during periods of reduced available water, providing political structure to ensure that decision makers took measures to distribute water equitably within and between canal systems. The Hohokam reached the technological threshold of their ability to manage Salt River discharge levels early in the Hohokam sequence, and their sociopolitical organization appears to have been effective, as little evidence in the archaeological record indicates violence or aggression.

Faced with similar problems of water allocation as demands on Salt River water increased far beyond average discharge levels, farmers of the late nineteenth century looked for and established methods to provide organizational control. It is highly probable that during periods of decreased effective moisture, such as during the Classic period, the prehistoric inhabitants exceeded the carrying capacity of the Salt River Valley as well. Disputes over water allocation could have arisen from the resulting stress among system participants, leading to the development of intrasystem social organization to assure equitable distribution of water and general distribution scheduling during these periods to avoid conflict. Historically, farmers often protected their rights to water allocation by arming themselves. Such actions and the general outcry for determination of water rights led to court rulings concerning water allocation.

If such conflicts arose prehistorically, what mechanisms were in place to supply adequate amounts of water to each village within an irrigation community the size of Canal System 2? Brush and rock dams may have diverted only a portion of the streamflow, and geologic features often result in the surfacing of water at various points along a river. Low-velocity flows of this type thus may have been adequate for small systems such as Canal Salado despite the presence of larger systems located upstream. Perhaps water allocators denied water rights to some irrigation settlements during periods of low discharge, resulting in the relocation of settlements and construction of small systems such as Canal Salado on the geologic floodplain.

At times, too much water in the Salt would have been as devastating as too little, and historic farmers responded by constructing diversion dams that could withstand high-velocity discharge.

Although the technology associated with the larger universe of the historic era provided a solution, the Hohokam had no choice but to endure the whims of their environment. By contrast, to manage the discharge of the river more effectively, the valley's historic inhabitants constructed a new, larger canal (the Arizona Canal) with a more durable head. The Arizona Canal was designed to more effectively distribute water to other canals, but it was not the answer to water shortages. To ensure a reliable source of water, SRVWUA and the Reclamation Service constructed dams along the Salt River. The prehistoric Hohokam did not have access to the larger social system or the technology that made large-scale impoundment possible, yet they practiced irrigation along the Salt for nearly a millennium.

CONCLUSIONS

Significant parallels exist between the prehistoric agriculturists and the early historic farmers, similarities directly influenced by environmental fluctuations. Through necessity, the historic populations developed an organizational structure that provided control for the construction, operation, maintenance, and allocation of irrigation water. Archaeologists posit that the Hohokam developed a similar structure. Although researchers generally accept that such a structure was in place within the separate canal systems, Hohokam social structure as it related to irrigation requirements also may have been developed for pansystem organization. Historic analogies that may be applicable in demonstrating why the Hohokam developed a pansystem organization can be drawn from the establishment of the large-scale canals in the vicinity of modern-day Mesa and the threat that those canals posed to the continued use of the canals in the Phoenix area. Use of canals upstream would reduce discharge levels available to farmers farther downstream, a problem compounded in years of low discharge levels and through the increase in cultivated acres. As a result of increased demands for Salt River water, the judicial system reviewed and enacted legislation governing its allocation (Kent 1910).

The Hohokam developed the technology necessary to survive and prosper in the Sonoran Desert through the application of irrigation methods. Irrigation agriculture was a widely practiced subsistence strategy in the alluvial valleys, and the Hohokam of the Salt and Gila basins made extensive use of it, creating the largest prehistoric system of irrigation in North America. The adoption of irrigation agriculture by historic Euroamerican populations was a logical response to the same unpredictable environment, although with the application of increased technology historic farmers had more control over environmental conditions than did their prehistoric counterparts.

CHAPTER 2

DIFFERENTIATING THE EFFECTS OF URBANIZATION, ALLUVIATION, AND TIME ON HOHOKAM SITES IN THE PHOENIX BASIN

David H. Greenwald
Mark L. Chenault

This chapter evaluates transformation processes associated with urbanization (resulting primarily from historic use of the area) and natural processes associated with deposition (resulting primarily from alluviation). Investigations in the Phoenix Sky Harbor Center served as the impetus and focus for this chapter, although the authors also derived information contained herein from other projects undertaken within the Phoenix metropolitan area.

Urbanization and alluviation have had differing effects on cultural resources in the Phoenix Basin, due in part to the location of the resources and in part to the amount of time that the resources were subjected to these processes (Schiffer 1987), so that time becomes a major component in the evaluation. This chapter examines the range of impacts, the nature of the damage to resources, and the ways that the above processes limit data recovery. The quality of information generated through archaeological studies in urban settings and the variables that contribute to site preservation also are discussed.

HISTORICAL BACKGROUND

Located in a broad alluvial area consisting of deep, well-developed soils interrupted by occasional uplifts of parent materials, the Phoenix Basin has received deposition from various streams and rivers (the Gila, Salt, Verde, and Agua Fria rivers, Queen Creek, Indian Bend Wash, New River, Skunk Creek, and Cave Creek). As discussed in the preceding chapter, Euroamerican settlers recognized the agricultural potential of the Phoenix Basin in the 1860s. Following the Civil War and the discovery of large mineral deposits in the territory, the federal government established military installations to suppress Apache and Yavapai aggressions against the small but growing Euroamerican population. To supply Fort McDowell in the lower Verde Valley with needed resources for personnel and livestock, the army awarded contracts for the production of hay and vegetables. Shortly after this, settlers in the Phoenix Basin undertook construction of irrigation canals to increase agricultural productivity. Prolonged agricultural use significantly modified land surfaces, making smaller archaeological sites less identifiable. Settlement activities also damaged larger sites, although settlers often avoided mounds, at least initially, or diverted ditches around them to avoid the effort required to level them. They also built farmhouses and hay barns on some mounds for protection from surface flooding.

By the 1930s, farmers had begun to cultivate much of the area adjacent to the floodplain of the Salt River. Historic canal systems were as extensive as the prehistoric ones had been, providing irrigation water to tracts of land on both sides of the river. Agriculture intensified, utilizing ever less-desirable tracts as demand for farm produce increased. Populations of the various townsites in the Phoenix Basin also increased, and with this increase came a demand for more housing. By the 1940s, several of the early farm parcels adjacent to the original Phoenix Townsite were subdivided into housing parcels. Over the next 20 years much of the Phoenix Sky Harbor Center project area became residential, with some commercial developments located along major arteries such as 16th,

20th, and 24th streets and Buckeye Road. The area remained in residential use until continued expansion of the adjacent Phoenix Sky Harbor Airport rendered the noise impacts too great.

DIFFERENTIATING THE EFFECTS OF URBANIZATION

Disturbance Processes

Anthropogenic processes that can disturb the archaeological record are many; most, however, involve some type of earth-moving activity. Trampling of surface remains and plowing can cause widespread disturbance (Schiffer 1987), and construction of buildings, roads, and canals requires the movement of large amounts of earth. Humans, of course, are not to blame for all disturbances to the archaeological record. Flooding, erosion, freeze-thaw action, and other natural forces also cause alteration of archaeological remains (Butzer 1982). However, this discussion is limited to the disturbance processes associated with human alteration of the landscape.

Economic development began in the Phoenix Sky Harbor Center area shortly after 1868 when farmers built the first historic irrigation canals on the north side of the Salt River in the vicinity of the Phoenix Settlement (Figure 1.1; Volume 1:Figure 4.3). Following the establishment of the earliest Euroamerican farms, construction and occupation of residential and commercial facilities disturbed archaeological remains in the area. Schiffer (1987; see also Schiffer and Gumerman 1977) has described stages of impact to archaeological sites caused by economic development that are appropriate in discussing the nature and effects of disturbances at Phoenix Sky Harbor Center sites. The stages, briefly described below, include planning, construction, and operation.

Planning stage impacts involve preparatory activity in the area to be developed, such as the building of access roads, surveying activity, and soil-test drilling. In studying an archaeological site and assessing the consequences of historic development, researchers may have difficulty in distinguishing impacts associated with the planning stage from those occurring during construction.

Schiffer further divided construction-stage impacts into three levels: primary, secondary, and tertiary. Primary impacts are those caused by any actual construction activity, including removing vegetation, leveling and blading the land surface, and digging for any purpose, such as obtaining construction materials (e.g., sand and gravel), installing utilities, or building a structure foundation. Secondary impacts are those caused by support activities, such as the use of access roads or the discard and burying of construction debris. Tertiary impacts would be those that are not a direct result of the construction activity but are connected to the presence of the activity; an example might be the collection of artifacts by construction personnel. Other researchers have referred to primary impacts as direct impacts, and Price (1977) and Schiffer and House (1977) have referred to secondary and tertiary impacts as indirect impacts.

Schiffer calls the third impact stage the operating stage. In the case of Phoenix Sky Harbor Center, with its preponderance of residential construction, this might better be thought of as the occupation stage. Occupation/operation-stage impacts result from the actual use of the constructed facilities, whether as a result of the function of the facilities, from other intended uses or associated uses, or because of changes in demographic trends and land use (Schiffer 1987:135–136).

Any of the impact stages may have occurred at an archaeological site in either recent or prehistoric times. Schiffer (1987:136–140) provides an example from Snaketown, where prehistoric building ("reclamation processes") disturbed extant cultural remains. Although this type of disturbance also may have occurred at Dutch Canal Ruin and Pueblo Salado, this chapter focuses on historic and modern activities that affected the prehistoric remains. For a discussion of the historic occupation of the Phoenix Sky Harbor Center area, see Volume 1, Chapter 4.

Disturbance of the Archaeological Record

Plowing

The earliest recognized historic disturbance in the project area was agricultural, principally plowing, which extended 40–60 cm below the present surface. However, little of an actual plow zone remains today, probably the result of blading and leveling of the area prior to commercial and residential construction. Aerial photographs taken in 1934 (Soil Conservation Service 1934) indicate some areas of farming in the vicinity. However, by 1964, when aerial photographs were taken again, no agricultural fields remained.

Roads

Streets functioned in three ways within the project area, with three levels of disturbance to cultural resources. Main arteries, such as Buckeye Road, 16th Street, and 20th Street, carried heavier traffic loads and also served as major utility corridors. Field investigations indicated that the main arteries had little effect on project resources because they lay on the periphery of the archaeological loci. Secondary streets, such as 18th, Mohave, and Grant, had functions similar to the main arteries but were designed to limit the flow of through traffic, and they carried less traffic overall. They served as primary utility corridors into subdivisions, and project investigators found that the utilities intruded into project resources that otherwise remained undisturbed by construction of the streets. Residential streets, the third level of streets in the project area, were numerous because of the design of the residential subdivisions. In general, residential streets lacked the deep ground disturbance zone of the secondary streets, but preparation of the roadbed did affect project resources, especially at Pueblo Salado.

The most extensive damage from roads occurred at the intersection of 18th and Cocopah streets, located above and adjacent to the compound of the Classic period site of Pueblo Salado. Disturbance from the construction of 18th Street extended fairly uniformly to a depth of approximately one meter below the modern ground surface, while prehistoric occupation surfaces throughout the site reached only to a depth of approximately 50 cm below the modern ground surface. As a result, no intact prehistoric surfaces remained in the former location of 18th Street, and in a few places excavators found nothing but the bases of the deepest wall footers. The depth of the disturbance caused by 18th Street generally was greater than that for Cocopah Street because of the function of the streets; in this particular location, however, the construction of both streets had destroyed parts of the site.

Blading and Leveling

Preparing the ground surface for residential development some 40 to 50 years ago was one of the most destructive of the disturbance processes, yet it helped to preserve other areas by contributing to the depth of the deposits that covered the resources. Blading apparently reached a depth below the level of the prehistoric occupation surface in some portions of the compound at Pueblo Salado, destroying prehistoric surfaces and features in these areas. The deeper wall footers and deep pits usually survived, but in some cases the blading and leveling extended deep enough to remove shallow building footings.

In most parts of the compound, however, blading and leveling remained above the depth of the occupation surfaces, truncating only the tops of the adobe walls. Although this construction activity did not destroy the occupation surfaces, it did move and scatter adobe construction material and cultural fill, complicating the task of defining walls amid the jumbled fill matrix.

House and Building Construction

Construction of houses and other buildings appeared to have affected sections of the Phoenix Sky Harbor Center sites to varying degrees. The bungalow-style house, common to Phoenix and ubiquitous in the neighborhoods in and around the project area, did not generally include a basement or deeply placed foundations. For this reason, the building of a house on top of prehistoric remains generally required only shallow footings (extending less than 50 cm below the modern surface). More destructive was the construction of septic systems in conjunction with those houses. Investigators identified many such historic features within the compound at Pueblo Salado. Outhouse pits associated with earlier houses also occurred, and excavators found what may have been one such feature in Area 8 of Dutch Canal Ruin. Historic pits resulted in spatially limited but deep impacts (some septic systems extended over 12 feet below the surface).

By contrast, the construction of more substantial buildings, those associated with commercial use of the project area, extensively disturbed the archaeological remains. In Area 4 of Dutch Canal Ruin, for example, field personnel found areas of ashy soil and some scattered burned rock fragments at a depth of more than a meter below the modern ground surface. They could identify no actual prehistoric features there, however, due to disturbance from construction of a commercial building in that location.

Landscaping of the house lots and the areas around the commercial buildings also had the potential for damaging the prehistoric remains. Historic and modern occupants dug pits for the planting of trees, constructed sidewalks, built shallow drainage ditches, and erected fences. Although landscaping within the project area was less destructive than construction activities, the cumulative impact of several decades of occupation and landscaping was evident.

A series of aerial photographs of the project area provided by the City of Phoenix aided in understanding the effects that construction of residential and commercial buildings had on the cultural resources. Photographs taken in 1964 indicate dense construction within the project area, a marked contrast to the agricultural fields visible on the 1934 photographs. Aerial views from 1975 show the same buildings. However, by 1986, a series of aerial photographs indicate that most of the structures had been removed as part of the development of Phoenix Sky Harbor Center.

Pipes and Trenches

Occupants of the project area during twentieth-century construction and occupation/operation stages dug trenches throughout the area. Many of those trenches contained utility lines leading from the main lines in the streets to individual houses and buildings. Project excavators found utility trenches throughout the project area, and many intruded through features.

Ditches

Ditches dating to historic and recent times have affected parts of the Phoenix Sky Harbor Center sites. For example, a small irrigation ditch ran east-west through the compound at Pueblo Salado, just south of Cocopah Street. The ditch is visible in both the 1964 and 1975 aerial photographs of the project area. Its construction intruded into several features in the compound and removed sections of several walls.

One ditch of historical importance in the project area was a re-engineered version of the old Dutch Ditch. Believed to have been built late in 1868 (Volume 1:Chapter 4), it ran east-west through the northern part of Area 8 of Dutch Canal Ruin. The ditch did not appear to have drastically affected the site. However, a small lateral to the south of the main ditch slightly damaged a vessel containing a cremation.

The Archaeological Record: What Remained?

Despite more than a century of disturbance, much of the archaeological record remained after the construction, use, and dismantling of houses, buildings, and roads from the modern occupation of the Phoenix Sky Harbor Center project area (see Volumes 2 and 3 for site descriptions). The cultural resources included various types of habitation features and a number of canal alignments. The impacts to the canals did not seriously alter their integrity and information potential because of the length of the preserved alignments.

The recovery of intact cultural remains at the two project sites, Dutch Canal Ruin and Pueblo Salado, allowed an extensive and detailed reconstruction of the culture history of these sites. Investigators recovered sufficient data to develop an understanding of site structure, activities performed at the sites, burial practices, and the relationships of the sites to other prehistoric villages in the Phoenix Basin.

Damage to the Archaeological Record: What Was Lost?

The archaeologist working in urban areas has a challenging task. Data collection is difficult, and archaeological analysis provides no method for determining the nature or full extent of the data lost to historic or modern disturbance. Individual features suffer damage from the construction of buildings, utility trenches, ditches, septic systems, and trash pits and from other subsurface intrusions. Where urban development has taken place, in most cases construction activity will have destroyed all but the deepest few centimeters of prehistoric structures. Archaeologists will thus lose most evidence of postabandonment formation processes, that is, whether prehistoric inhabitants abandoned a structure and allowed it to deteriorate or actively dismantled it. At sites spared from such large-scale urbanization, excavators may be able to reconstruct site formation processes well enough to achieve some understanding of use and abandonment modes. For example, if structures collapsed in place through natural processes, investigators may be able to estimate original wall heights. Project archaeologists unfortunately distinguished little useful data on use and abandonment pertaining to the Classic and post-Classic period components in the Phoenix Sky Harbor Center.

Evidence for the horizontal extent of the compound at Pueblo Salado was missing (Volume 3:Figure 2.1). For example, in the far southeastern corner of the compound, a very large (ca. 9 × 12 m) historic pit had destroyed portions of a pair of parallel adobe walls, and 18th Street had cut through those same walls to the west. Only two short sections of the walls remained, providing no indication of their former extent. In the southwestern corner of the compound, blading had extended below the level of the plow zone and the typical level of historic disturbance; construction activity thus may have completely removed many walls and portions of walls in this area. As a result, project investigators faced a difficult task in defining the southern boundary of the compound and determining the relationships of features within the southern portion of the compound.

Damage to project resources also interferes with the analysis of site structure. Defining site structure requires an understanding of the spatial relationships of features; moreover, archaeologists rely on spatial relationships in defining use areas and identifying social groupings. Perhaps the historic or modern activity most destructive to site structure was the creation of 60-foot-wide transects during the construction of 18th and Cocopah streets, which obscured some relationships between features and blocks of features at Pueblo Salado. Impact on site structure was less of a problem at Dutch Canal Ruin because of the greater depth of the alluvium covering this site. Destruction of trash mounds by historic activities also limits the extent to which archaeologists can examine topics such as site structure and behavior (e.g., discard patterns and resources used).

Some areas within the Phoenix Sky Harbor Center endured less severe disturbance, providing enough data that project investigators could define features and reconstruct their attributes. At Pueblo Salado, for example, utility trenches and the construction of 18th Street had removed the western portion of Feature 25, an adobe compound room (Volume 3:Figure 2.15). Nevertheless, field personnel used the preserved portion of the structure to reconstruct the approximate dimensions of the room and to define its orientation and function. In other instances, intensive disturbance from street construction and commercial property development had damaged cultural remains to the extent that reconstruction was not possible. In some portions of Area 4 at Dutch Canal Ruin, nearly all evidence of features had been destroyed.

On a smaller scale, disturbance of the spatial arrangement of artifacts within the fill of features and on occupation surfaces also means a loss to the archaeological record. Spatial analysis can be complicated by the loss of artifact patterning, making definition of activity areas problematic. Artifact counts will also be incomplete or skewed if materials are removed from their original context and redeposited elsewhere. In the case of Hohokam sites such as Dutch Canal Ruin and Pueblo Salado, the leveling of the surface will truncate trash mound features, spreading the deposits across a much larger area than they originally occupied. In addition, earth-moving activities can obscure patterns of deposition and the spatial association of refuse features and architectural features. The resultant removal, churning, and redeposition of cultural deposits can complicate the dating of those deposits.

DIFFERENTIATING THE EFFECTS OF ALLUVIATION

Archaeologists often consider the accumulation of alluvial materials during the course of reconstructing archaeological events, more often than not from the perspective of the effects of alluviation on a site during occupation, around the time of abandonment, and over the span of time between abandonment and investigation of the site. One aspect that investigators seldom consider is the association of presettlement alluviation with site transformation processes. For example, what physical conditions did occupants encounter when selecting a habitation site? What measures did they undertake to establish their residence? How did people adapt to and modify the physical environment to subsist? To answer such questions, researchers must both examine the archaeological evidence and evaluate variations that occurred at sites in different settings.

Alluvial Processes

Variation in the accumulation of alluvial deposits depends on where a site lies (usually in terms of distance and topographic setting) in relation to drainages. The topographic relief of the central Phoenix area is relatively low, making the boundary between the geologic floodplain and the lower bajada difficult to discern. As a result, Salt River flooding can inundate broad expanses north of the Salt River in the area now occupied by central Phoenix.

The drainage of the Salt River is from east to west, while secondary drainages flow nearly north-south. Alluvial materials along the Salt River channel consist primarily of sands, silts, and clays transported from upstream watersheds. The valley margins receive considerably less material from the Salt River and more from secondary drainages and from the uplift parent materials, sources of colluvial materials common to the Basin and Range physiographic province. The annual and seasonal discharges of the Salt River have exhibited great fluctuations, as suggested by a recent streamflow reconstruction that focused on the period from A.D. 740 to 1370 (Graybill 1989; Nials, Gregory, and Graybill 1989). Nevertheless, through time the Salt River would have exceeded every secondary drainage in the Phoenix Basin in the amount of alluvial materials it deposited across the valley floor.

Researchers have not determined the maximum extent of alluvial deposition by the Salt River in what is now central Phoenix. As an example of the potential for flooding and alluviation, however, the flood of 1891 extended north of the river nearly 2.5 miles, to the vicinity of Van Buren Street at 24th Street (Volume 1:Figure 3.2). Larger floods may have deposited alluvial materials beyond this distance, an inference supported by the recent streamflow retrodiction study by Graybill (1989:33–38). His analysis of tree-ring records of the Salt River watershed indicated that floods of greater magnitude than the 1891 flood occurred after A.D. 800. Therefore, deep alluvial deposits originating from the Salt River probably overlie remnant terrace materials at the interface of the lower bajada and the geological floodplain. Alluvial materials beyond this general boundary have come from the secondary and tertiary drainages fed by the uplands to the north. These drainages dump their bed load as their velocity decreases upon entering the broad plain of the Phoenix Basin. Data from several archaeological investigations and soil surveys (Adams 1974; Hartman 1977) indicate a general trend toward shallower deposits of alluvial materials with increased distance north of the Salt River between 7th Street on the west and 32nd Street on the east (Figure 2.1). The illustration includes alluvial materials that predated site occupation to illustrate the depositional history of the Phoenix Basin during the Holocene epoch.

Besides examining the depth of deposit, archaeologists may also trace depositional history by analyzing the soil types within alluvial accumulations. The Phoenix Basin contains calcium carbonate horizons (in varying degrees of development) that are commonly associated with the lower bajada. By contrast, carbonates associated with soils on the geologic floodplain are weakly developed, if present at all. Here, alluvial materials may be several meters thick but often overlie cobble, gravel, and sand substrate. Investigators commonly found thick deposits of alluvium with little or no carbonate development at Dutch Canal Ruin and also identified a gravel and cobble substrate under the compound at Pueblo Salado. Previous research documented a calcium carbonate, or caliche, horizon at the site of Los Solares, approximately 1500 m north of Dutch Canal Ruin (Figure 2.1). Moreover, data indicate a considerable decrease in alluvial materials, to thicknesses ranging between 0.85 m and 1.10 m (Rice 1987), above the calcium carbonate horizon at Los Solares compared to alluvial deposits at Dutch Canal Ruin. Pueblo Patricio, at the interface of the geologic floodplain and the lower bajada, exhibited a range in the depth of alluvial materials covering the calcium carbonates of 0.85 m to 1.20 m (Cable and Allen 1982:31–71; Cable and Doyel 1984a, 1985a),

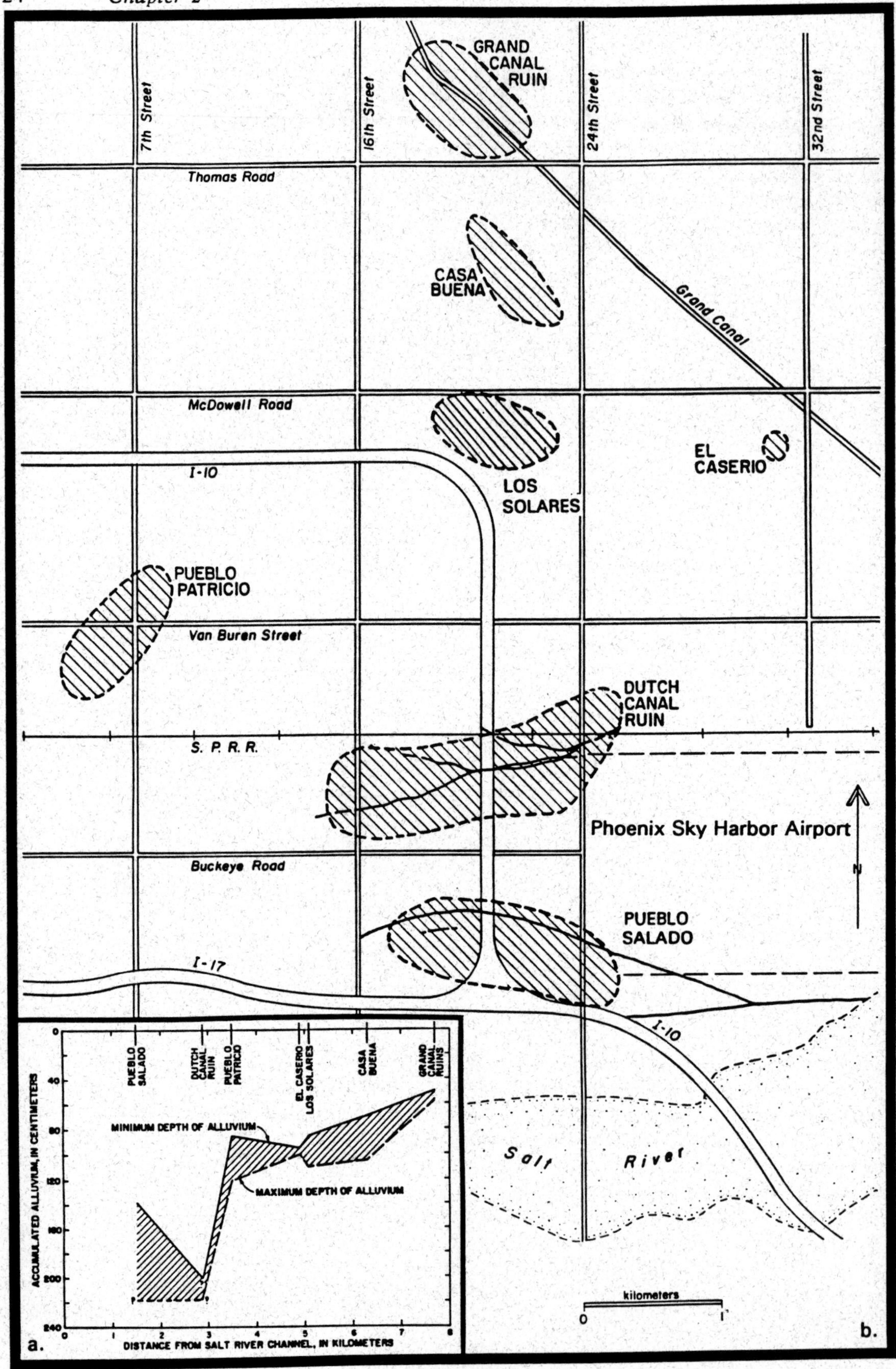

Figure 2.1. (a) Reconstruction of accumulated alluvial materials north of the Salt River; (b) distribution of archaeological sites used to reconstruct alluvial accumulations.

similar to that at Los Solares. Researchers also reported a similar but slightly narrower range (0.95–1.00 m) for the site of El Caserío (Huckleberry 1989).

The sites of Casa Buena and Grand Canal Ruins lay farther north of the Salt River. Casa Buena had a broader range in the depth of accumulated alluvial materials above the calcium carbonate horizon—0.70 m to 1.05 m (Howard 1988a)—at least partly due to the extended linear pattern of the site and its location on two different soils. The southern portion of the site was situated on Avondale clay loam, generally a deep soil, whereas the extreme northern portion of the site was on Gilman loam, a moderate to shallow soil. The northernmost site under consideration here is the Grand Canal Ruins, located on Estrella loam. Huckleberry and Kwiatkowski (1989:25–28) reported alluvial materials ranging from 0.50 m to 0.58 m above the caliche horizon for the site.

Although the conclusion that less alluvium is deposited with increased distance from the Salt River may seem obvious, its implications are important to understanding and evaluating the effects of transformation processes on cultural resources in the Phoenix Basin. Specifically, what factors in the Phoenix Basin have protected the resources, and to what extent does the protection afforded the resources change due to location, physiographic relations, alluviation, and age of the resources?

Variability of Sites and Physiographic Features

As described above, depth of alluvium varies with site location and relation to physical features. Within the sites of the Phoenix Sky Harbor Center, investigators identified two extremes.

The deeper alluvial deposits occurred at Dutch Canal Ruin, which exhibited rock-free alluvium in an area of apparent erosional stability where alluvial deposits continued to accumulate during and after site occupation. Because of the presence of pre-Classic, Classic, and post-Classic period remains, Dutch Canal Ruin offered an excellent opportunity to examine how alluviation had protected resources of different temporal association. Because the temporal components lay only 100–150 m apart, differences in depth of alluvium should have been a function not of geophysical setting but of the age of the component.

Elevations from Dutch Canal Ruin loci indicate the amount of deposition that occurred during the pre-Classic, Classic, and post-Classic periods (roughly A.D. 900-1450 or later). The occupation levels of the pre-Classic period components occurred at 3.95–5.11 meters below datum (mbd), or 0.77–1.04 m below the modern surface; those of the Classic and post-Classic periods lay at about 3.80 mbd, or 0.40 m below the modern surface. This additional 0.37–0.64 m of accumulated alluvium was sufficient to protect the pre-Classic components from the effects of historic and modern residential use but insufficient to protect the Classic and post-Classic period components. As a result, historic and modern use had truncated the later components but had only minimally disturbed the pre-Classic period components, primarily through intrusive pits and deep utility lines.

Shallower deposits occurred in Area 8/9 at Pueblo Salado. Located closer to the center of the floodplain than Dutch Canal Ruin and immediately north of the modern channel of the Salt River, Pueblo Salado lay, in effect, on an "island" in the floodplain. As at Dutch Canal Ruin, alluviation at Pueblo Salado had occurred rapidly during the occupation of the site. In general, the earlier remains, those dating to the Soho phase, were buried by deep deposits, indicating that the area had been subjected to varying degrees of flooding during and after occupation. Deeply buried features associated with the use of the compound indicated flooding of the site area during the Civano phase occupation as well.

The Hohokam built the compound, Area 8/9, on a topographic rise (Figure 2.2), the highest elevation within the project area south of Buckeye Road. As a result, builders made their compound less likely to be inundated by flooding and subject to less alluvial deposition. Other factors not yet fully defined, such as the construction of the compound wall, may also have reduced the incidence of alluvial materials in this area. One unique feature of Pueblo Salado was the dual wall along the north and south sides of the compound. Although investigators have not yet determined the function of this wall, residents of the compound may have constructed it in response to increased threats of flooding, thus further protecting the enclosed habitation area. Beyond the compound walls and at lower elevations, excavators found deeply buried features associated with Area 8/9. The accumulated deposits over these features indicated the amount of alluviation both during and after occupation.

Project analysts distinguished three types of deeply buried features: (1) adobe-lined pits, used for the mixing of adobe mud for construction of the compound and house walls; (2) borrow pits, irregular depressions created by the removal of materials, presumably for construction purposes, and used in conjunction with the adobe-lined pits; and (3) agricultural fields. Each of these feature types lay under deep accumulations of alluvial materials, but project investigators have demonstrated their temporal associations with the construction and occupation of the compound at Pueblo Salado (Chapter 5).

The earliest adobe-lined pits and borrow pits lay at elevations of 1.00–1.10 m below the modern ground surface, or 5.67 mbd. They had been buried by the accumulations of alluvial materials associated with the agricultural fields, the uppermost levels of which lay at 5.17 mbd, beneath 20 cm of postabandonment alluvium and 40 cm of historic and modern disturbance (Chapter 5). In contrast, the earliest Civano phase house floors lay at a maximum of 0.50 m below the modern surface, or 4.87 mbd. The amount of deposition at Pueblo Salado ranged from 0.68 m to 1.10 m, except within the compound. The difference in elevation (0.80 m) between the earliest adobe-lined pits and the earliest Civano phase house floors at least partly reflected the difference between the naturally elevated area on which the habitation area was built and the lower, and probably more typical, elevation of the surrounding areas (Figure 2.2).

The construction of the initial compound and the digging of the earliest adobe-lined pits preceded the abandonment of the compound by approximately 200 years, an occupation range of circa A.D. 1300–1500. From the beginning of the Civano phase until about A.D. 1900, the area adjacent to the compound at Pueblo Salado accumulated about 1.00 m of alluvial deposits. Since the early part of this century deposition within the project area has been limited because of the construction of impound dams along the Salt and Verde rivers (Chapter 1).

In summary, alluviation protected cultural features in the project area except in the elevated compound area of Pueblo Salado. The compound, due to its position on the topographic rise, received the direct impacts of historic and modern use of the area.

At the other sites included in this study, alluviation buried cultural features and deposits to varying depths, depending on their temporal and stratigraphic associations. Because of the shallower accumulation of alluvial materials at these sites, later deposits were susceptible to historic and modern disturbance. Pre-Classic period features, often dug into the caliche substrate, thus received greater protection than features affiliated with the Classic and post-Classic periods.

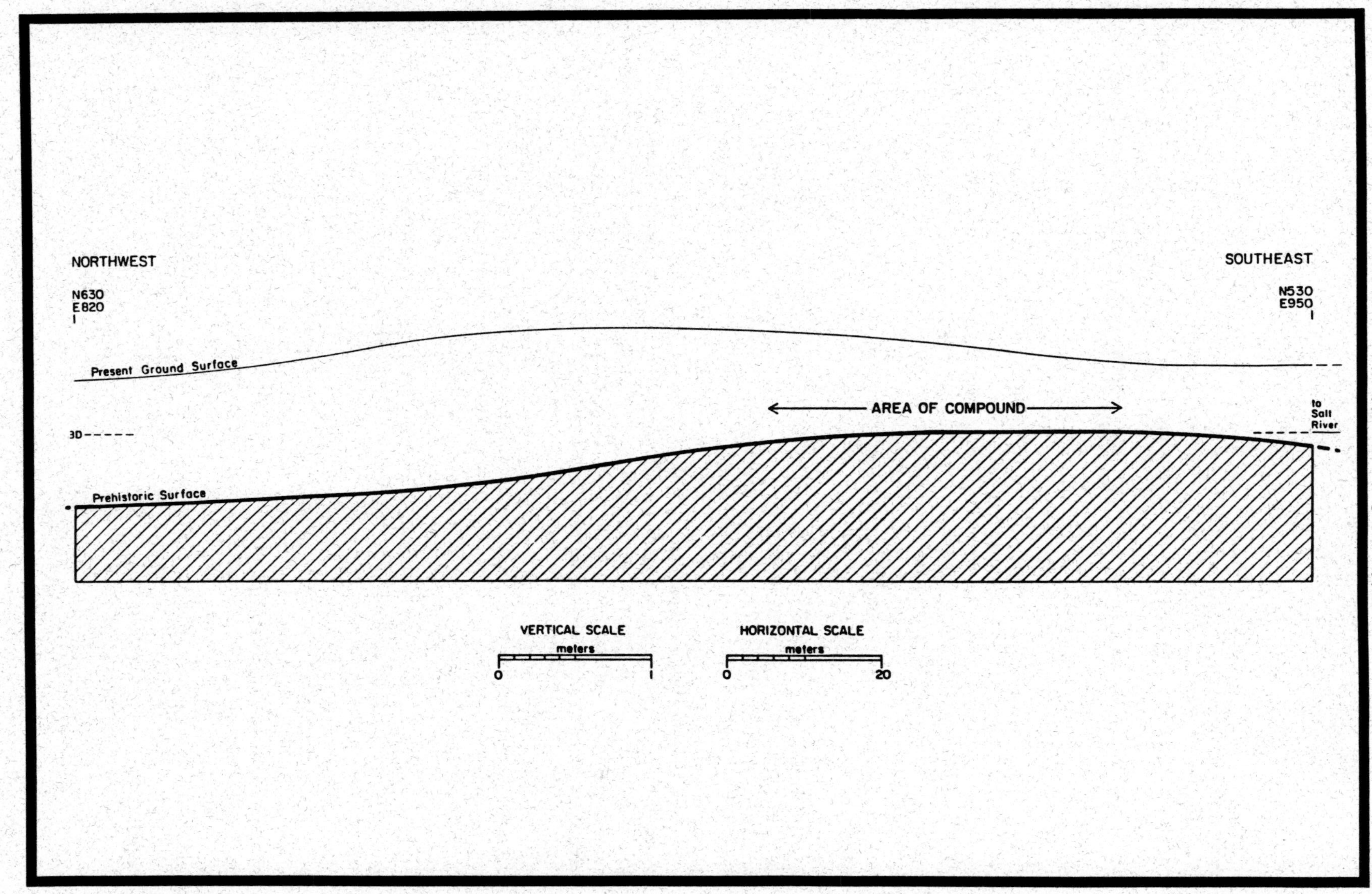

Figure 2.2. Cross section of Area 8/9 showing the prehistoric and present ground surfaces in the area of the compound at Pueblo Salado.

CONCLUSIONS

The main conclusions to be drawn from this study may seem obvious, but they deserve to be stated nevertheless. Despite decades of commercial and residential development and use, the Phoenix Sky Harbor Center project area still contained numerous intact or partially intact prehistoric cultural remains, and much archaeological information remained. The analysis of data recovered during the course of fieldwork allowed us to reconstruct lifeways at Dutch Canal Ruin and Pueblo Salado and will contribute to the existing data base for the Hohokam in the Phoenix Basin.

Even though the modern ground surface prior to excavation yielded only a few scattered artifact fragments, field personnel excavated 68 habitation features at Dutch Canal Ruin and 155 at Pueblo Salado. Cultural resource managers should consider the location of a project area in relation to physical features as well as historic and modern impacts to sites when assessing the preservation and data potential of sites located in urban settings. Surface indications alone are unreliable in assessing the potential for subsurface remains in urban areas, although they are helpful when used with other sources of information. When available, archival records in the form of historic maps and early archaeological records provide important information for assessing the potential for archaeological remains. Until researchers identify more reliable methods, they must continue extensive subsurface testing in urban settings to determine the extent of disturbance and the condition of the remains.

Testing at the Phoenix Sky Harbor Center ranged from 1% of the total available project area, in those areas where archival information and surface indications suggested a low probability of subsurface cultural remains, to 2% in areas considered to have a moderate to high probability of occurrence. For the purposes of determining the presence of subsurface features, a 1% sample provided reasonable results. However, experience in the Phoenix Sky Harbor Center indicates that a 1% sample should be augmented with judgmental trenching when defining the spatial extent of features and attempting to estimate the number of features present. Given the early records associated with Dutch Canal Ruin, for example, project investigators expected to find a greater number and variety of features than were actually present. Dutch Canal Ruin appeared to be a large habitation site because of the numerous mounds illustrated on the Midvale maps (n.d., 1934). In this case, close scrutiny of the archival records, along with the testing results, revealed that a misconception concerning Dutch Canal Ruin had developed through the use and reuse of archival records. The present project has clarified the record, and the misinterpretation should not be used to argue against the use of archival data in future undertakings. These "mounds" may have represented horno locations or perhaps artifact concentrations such as Area 8.

A number of factors are instrumental in the preservation of archaeological remains. Among those are the passage of time, the degree of deposition, and the type of modern activity that disturbed the remains. At the Phoenix Sky Harbor Center sites, investigators found that the older prehistoric remains stood a better chance of being preserved than the more recent ones because the older features occurred at a greater depth and were protected by thicker layers of alluvial deposits. The later remains, being closer to the surface, were subjected to greater degrees of damage. Again, this conclusion may seem obvious given the explanation of alluvial processes in the preceding section. However, it is the opposite of what archaeologists may expect given the norm, in which the processes of "deterioration, decay, alteration, and modification" (Schiffer 1987:143) increase with time. This inversion of preservation at the project sites did not hold true for all aspects of data collection. For example, the results of the radiocarbon assays from the pre-Classic period components were extremely poor. At present, project analysts attribute the deterioration of sample materials to the extended periods of saturation by floodwaters and the introduction of organic contaminants into the wood charcoal. Samples from the Classic and post-Classic period components, by contrast, fell within the expected temporal range and apparently were less affected by flood-related contamination.

The type of disturbance processes at work is another factor affecting preservation of archaeological remains. Our excavations suggested that disturbance from residential construction was, as might be expected, less destructive than disturbance from the construction of commercial buildings. Nevertheless, residential construction tends to be more widespread, covering the landscape in a more systematic manner. Moreover, although a commercial building may occupy a large area, the spatial extent of its impact may be proportionately less than that of residential areas, a situation also observed at Pueblo Grande (Cable 1988). A commercial development area often includes large parking and other nonbuilding areas, which generally contain only underground utilities to disturb subsurface remains.

From an archaeological perspective, it is unfortunate that any site has been damaged by historic or modern activities, but the presence of such facilities on top of prehistoric sites can protect portions of those sites from further damage. In a sense, historic development of the Phoenix Basin has prolonged the preservation of some resources. Modern developments, in general, result in greater disturbance to the landscape than do earlier historic developments. Transportation corridors are greater in size, buildings (both residential and commercial) are more substantial, and underground utility lines are more extensive. Archaeologists and land planners therefore should be aware of the research potential of sites that underwent construction in prior years. The experience of investigators in both the Phoenix Sky Harbor Center Project and projects in other urban areas dictates that project managers consider reducing the level of archaeological investigations in those areas that potentially have been significantly disturbed by modern activities such as road construction, utility installation, and deep subgrade disturbance when less disturbed areas are available nearby.

Archaeologists should not lose sight of the prehistoric record potentially still contained in urban areas of the Phoenix Basin. Although this chapter has focused on the portion of the Phoenix Basin north of the Salt River in east-central Phoenix, other researchers should examine formation processes associated with urbanization and alluviation throughout the region. Assessment of the effects of urbanization and alluviation on cultural resources can be applied to areas of the Phoenix Basin, and to other areas of Hohokam culture such as the Tucson Basin and Safford Valley, to aid future land planners and agencies during initial planning stages.

CHAPTER 3

IRRIGATION SETTLEMENTS: AUTONOMY WITHIN CANAL SYSTEMS

David H. Greenwald
M. Zyniecki

Small settlements associated with canal irrigation were a major focus of the Phoenix Sky Harbor Center investigations. This chapter reviews the apparent association of the habitation remains with canals and fields and how those remains may have been integrated into larger, more complex social networks. Within the Phoenix Basin, Hohokam settlement systems often took the form of linear canal networks that occupied spatially discrete areas of the valley floor. Researchers have defined canals and related habitation areas that occupied a common geographical area as canal systems and interrelated sites that share the canal system as irrigation communities (Doyel 1974, 1980; Howard 1987:211). Howard and Wilcox (1988:911) expanded the definition of irrigation communities to include all sites along a canal system that share a common headgate or series of headgates. Gregory (1991:170–174) defined irrigation communities formed during the Classic period based on the distribution of platform mounds and identified at least 16 such communities.

To address the question of autonomous irrigation settlements within the Phoenix Sky Harbor Center, investigators had to first define the level of participation of Dutch Canal Ruin and Pueblo Salado in the sociopolitical organization of canal systems. The relationship between the habitation features and canals at Dutch Canal Ruin appeared to conform to previous models for the intrasite role of the field house and how it may have functioned in relation to larger, permanently occupied sites within Canal System 2 (Volume 2:Chapter 19). Pueblo Salado, however, exhibited more autonomy in its establishment and growth than did other sites in Canal System 2 or other canal systems in the Phoenix Basin. This chapter presents evidence supporting an autonomous settlement and independent canal system both for Pueblo Salado and for other as yet undocumented sites.

IRRIGATION SETTLEMENTS IN THE PHOENIX SKY HARBOR CENTER

An irrigation settlement (Volume 1:Chapter 1) consists of an individual habitation locale that may have functioned as a field house, farmstead, hamlet, or village. Irrigation settlements are components of irrigation communities. The pre-Classic period occupation of Dutch Canal Ruin represented seasonal use of the area by residents from other communities within Canal System 2. For the most part, the canals at Dutch Canal Ruin served as extensions of that system, representing the southern extent of land-use practices associated with farming in Canal System 2 during the pre-Classic period. Pueblo Salado, by contrast, was physically separated from the canal network defined as Canal System 2 by a channel of the braided Salt River (Turney's Gully; see Figure 3.1). Although early canal maps (e.g., Figures 1.2, 3.2) show a crosscut canal extending from Canal Patricio (which headed near the Park of Four Waters area) in a south-southwesterly direction tying Canal Salado with the main network of Canal System 2, recent investigations at the Phoenix Sky Harbor International Airport did not locate a canal in that area (Greenwald and Zyniecki 1993). Investigators identified Turney's Gully, however. Artifacts within its flow deposits suggested that this channel had been open during or after the prehistoric occupation of the area; the geomorphology of the floodplain indicated that the channel probably existed during the occupation of Pueblo Salado when residents of the site were using the Classic period canals in this area. Its presence may have presented the canal builders with an engineering problem requiring more labor than the irrigation water was worth. A canal constructed across an active river channel would have

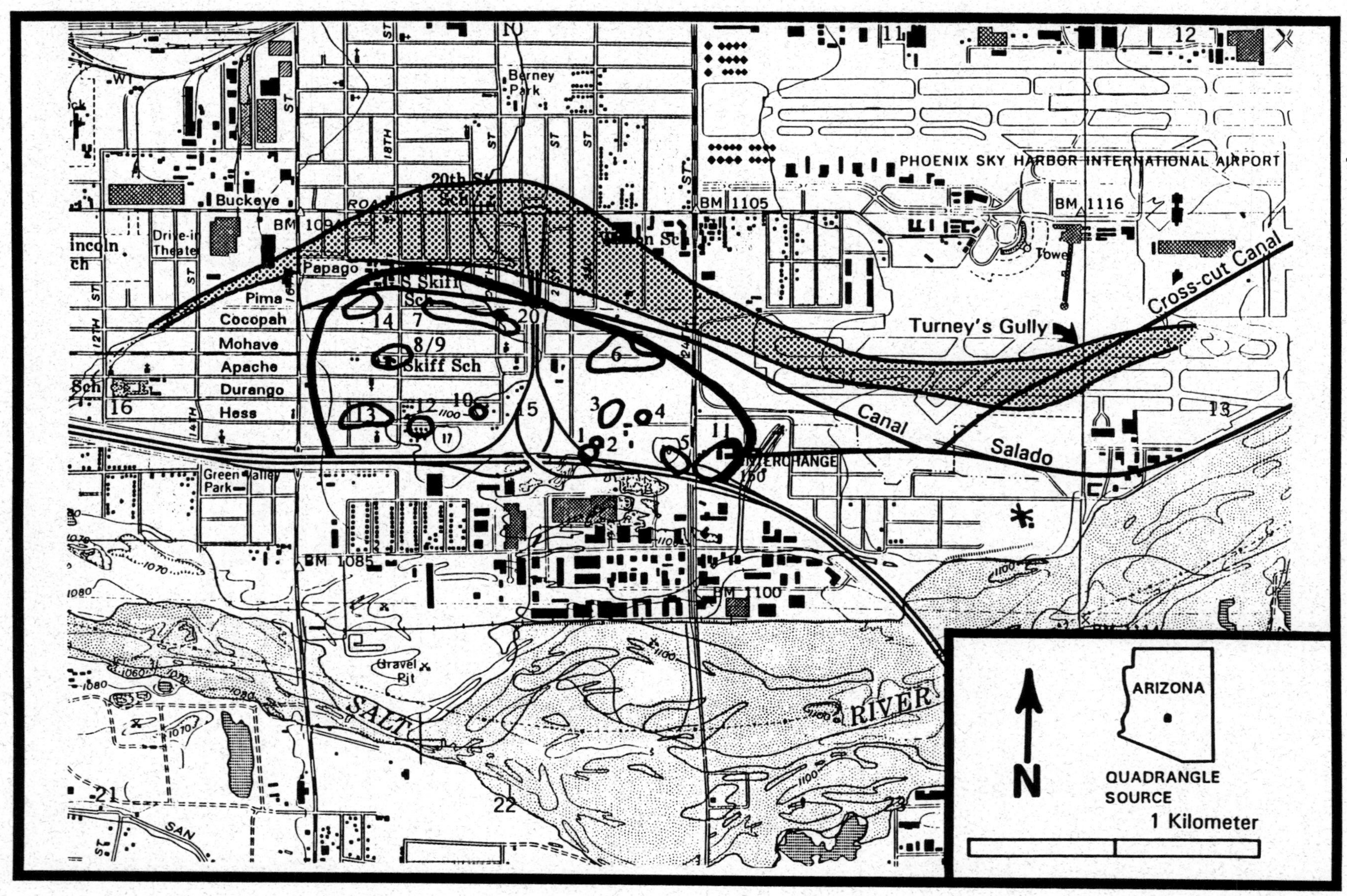

Figure 3.1. Plan map showing Pueblo Salado (AZ T:12:47, ASM) in relation to Canal Salado and the Salt River channels. Base map is Phoenix, Arizona, 7.5 minute USGS topographic map, photorevised 1982.

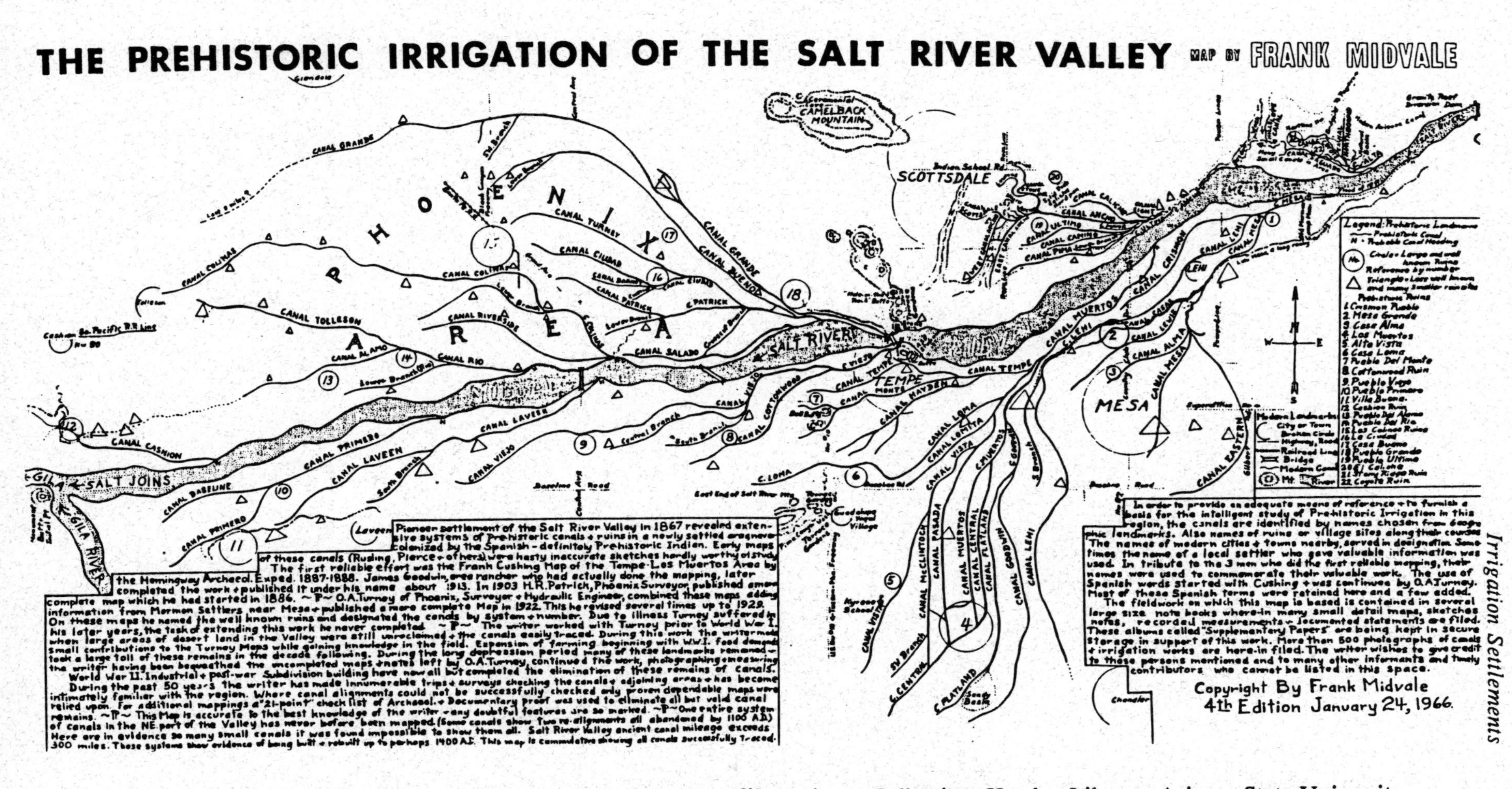

Figure 3.2. Midvale's 1966 map of Salt River Valley canals (copy on file, Arizona Collection, Hayden Library, Arizona State University, Tempe).

been susceptible to erosion each time water flowed in the channel, sufficient reason to question the early canal maps, the existence of the crosscut canal, and the inclusion of Pueblo Salado in Canal System 2 and its irrigation communities (Howard 1987:211; Howard and Wilcox 1988:911; Howard, personal communication 1991).

Pueblo Salado was constructed on an alluvial island within the Salt River floodplain. The earliest occupation of the site, consisting of a series of farmstead and hamlet settlements scattered along the Canal Salado system, appeared to have occurred during the Soho phase of the Classic period. Later, during the Civano phase, population aggregation resulted in the abandonment of the small, scattered irrigation settlements and the construction of a compound between the alignments of Canal Salado and its north branch on an elevated topographic feature. Later the inhabitants enlarged the compound and built at least five courtyard groups within the expanded, double-walled portion. Occupation of the compound continued into the Polvorón phase of the post-Classic period, although at population levels dramatically reduced compared to those of the previous occupation.

CANAL SYSTEM ORGANIZATION

During the Classic period within the Phoenix Basin, the irrigation communities identified by Gregory (1991) consisted of various configurations, including both single- and multiple-village systems. As noted above, these communities encompassed settlements such as field houses, farmsteads, hamlets, and villages, in addition to dry farming loci and other features separate from particular settlements.

An easily recognizable pattern for irrigation communities of the Classic period is the relatively even spacing of 5 km between platform mound villages within systems that contained more than one such village (Crown 1987; Gregory and Nials 1985; Midvale 1968). Gregory (1991) also noted that in single village systems, the platform mound village is 5 km from the heads of the canals, the largest villages are near the end of the canal system, and settlements are usually identified on the upslope side of the canals. Occupation zones form another pattern within irrigation communities (Cable and Mitchell 1991; Mitchell 1989b; Turney 1929; Wilcox 1984), exhibited in the tendency toward a north-south alignment of major sites across east-west-oriented major canals. The irrigation communities would have been the sociopolitical unit responsible for the operation of the canals (Gregory and Nials 1985; Gregory 1991), while the occupation zones would have operated at a level between an irrigation community and the canal system.

Researchers have proposed various systems for the integration and operation of canals within the Phoenix Basin. Gregory and Nials (1985) suggested a three-level hierarchy: irrigation communities at the lowest level; canal systems that integrate a number of irrigation communities, such as Turney's (1929) Canal System 2, at the next level; and a regional system comprising a number of canal systems at the highest level of complexity. Others (Rice 1987; Teague 1984, 1985; Upham and Rice 1980; Wilcox and Shenk 1977; Wilcox and Sternberg 1983) have suggested that the Hohokam had some version of a chiefdom system. Mitchell (1989b), however, argues against a chiefdom system and suggests that, while some system of authority seems to have existed, no single person appears to have been in control. A lineage system may have been the mechanism used to control and manage the canal systems, with lineage heads, whether religious or secular, residing at major sites within occupation zones (Mitchell 1989b; see also Bostwick 1992). Such a system may have established mutual cooperation through affinal connections, religious constructs, and resource interdependence, with village heads linked through direct affinal ties or marriage networks.

Given the lack of evidence for a chiefdom system, a lineage system seems the most likely mechanism for integrating the settlements within an irrigation community and occupation zone. For a multiple-village canal system to have functioned, the Hohokam would have had to establish some form of water regulation to ensure parity among water users. Archaeologists have not determined whether prehistoric irrigators had relied on a system to regulate water usage between canal systems, as historic Euroamerican farmers did (Chapter 1); some such mechanism may have operated at least during periods of reduced streamflow or other periods of system stress.

PUEBLO SALADO: EVIDENCE FOR AUTONOMY

Three lines of evidence support the inference that Pueblo Salado functioned as an autonomous settlement within the geographic boundaries of Canal System 2. These are data produced through temper sourcing analysis of the site's ceramic assemblage, the relationship of Pueblo Salado to the physical features of the Salt River floodplain, and the isolation of Canal Salado and its north branch in relation to the rest of Canal System 2.

Ceramic Evidence

Recent advances in analytical technology have enabled specialists to closely scrutinize Hohokam ceramics from the Phoenix Basin with respect to the types of temper material used in ceramic production. Analysts have identified nine mutually distinguishable rock types used as sources for ceramic temper in the Phoenix Basin (Abbott and Schaller 1991; Abbott, Schaller, and Birnie 1991; Schaller 1993). The rock types possess unique qualities, and investigators have linked their geographical distributions to specific locations within the Phoenix Basin. By identifying the source of the temper agent(s) within ceramics, analysts can demonstrate general trends in acquisition, which may be related to methods of exchange and the sociopolitical organization of the Hohokam within Canal System 2 and the Phoenix Basin more generally. Petrographic (Schaller 1991) and microscopic analysis of the temper agents (derived from sherd data) of ceramics from Pueblo Salado (Chapter 11) revealed that through time between 14% and 22% of the plainware ceramics were tempered with South Mountain granodiorite, alone or in combination with Estrella gneiss (Table 11.7), sources of which lie south of the Salt River. Given the location of Pueblo Salado on the north side of the river and within the geographic boundaries of Canal System 2, the frequency of South Mountain granodiorite was surprisingly high when compared with that in ceramic assemblages from other sites north of the river. If Pueblo Salado participated in the social, political, and economic structure of Canal System 2, that participation is not well supported by the ceramic assemblage. In fact, the ceramic assemblage from Pueblo Salado more closely resembles the small sample recovered from Pueblo Viejo, a site located on the south side of the river. Ceramic temper analysis indicated that the overwhelming majority of the sherds from this site contained South Mountain granodiorite (Abbott, personal communication 1992).

Diachronic review of the data on ceramic temper in plainware indicated that temper materials from sources south of the Salt River increased between the Soho and Civano phases (from 15.3% to 23.8%, respectively) but decreased during the Polvorón phase (to 18.9%) (Chapter 11). Temper materials from sources north of the Salt River showed a steady decline through time from 21.8% during the Soho phase to 11.8% during the Civano phase to 4.9% during the Polvorón phase (Chapter 11). This decline suggests a shift away from northern interaction spheres, possibly culminating in a strong association with groups living south of the Salt River, especially in the South Mountain area. Production of ceramics that used temper agents from unknown sources, at least some of which presumably were locally available at Pueblo Salado, increased from the Soho phase (62.9%) to the Civano phase (64.5%) and became higher still during the Polvorón phase (76.1%). Because investigators have not yet determined the source(s) of these materials, analysts can only speculate that residents of Pueblo Salado depended on outside sources such as villages near South Mountain for the acquisition of pottery. Whether the Pueblo Saladoans produced their own pottery is not clear, but they almost certainly acquired only minimal numbers of vessels from northern production areas. A similar review of the redware from Pueblo Salado indicated an even contribution of temper materials from south of the river during the Soho (48.7%) and Civano (47.3%) phases and an increase during the Polvorón phase (56.0%). Redware from northern sources was rare during the Soho (1.3%), Civano (0.3%), and Polvorón phases (1.0%). However, the number of redware ceramics from the indeterminate (perhaps local) source areas remained constant during the Soho (50.0%) and Civano (52.4%) phases but decreased during the Polvorón phase (43.0%).

The first occupation of Pueblo Salado occurred during the Soho phase. The percentages of sherds that can be traced to northern and southern production areas suggest that the initial settlers of Pueblo Salado were a group or groups that had previously been part of a settlement system south of the Salt River. Although the Pueblo Salado site can be included within the geographic boundaries of Canal System 2, the data from the ceramic assemblage discount a strong association with other settlements in that system and supports the concept of autonomy. Residents of Pueblo Salado continued to participate in social, political, and economic spheres south of the river, while the lower frequencies of northern temper material suggest limited interaction with groups north of the river. Through time, the level of interaction with groups south of the river appears to have increased, especially during the Polvorón phase. Pueblo Salado thus appears to have participated in the Canal System 2 exchange system while maintaining close interaction with groups south of the Salt River. Continued temper analysis may yield more specific information regarding exchange and association in the Phoenix Basin. For now, the data suggest that Pueblo Salado may have been closely aligned with sites such as Pueblo Viejo, Las Canopas/Cottonwood Ruin, Pueblo del Monte, Los Hornos/Casa de Loma, Alta Vista, or Los Muertos, all located within Canal System 1 (Figures 1.2 and 3.2).

The Geomorphic Setting of Pueblo Salado

Physical evidence regarding the sociopolitical and economic structure of settlement systems is often restricted to spatial patterns, exchange networks, and geophysical constraints. The following examination of the question of the autonomy of Pueblo Salado within the nearby canal systems considers the location of the site on the floodplain, its limited area for expansion due to the physical constraints of the floodplain, and its spatial relationship to Canal Systems 1 and 2.

The location of Pueblo Salado on the floodplain of the Salt River (Figure 3.1) may seem aberrant because most known site locations along the Salt and Gila rivers lie well above the geologic floodplain on the lower reaches of the bajada slope. The Hohokam appear to have preferred the lower bajada for their permanent habitations to reduce the threat of flooding to the inhabitants, their villages, and their food supplies, canals, and fields. Canal intakes and sections of the canal systems located on the floodplain would have been much more susceptible to the erosional and depositional effects of floods. Often no physical distinction is apparent between the floodplain and the bajada in the Phoenix Basin, but they can be identified geomorphically. Archaeologists know of few permanently occupied sites established in the Phoenix Basin below the boundary between the floodplain and the bajada, and those that have been identified appear to be small and exhibit limited complexity when compared to those on the lower bajada (e.g., Bostwick and Rice 1987; Zyniecki, Motsinger, and Greenwald 1990).

By contrast, the Hohokam constructed Pueblo Salado adjacent to the main channel of the Salt River immediately north of an area that historically contained three or four eroded channels (Figure 3.1). Investigators considered the main channel of the river to be the extreme southern boundary of the site, as no field investigations have been conducted in that area to provide other information. A large, apparently continuous secondary channel of the Salt River ran north of the main habitation area, forming the northern boundary of the site. Researchers have traced this channel, referred to as Turney's Gully because Turney (1924) shows its western extent on his map of prehistoric canals, through the project area and into the Phoenix Sky Harbor International Airport through the use of aerial photographs (Soil Conservation Service 1934) and soil maps (Adams 1974). Recent investigations in the airport (Greenwald and Zyniecki 1993) confirmed the presence of this channel immediately south of the south runway, which corresponds with the soil units believed to represent Turney's Gully. In effect, the configuration of the main channel of the Salt River and Turney's Gully produced an alluvial island upon which Pueblo Salado and the Canal Salado system were built. Due to the generally isolated nature of this alluvial island, the inhabitants of Pueblo Salado probably were more independent in their use and management of canal water than were residents of village sites located within the larger, more complex canal systems that seem to have required some form of water regulation or allocation to achieve parity among participating settlements. In turn, the Pueblo Saladoans may have maintained greater autonomy in their daily activities and interaction with larger systems. They may thus have functioned as members of a microsystem or perhaps as members of a subsystem of the larger, more integrated canal systems. The microsystem concept is preferred here because it denotes a small independent system.

Because the Canal Salado system would have been limited in its areal extent due to the physical limitations of the alluvial island that it occupied, the number of residents that could be supported by the Canal Salado system would have been similarly restricted (see Chapter 6 for additional discussion). If the population of Pueblo Salado reached the maximum carrying capacity of this small system, reconstructions of population size, the extent of arable land, and the irrigation capacity of the canal system may provide important insights for estimating the carrying capacity of the larger canal systems within the Phoenix Basin. Factors that cannot be addressed adequately are related to the susceptibility of the Canal Salado system to flooding and the long-term effects of catastrophic flooding on other canal systems (Nials, Gregory, and Graybill 1989).

Other Autonomous Canal Systems

Early canal maps (Midvale 1966, n.d.; Turney 1929) show several canals that appear to be physically isolated from others while having their own intakes on the Salt River. Among these are Canal Riverside, the Canal Rio system, the Canal Saguaro/Alta system, Canal Laveen, Canal Primero, Canal Cashion (all on Figure 3.2), and Canal Liberty (Midvale n.d.). Most of these canals (systems) could be traced only a short distance, usually because of topographic restrictions, as with Canal Salado. In particular, Canal Cashion was restricted by the Agua Fria River channel, Canal Primero by the Gila River channel, Canal Laveen by the upper bajada of South Mountain, and the Canal Saguaro/Alta system by Desert Wash. Topographic features that may have limited the lengths of the other canals are not apparent.

Although canal length, or the physical limitations of the systems, and independent intakes may be useful aspects in identifying independent systems, they should not be used as the sole criteria when classifying a particular canal as an autonomous system. Ceramic temper analysis has the potential to demonstrate acquisition strategies and production areas of site assemblages. Trends within ceramic assemblages may be used to reconstruct interaction spheres and demonstrate the sociopolitical structure of the Hohokam.

CONCLUSIONS

Review of the association of the project sites with the larger, integrated canal system structure of the Salt River Valley indicates that the pre-Classic period field house and farmstead occupations at Dutch Canal Ruin were extensions of larger settlements within Canal System 2. By contrast, the ceramic evidence from Pueblo Salado and its geomorphic setting provide sufficient reason to consider this site an autonomous settlement. Absolute autonomy, however, cannot be demonstrated, although Hohokam scholars can argue that Pueblo Salado exhibited a greater level of autonomy than most other investigated sites in Canal System 2 based on its physical setting, ceramic assemblage, and independent canal system. Pueblo Salado was physically isolated from Canal System 2 by the braided stream channels of the Salt River, and it contained a much higher proportion of ceramics from southern than from northern production areas. It also appears to have been on a separate canal system, containing its own headgate or intake.

Investigators have suggested that the Classic and post-Classic period occupations of Pueblo Salado and its associated canal system represent either a separate microsystem or a subsystem of a larger canal system. Two basic assumptions may pertain to Hohokam sociopolitical structure within canal systems: (1) a hierarchical system appears to have existed and (2) water regulation and an allocation system appear to have regulated the equitable distribution of water. With an increase in the size and complexity of the canal systems, the need for management and regulation of water would have increased. Archaeologists previously associated the Canal Salado system with the La Ciudad subdistrict of Canal System 2 (Cable and Mitchell 1991; Mitchell 1989b; Wilcox 1984). However, because Canal Salado may not have been directly tied to Canal System 2 and therefore may not have participated directly with that system in the regulation and management of irrigation water, Pueblo Salado may have retained a greater amount of autonomy than villages that were more dependent on system-wide water management and regulation.

Recent advances in ceramic temper studies (Abbott 1992; Abbott and Schaller 1991; Schaller 1993), in particular the ability to identify specific geographical temper sources in the Salt River Valley, have now made it possible for analysts to demonstrate the interaction spheres in which sites and canal systems participated. Temper studies of plainware and redware have shown considerable variability in the temper constituents, and investigators have linked this variability to specific geological and geographical sources within the Salt River Valley. The apparent lack of homogeneity within the Hohokam Plain Ware and Red Ware types functions as an important component in defining production areas, interaction spheres, and intraregional exchange. Ceramic temper analyses also provide reason to believe that Pueblo Salado was far more involved with the western extent of Canal System 1 (the Las Canopas/Pueblo Viejo subdistrict) than with Canal System 2, even though Pueblo Salado lay across the Salt River from Canal System 1. Coupled with geomorphic and topographic data, the ceramic assemblage from sites and canal systems becomes an important component in defining the level of independence or association of a site or canal system. The information value of the ceramic temper studies in regard to the level of autonomy experienced by the canal systems identified above will need to wait for future research opportunities. The value of sourcing ceramic temper, however, has been demonstrated by the investigations at Phoenix Sky Harbor Center, which show that physical association does not necessarily mean physical interaction within canal systems. Although the concept of the canal system is useful, the technology now exists to further test that model and re-evaluate previous concepts, as investigators did with Pueblo Salado. In summary, given the geographic location, the physical restrictions of the surrounding topography, and the apparent interaction with groups from Canal System 1, the Canal Salado system may well have functioned as a microsystem that interacted as an independent entity with adjacent and larger canal systems.

CHAPTER 4

THE FIELD HOUSE AND THE SUBFAMILY CORPORATE GROUP

David H. Greenwald

Archaeologists in the Hohokam region as well as throughout the Southwest and Mesoamerica have identified field houses as seasonal or temporary structures related to subsistence pursuits. These small structures are a common feature at sites in the Hohokam area. Although researchers have defined a variety of functions for these structures (Crown 1985), this chapter focuses specifically on structures related to agricultural activities. Field houses functioned as shelters (Mindeleff 1891; Ward 1978; Wilcox 1978), storage structures (Wilcox 1978; Woodbury 1961a), expressions of field ownership (Adams 1978:105), or locations where families retreated from village pressures (Ellis 1978:59). This chapter examines the social implications of field houses and the activities conducted at these loci by groups comprising subfamily units, periods of occupation, and use by economic units or corporate groups. Because of the ephemeral nature of field house sites, the discussion uses historic and cultural analogies.

THE FUNCTION OF FIELD HOUSES

The field house concept was first employed in Arizona by Woodbury (1961a:14–15) in the Point of Pines area in the east-central part of the state. Woodbury viewed the field house as a component of the agricultural system, used as a temporary habitation during the growing season and for storage during other times of the year. The field houses that Woodbury identified at Point of Pines consisted of one- or two-room masonry structures in or near agricultural fields. Wilcox (1978), in reviewing Woodbury's original description, defined field houses as "architectural facilities built on sites adjacent to or on fields or gardens; they are inhabited on a temporary basis during the growing season. While they may be used for storage, this is not a necessary criterion in the definition" (Wilcox 1978:26). Archaeologists have generally accepted and applied this definition in the Hohokam area and have assumed that subfamily groups rather than entire families used field houses (Crown 1983:5). In this study, the extended family rather than the nuclear family is considered the family unit.

Given the various lines of data gathered in the Phoenix Sky Harbor Center as a result of this project and investigations in the portion of the Squaw Peak Parkway running through the Phoenix Sky Harbor Center (Greenwald and Ciolek-Torrello 1988a), project investigators have interpreted as field houses the small, pre-Classic period structures at Dutch Canal Ruin that are spatially associated with the canals (Figure 4.1; see also Volume 1:Chapter 1). Farmsteads, as defined in Volume 1, Chapter 1, similarly are seasonally occupied sites, composed of multiple field houses. Although still related to agriculture (Woodbury 1961a), they exhibit greater diversity in associated subsistence activities. As with field houses, farmsteads are functional extensions of larger, permanently occupied sites (Gregory 1991:163). Investigations at Dutch Canal Ruin yielded 14 field houses, with another 21 features recorded in profile that excavators considered possible field houses but did not confirm through further examination. Investigations in the Squaw Peak Parkway identified 5 field houses, with 4 others remaining unconfirmed. The number and variety of features within excavated project field houses were limited (Table 4.1). Even postholes occurred infrequently within these structures. The informal design of the structures, the relative lack of features, and the limited number of artifacts (65 items or less from any structure) supported the premise of restricted use and a reduced number of associated activities. Botanical remains indicated that corn and plants

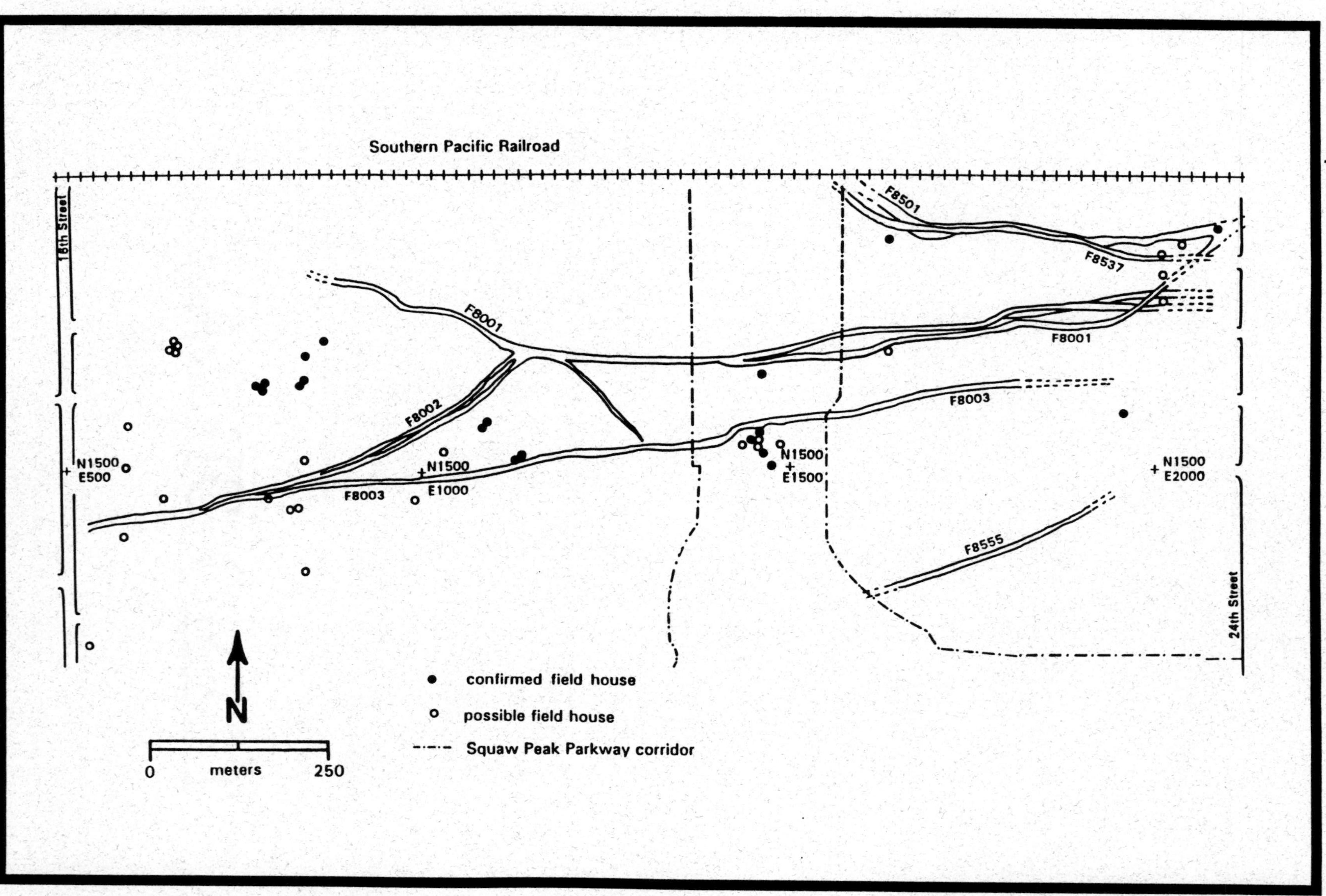

Figure 4.1. Field house sites at Dutch Canal Ruin.

Table 4.1. Attributes of Fully Excavated Field Houses at Dutch Canal Ruin

Attribute	**Feature Number**											
	1-1	1-2	2-2	3-1	3-2	3-6	3-9	3-14	3-16	3-19	5-1	7-1
Dimensions (m)												
Length	(3.80)	4.48	3.10	2.76	4.20	2.08	2.95	?	(3.05)	3.30	1.97	2.44
Width	(2.90)	4.28	2.85	2.32	4.00	1.96	2.77	?	(2.03)	(2.40)	1.78	2.38
Feature Types												
Entry	—	+	+	+	+	—	+	—	+	+	—	?
Hearth/Fire Pit	—	—	1	1	1	—	1	2	—	—	1	—
Posthole	—	9	—	—	1	—	—	1	—	—	—	—
Oxidized/Burned Area	—	2	2	1	1	—	—	—	—	1	—	1
Material Culture Remains[1]												
Ceramics	—	2	5	4	8	—	5	—	—	1	4	5
Flaked Stone	—	5	50	29	3	—	32	—	4	5	1	47
Ground Stone	2	3	2	2	—	3	13	—	1	—	2	1
Other	—	5c	—	—	9b, 4s	3fcr	2b, 2fcr	—	22r	—	—	12b
Botanical Remains[2]												
Arboreal												
Alnus			+									
Juniperus			+		+						+	
Pinus	+	+	+		+					+	+	
Prosopis		+	+		+						+	
Quercus			+		+					+		
Non-Arboreal												
Agave											+	
Anacardiaceae										+		
Cheno-ams	+	+	+	+	+				+	+	+	
Compositae												
Artemisia		+	+	+	+					+	+	
Low-spine	+	+	+	+	+				+	+	+	
High-spine	+	+	+	+	+				+	+	+	
Tubuliflorae		+								+		
Liguliflorae			+		+				+			
Cruciferae			+		+							
Cylindropuntia									+			
Cyperaceae			+									
Ephedra torreyana-type		+	+									
Eriogonum		+			+					+		
Erodium		+										
Euphorbia	+	+	+	+	+				+	+	+	
Gramineae	+	+	+		+					+	+	
Larrea	+		+	+	+					+	+	
Leptochloa-type				+								
Liliaceae			+									
Nyctaginaceae	+	+	+	+	+				+			
Onagraceae	+		+								+	
Plantago		+								+		
Solanaceae		+										
Sphaeralcea	+	+	+		+				+			
Tribulus					+							
Zea			+	+	+		+			+	+	
Spores												
Selaginella densa			+									
Trilete	+								+			

( ) = estimated dimensions; c = cobble; fcr = fire-cracked rock; b = faunal bone; s = shell; r = rock
[1]floor contact and floor fill proveniences; [2]data included only for features yielding processed and analyzed samples

found in association with agricultural activities were common (Table 4.1), while remains from cacti, agave, and most legumes were either absent or low in frequency. Mesquite (*Prosopis*) was present at much higher levels in the project field houses than in those investigated in the Squaw Peak Parkway (Ruppé, Cummings, and Greenwald 1988). This difference may be a result of changing procurement strategies between the late Pioneer period, to which previous investigators assigned most features in the Squaw Peak Parkway at Dutch Canal Ruin, and the Colonial period, to which project analysts attributed most of the field houses elsewhere at Dutch Canal Ruin.

The temporal and spatial distribution of field house locales at Dutch Canal Ruin correspond with the canals that transect the Phoenix Sky Harbor Center (Figure 4.1). Given these associations, investigators concluded that field house use was directly associated with the use of the canals. As features related to agriculture, project field houses would have been used during periods of planting, cultivating, and harvesting (Cable and Doyel 1985b; Greenwald and Ciolek-Torrello 1988a) and during periods of canal maintenance and water regulation. The botanical remains recovered through excavation indicated that corn agriculture was a primary focus of subsistence activities, although a variety of wild plant resources were also present (Volume 2:Chapters 15 and 16). Faunal resources, perhaps the result of garden hunting—an opportunistic strategy—were represented by small rodents and mammals (Volume 2:Chapter 17; Szuter 1991). In general, hunting and wild plant procurement were secondary to activities relating to farming and irrigation. The Hohokam also used field houses from the site of Pueblo Patricio, nearby to the northwest, in conjunction with mesquite exploitation (Cable and Doyel 1985c), possibly because of the proximity of that site to a mesquite bosque (Barney 1933; see also Volume 1:Figure 4.3 [stippled area at west-central edge of map]). The functions of the field houses in the Phoenix Sky Harbor Center do not encompass the full range of activities represented by field houses elsewhere in the Phoenix Basin (Cable and Doyel 1984b, 1985c), which seems to reflect function-specific activities and the availability of nonagricultural resources on the geologic floodplain at Dutch Canal Ruin. Moreover, field houses at various sites in the Phoenix Basin exhibit significant variation in morphology that may reflect differences in function (Figure 4.2).

Specific uses of field houses at Dutch Canal Ruin are difficult to reconstruct, and investigators have had to speculate based on the presence or absence of evidence to place these structures within functional parameters. One reasonable assumption is use as shelter from the sun or for sleeping quarters. Ethnographers have documented such uses for historic Native American groups (Ellis 1978; Mindeleff 1891), although the morphology of the structures observed varies considerably. At project field houses interior features were uncommon (Table 4.1), but half of the investigated structures contained hearths or fire pits, suggesting that occupants conducted activities relating to heating or food preparation within those structures.

The Hohokam also may have used field houses as storage structures, a function observed historically for small structures associated with fields; these structures protected the harvested crops from animals and raiding groups (Hammond and Rey 1928:325; Mindeleff 1900). Additionally, field houses may have provided base camps from which the Hohokam gathered wild resources. Although analysts did recover evidence of this type of activity in the botanical record, wild resources appear to been considerably less important than agriculture at Dutch Canal Ruin (Volume 2:Chapters 15, 16, and 17; see also Table 4.1).

The spatial associations of the field houses with the canals and the similar temporal affiliations of these features imply that Hohokam farmers used the structures in conjunction with the canals. Canals that exhibited the need for increased maintenance due to rapid infilling or erosion may have required greater attention during periods of use. A "ditch tender" or *zanjero*—who regulated discharge levels and attended to bank erosion or siltation problems—may have used the nearby

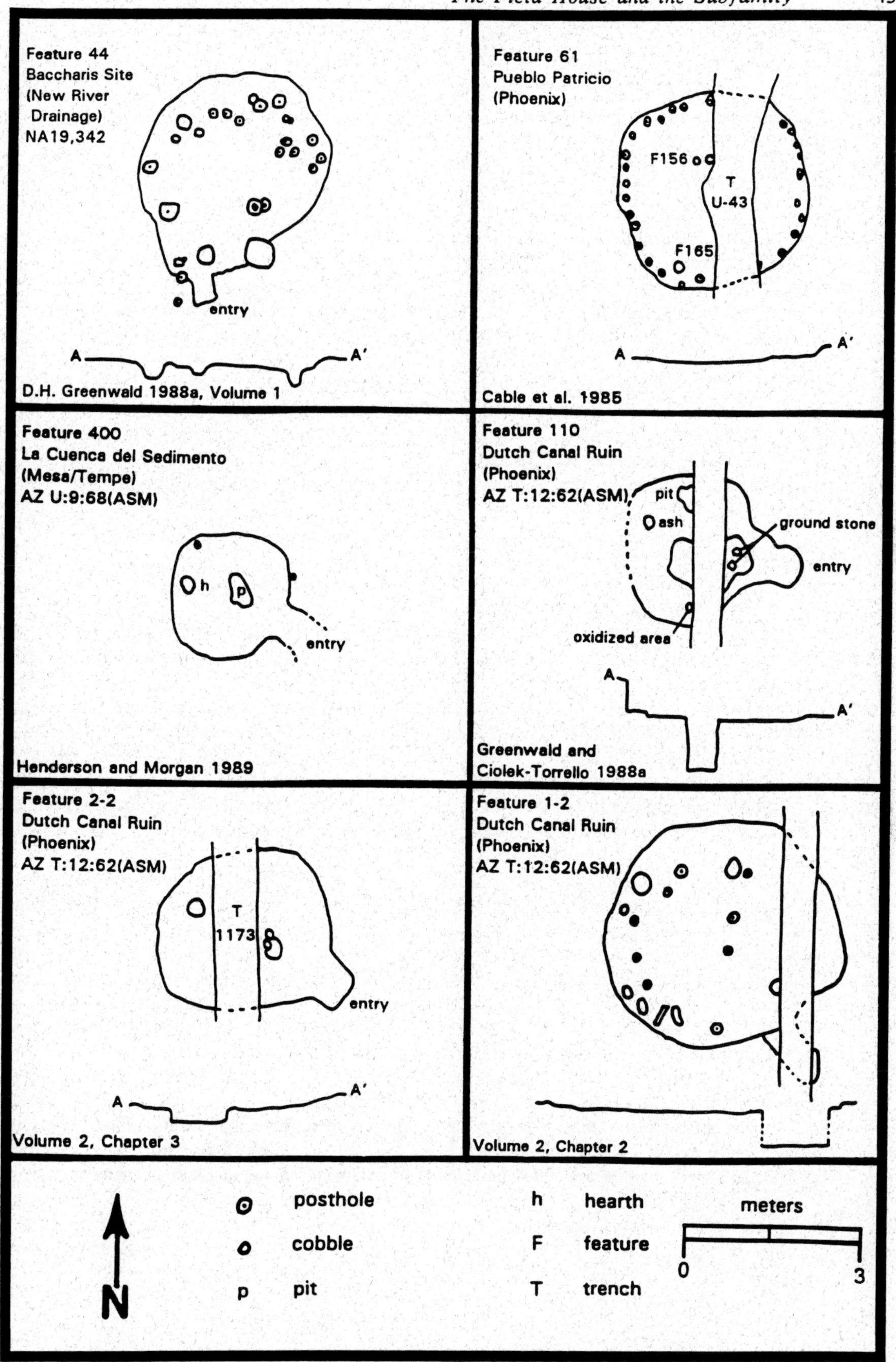

Figure 4.2. Examples of field houses from various sites in the Phoenix Basin.

structures periodically for temporary quarters. These individuals, who may have been the farmers themselves, also may have regulated the amount of water dispersed to fields through distribution or lateral canals, making use of the temporary shelters when performing this task. Evidence for these activities, however, cannot be identified specifically in the archaeological record.

Another possible function of the field house is to denote ownership of a field area (Adams 1978:105). Farmers probably built field houses not within any single field but along field margins or between fields (Fish and Fish 1978:52; Woodbury and Neely 1972:121), as observed among present-day subsistence farmers. Information from interviews with historic and modern groups indicates that the field house structure can denote or assert ownership (Fish and Fish 1978:52–53) and thus prevent ownership disputes from arising during periods when fields are not in use.

In summary, field houses in the Phoenix Sky Harbor Center may have functioned as shelters during periods when farmers tended fields and maintained and regulated canals, as temporary storage facilities for harvested goods, or to denote ownership or access rights. The small structures used in conjunction with agricultural activities at Dutch Canal Ruin exhibited limited use by residents, producing few associated artifacts, features, and activity areas and a limited variety of botanical remains.

SOCIAL IMPLICATIONS

Determining the social composition of prehistoric groups is difficult. Some researchers have used ethnographic analogy to develop formulae for estimating prehistoric population size based on available living area, generally defined as the interior space of structures (Cook 1972; Cook and Heizer 1965; Hill 1970; Naroll 1962). The applicability of such an approach to the Hohokam field house structure is questionable because of the specific function of these features and their apparent intermittent use. Furthermore, the climatic conditions of the Phoenix Basin during the growing season would have allowed many of the daily habitation activities, such as food preparation and sleeping, to take place outside the structure. Attempts to define extramural activities produced few remains associated with such activities, indicating the ephemeral nature of the field house locales at Dutch Canal Ruin.

Using ethnographic data, Moore (1978:10) identified the residential composition of Puebloan field house sites as varying among "the individual farmer, one or more children or elderly men, and particularly during the harvest, whole families and selected friends." Ellis (1978:63) similarly identified the range of field house occupants from individuals (bachelors) to entire families, stating that field houses "were sparsely equipped with household goods" and were "far from spacious." In both cases, only a few people would have used a given field house. In the Hohokam area, a number of wild resources that are economically important to desert farmers such as mesquite beans and pods, saguaro fruits, and seed grains mature during the farming season. These resources may have been distant from field house sites, requiring the economic unit to split labor efforts among its members, so that the group could exploit both wild plants and agricultural crops. In general, ethnographic data indicate that the harvest season required larger numbers of individuals, sometimes including friends (Moore 1978:10). The size of the social unit that used field houses, then, appears to have been related to the labor requirements associated with farming practices and with maintenance and regulation of the canals. Ethnographic analogies indicate that the greatest labor demands came at harvest time; therefore, field house sites probably experienced the greatest use during that period.

THE SUBFAMILY CORPORATE GROUP: AN ECONOMIC UNIT

The subfamily unit, as the term implies, consists of family members organized to perform specific or related tasks. As an economic unit, this group may comprise as few as two individuals. The following scenario for a single-crop agricultural cycle, based on ethnographic analogy, may describe the annual activities and the economic units represented by the field house remains at Dutch Canal Ruin. Although double-cropping is a possibility, an early spring planting would have coincided with the period of highest river discharge (Graybill 1989). Water levels at this time would have interfered with the maintenance of diversion dams and canal heads.

The planting season would have required preparation of the soil and sowing of seeds. Preplanting might have required field clearing, when farmers would have burned or otherwise removed weedy species that had invaded the fields during the off-season or fallow periods and also would have repaired field borders or irrigation features. Labor demands to accomplish this might have varied considerably, depending on the size of the fields, the amount of preparation required, and whether the farmers were bringing new fields under cultivation. The single-crop agricultural cycle may have been initiated each year following the spring runoff, when the threat of flooding was reduced and streamflow in the rivers had returned to manageable discharge levels. This period generally would have corresponded to the month of April (Nials and Gregory 1989:Figures 2.3 and 2.7) and would have been followed by a general decrease in available moisture until the monsoon season that may not have begun until July. The initial requirements of the planting season might have been labor intensive, at least until farmers had made all repairs to the canals and irrigation systems, readied fields, and planted crops. At this time, the economic unit may have consisted of several individuals or perhaps entire families.

During the cultivation season, following the planting season in the sequence of labor requirements, farmers would have tended fields by applying irrigation water and weeding and would have protected seedlings from birds, rodents, and small mammals. Labor requirements would have been greatly reduced from the planting season, although the cultivation season may have exceeded four months. Daily activities would have included regulating the amount of irrigation water applied to the field, repairing ditches and field borders, and general caretaking. Because the labor demand would have been significantly less than during the preceding phase, only one or two individuals may have resided at the field house, possibly with frequent visits from other family members. Even those who were responsible for daily activities at the fields might have returned to the main village each night, as in the case of many of the Hopi farmers, despite the distance they would have had to travel (Bradfield 1971:38; Bryan 1954:46). Because some crops could have been harvested at varying stages of maturity (including economically important wild species), frequent trips to the village may have been necessary to transport "green harvests" before they spoiled. During periods when certain species could be harvested, family members might have made day trips to the fields to participate in the harvesting activities. These patterns suggest that corporate levels would have fluctuated between periods of cultivation and harvest, but labor requirements during the cultivation season would have been generally low.

Harvests are generally labor intensive, regardless of the level of technology used by farming societies. Efforts are focused on recovering as high a yield as possible in as short a time as possible, reducing crop loss from factors such as wildlife or spoilage. At this point, the Hohokam farmers probably would have increased the number of members in the economic unit to maximize efforts, and non-family members may have been included in these tasks (Castetter and Bell 1942; Moore 1978:10). Observations among the Pima, Papago, and Hopi support the notion that the number of persons engaged in activities at field house sites during harvests may have exceeded the number in the subfamily unit. However, the evidence from the Dutch Canal Ruin field houses consistently indicated that occupation levels and associated activity levels were low, interpreted to mean limited numbers of individuals.

CONCLUSIONS

While considering the role of the field house in the Phoenix Sky Harbor Center, this chapter has examined group size associated with field house sites and how group size may have changed with the types of activities that were performed. The agricultural cycle generally dictated (in logical order) planting, cultivating, and harvesting. Associated activities would have included preplanting preparation of fields, canals, and associated features. Harvesting periods also may have varied and labor requirements may have changed accordingly.

Field houses in the Phoenix Sky Harbor Center appeared to have served functions specific to agriculture. Although investigators noted some exceptions (e.g., activities in Area 4 and possible postcanal use in Area 5), the botanical record supported the supposition that the Hohokam focused primarily on agriculture (Table 4.1) at field house sites. Faunal resources did not represent a significant contribution to the subsistence base at Dutch Canal Ruin until the Classic and post-Classic period occupations, when the Hohokam established permanent settlements. Similarly, agave was not present in the botanical record until this time.

Few field houses in the Phoenix Sky Harbor Center appeared to have the capacity to house entire family units or to have been designed to do so. As previously mentioned, field personnel identified few features within the field houses and had difficulty in recognizing extramural activity areas. Although based somewhat on negative evidence, these patterns suggest that activities associated with project field houses were limited. This interpretation is supported by the limited variety and frequency of botanical and faunal remains. Certainly various stages of the agricultural cycle would have been labor intensive. However, data recovered from the project field houses do not necessarily support investigators' supposition of periods of increased activity or increased numbers of individuals, perhaps because of the proximity of the field houses to the permanent settlement.

It may be posited that field houses within a certain distance of the permanent settlement will reflect a limited range of activities, fewer artifact and feature types, and reduced variability among exploited flora and fauna species than field houses farther away. This reasoning is based on the expectation that occupants would have spent more time (i.e., conducted more activities) at field houses when travel requirements increased due to distance (Chisholm 1962). The limited extent of use of the field house loci at Dutch Canal Ruin (Volume 2:Chapter 19) may thus indicate that the project structures were located within an easily traversed distance, enabling farmers to avoid lengthy stays at the field houses. Project field houses were nearer to the La Ciudad/Los Solares occupation zone (Figure 4.3) than to other settlements that may have made use of the project area (Figure 4.3); if, as conjectured, the extent of use of field houses is directly related to the distance that they are removed from the parent settlement, the farmers of the pre-Classic periods at Dutch Canal Ruin may have been permanent residents of La Ciudad/Los Solares.

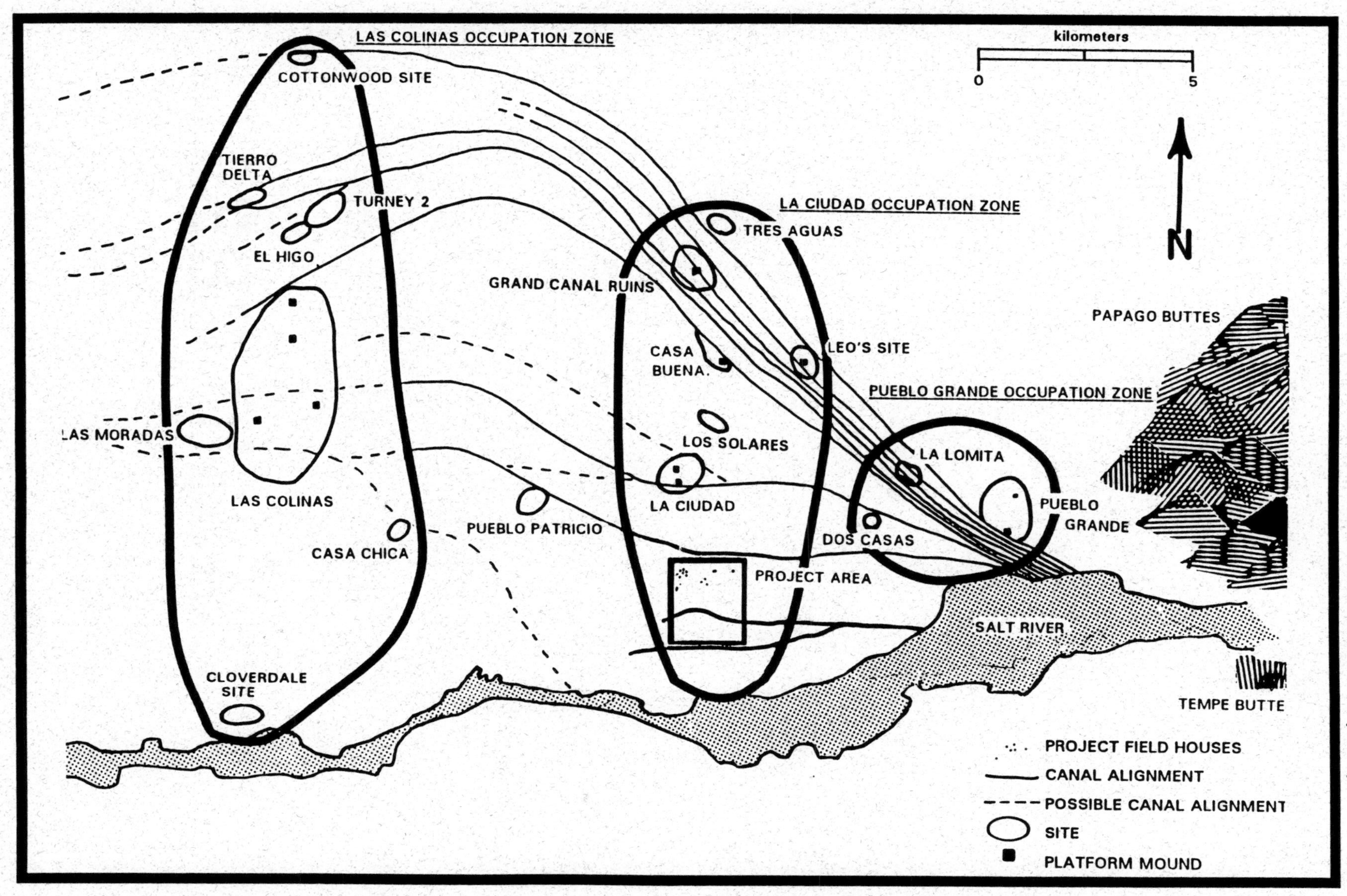

Figure 4.3. Distribution of project field houses in relation to the three occupation zones of Canal System 2.

CHAPTER 5

AGRICULTURAL FIELDS: GEOMORPHIC, CHEMICAL, BOTANICAL, AND ARCHAEOLOGICAL EVIDENCE

David H. Greenwald
Kirk C. Anderson
Scott Kwiatkowski
Linda Scott Cummings

During the testing phase in the Phoenix Sky Harbor Center, project geomorphologists Fred Nials and Kirk Anderson observed evidence of pedogenic processes in the form of oxidized soils that were unusual given the location of the project area on the geologic floodplain (Volume 1:Chapter 8). The soils in question lay immediately north of Area 8/9 at Pueblo Salado (Figure 5.1) and consisted of a red "oxidized" zone positioned between two distinct culture-bearing stratigraphic units. Analysts concluded that the oxidized soils had developed through other than the usual pedogenic processes, most likely agricultural activities.

At the close of the testing phase, all data gathered regarding these soils were from backhoe trench profiles. Investigators adopted a working hypothesis that the oxidized soils had resulted from the repeated application of irrigation water, which had subjected the soils to alternating wet and dry conditions. They inferred that features observed below the oxidized soils were field ditches or rills built by Hohokam farmers to bring water into fields. Later, during the data recovery phase, the investigating team determined that these basin-shaped features, often present opposite each other in the trench profiles, were instead adobe-lined pits probably used for the mixing of adobe for construction. Trench profiles (e.g., Trench 3015) contained several adobe-lined pits (Figure 5.2), indicating that residents of this area had extensively used the pits to produce construction materials for the compound and its interior structures.

The process of applying irrigation water to field areas can result in pedogenic conditions similar to those that form oxidized soils naturally over several thousand years (Nials, personal communication 1990). In effect, irrigation activities can significantly change soil properties through the introduction of finer sediments and minerals (Huckleberry 1992). At Pueblo Salado, field personnel identified four suspected agricultural fields during the testing phase and recognized an additional area southwest of the intersection of Pima and 18th streets during data recovery (Figure 5.1). At Dutch Canal Ruin, a linear area south of the South Main Canal contained elevated amounts of clay, another possible indication of irrigation-impacted soils. Investigators did not identify oxidized or reddish soils here, and historic activities (principally plowing, root growth, and rodent and insect turbation) had disturbed the upper extent. Analysts could not determine the temporal affiliation of these clay-rich deposits because they had been truncated by historic farming.

To date, with the exception of the work at Pueblo Salado, archaeologists have not identified the specific morphological attributes of buried agricultural fields in the Phoenix Basin. Investigations in the Tempe Outer Loop Freeway at Site AZ U:9:71(ASM) identified water-laid deposits thought to represent a buried field (Gardiner, Masse, and Halbirt 1987:43–44). Although botanical analyses revealed the presence of maize and agave remains in these sediments, their linear configuration and their restriction between two parallel canals suggested that they may have been associated with canal overbank sedimentation. The organic matter in these deposits, including the maize and agave remains, may also have been transported by the waters carrying the sediments.

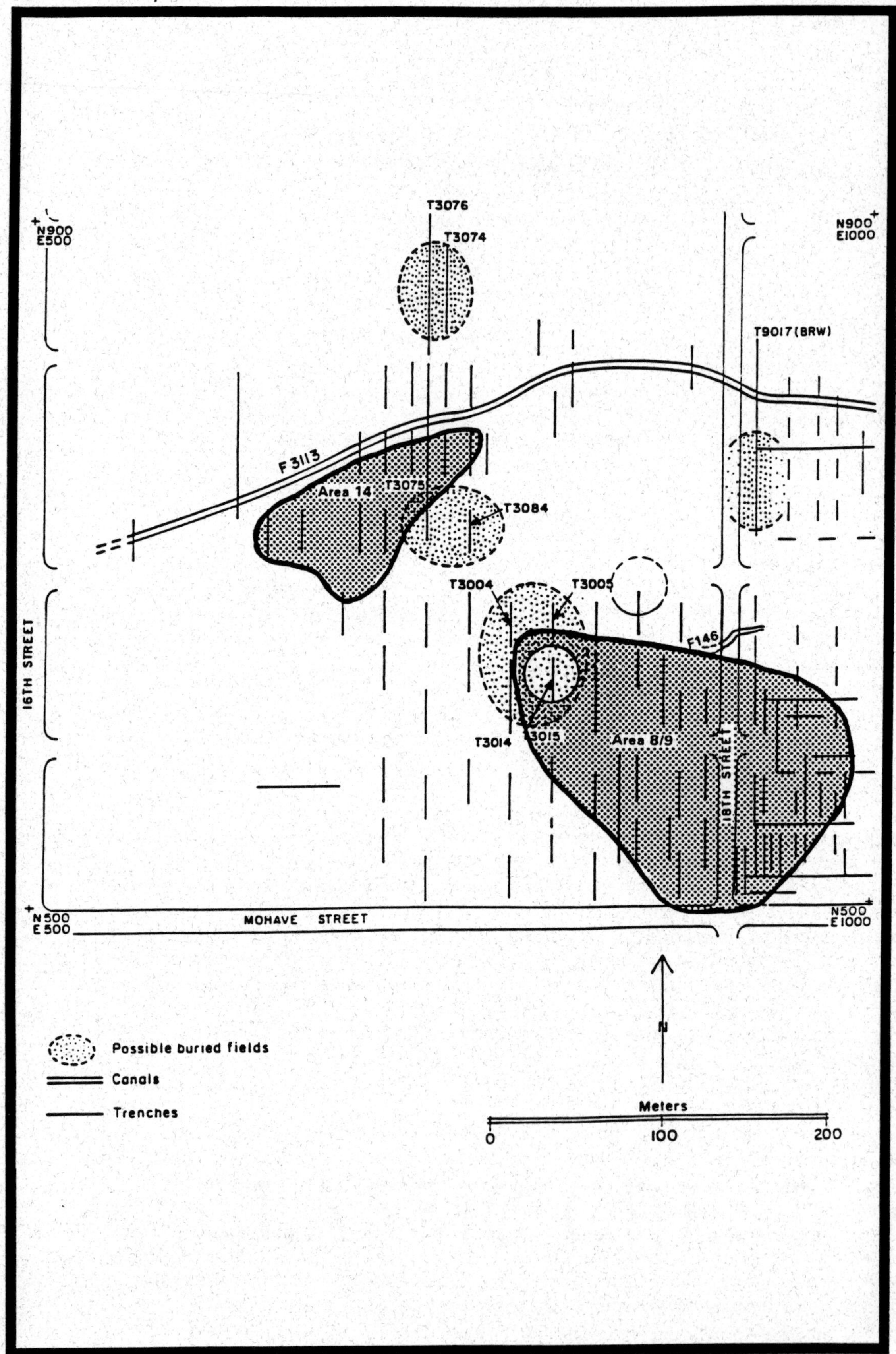

Figure 5.1. Plan map showing oxidized soils in relation to Area 8/9 and Area 14 at Pueblo Salado.

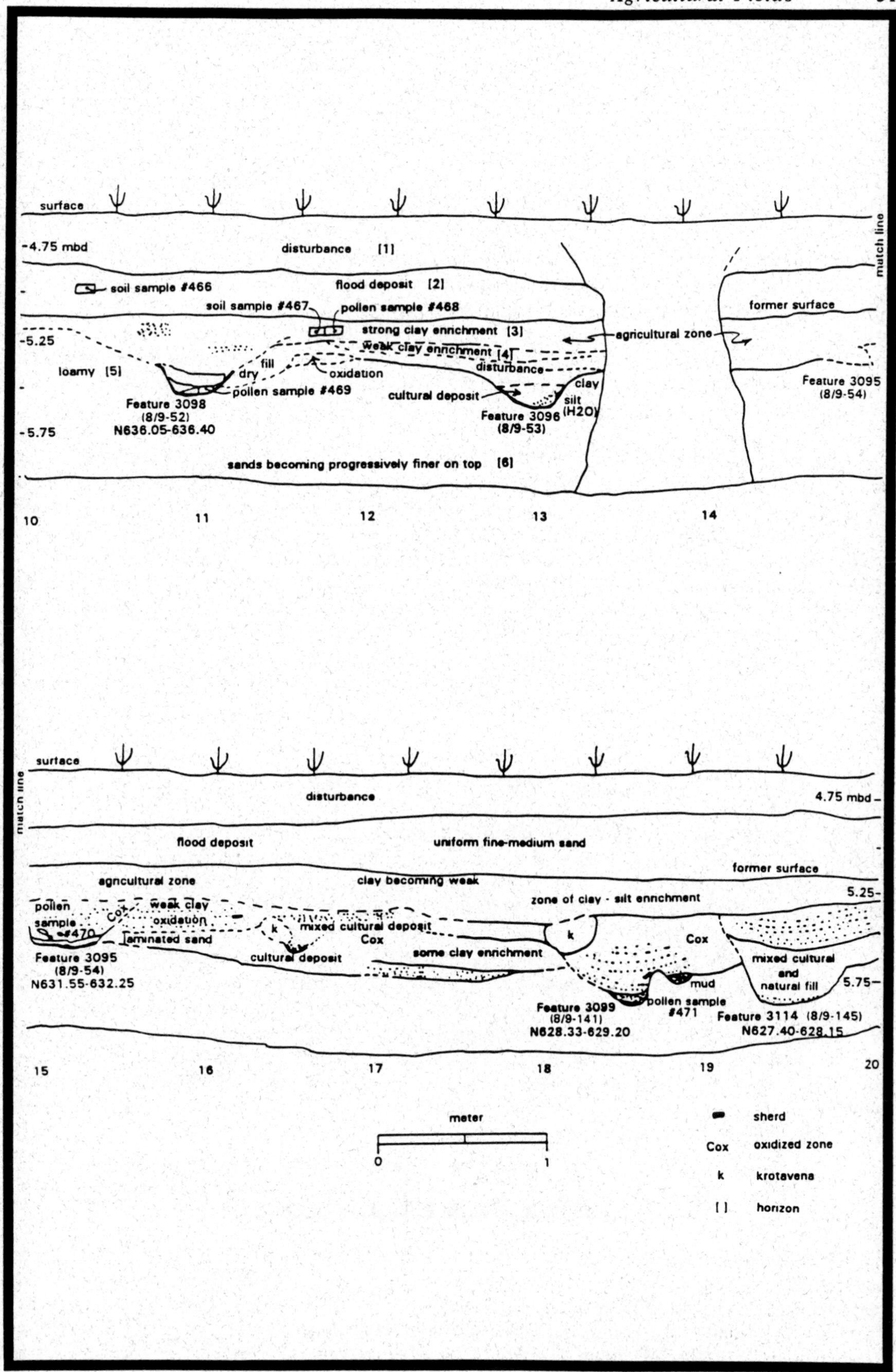

Figure 5.2. Profile of Trench 3015, showing relationship of oxidized soils, cultural features (adobe-lined pits), postabandonment flood deposits, and historic plow zone as determined during testing and indicating horizons sampled and investigated.

Huckleberry (1992), approaching the problem of prehistoric field preservation from a broader perspective, suggested that irrigation-affected soils are not more widely recognized because "sampling has been too limited to recognize large-scale soil patterns" and that "the cumulative irrigation at most places was inadequate to impart a distinct, physical signature" (Huckleberry 1992:244). In response, Fred Nials, Gary Huckleberry, Todd Bostwick, Judy Brunson, Robert Euler, and the authors visited Pueblo Salado to examine naturally and culturally altered pedogenic properties. Because pit features lay below the oxidized soils (which were unaffected by either fire or heat), the research group inferred that the alteration of the soils had occurred in a narrow geologic time span of less than a thousand years. They hypothesized that the pedogenic processes observed at Pueblo Salado could have resulted from alternating wet and dry conditions that had produced an oxidized soil zone relatively quickly, a phenomenon that was probably related to the application of irrigation water to agricultural fields. The authors thus undertook a multidisciplinary study of prehistoric field preservation.

Investigators examined five lines of evidence to test their hypotheses: (1) geomorphic conditions and processes; (2) soil chemistry and particle-size analysis; (3) macrobotanical and flotation analyses; (4) pollen and phytolith analyses; and (5) archaeological associations. Geomorphic, chemical, and particle-size analyses provided the strongest evidence in support of agricultural fields. The archaeological data provided important temporal information defining the period when the geomorphic processes could have occurred. The botanical data provided varying degrees of information that allowed evaluation of the hypotheses. In sum, the evidence generally supported the inference that the five areas of oxidized soil (Figure 5.1) were the result of irrigation of prehistoric fields. The following discussions present the results of each analysis and the archaeological data that provided the temporal parameters for this study. Appendix D contains a glossary of geomorphic terms.

GEOMORPHIC EVIDENCE

The agricultural fields at Pueblo Salado are unique in the apparent quality of their preservation compared to other such features within the Phoenix Basin. These prehistoric fields were medium-scale surface features, that is, in size more like today's gardens than extensive fields. After their burial under alluvial deposits they exhibited no obvious expression of the prehistoric cultural activity that had created them. Agricultural fields are usually difficult to identify if they lack associated cultural features such as stone lines, terracing, or stone bordering. In addition, unless such fields are preserved through burial, they will undergo natural soil-forming processes, developing soil characteristics unlike those imparted by the agricultural activity. In short, they will lose their defining physical and chemical signatures and become impossible to identify by these characteristics and difficult to identify botanically. The following section discusses geomorphic and pedogenic evidence from the buried soils and tentatively supports the notion that they were prehistoric agricultural fields. The observed alteration of the substrate, although of limited areal extent, has implications for future studies of Hohokam farming techniques.

Geomorphic Conditions and Processes

The unusual oxidized soils at Pueblo Salado apparently represented prehistoric fields buried and preserved by flood sands from the Salt River. Being located within the geomorphic floodplain of the Salt, the fields would have been subjected to periodic flooding, perhaps by the 10-year or even the 100-year flood event (Volume 1). The sediments that buried the fields could have isolated them from further soil formation, which would have preserved the physical, chemical, and botanical signatures of the fields imparted by Hohokam farming several hundred years ago. Indeed, the fields under investigation had been completely buried to depths of 50–70 cm.

The investigated fields were located near the north branch of Canal Salado and probably had received irrigation water from that channel. Pueblo Salado seems to have been a relatively isolated habitation zone, separate from much of Canal System 2 (Chapter 3; Volume 3:Chapter 1). Canal Salado thus may represent a relatively independent irrigation system, within the geographic boundaries of Canal System 2 but functioning separately from the larger system. The amount of irrigable land on the Pueblo Salado "island," which is bordered on the north, south, and west by floodplain channels of the Salt River, seems minimal in comparison to Canal System 2. Any estimate of the capacity of the irrigation system serving the lands surrounding Pueblo Salado must necessarily consider the size of this area, the restrictions on canal engineering, and the resulting restrictions on crop production. As the population of Pueblo Salado increased, the potential for expansion of the fields on the alluvial island would also have been limited. Therefore, the Pueblo Saladoans probably utilized other fields as well, perhaps expanding onto the fertile floodplain of the Salt River channel. Notwithstanding the dangers of catastrophic floods destroying crops, the floodplain would have been prime farmland, albeit risky at best. The concept of locating fields in various environmental settings is not new. The Hopi, for example, make use of a variety of locations to ensure the productivity of at least some fields (Hack 1942).

Field Methods

Backhoe trenches at Pueblo Salado excavated to the north of Area 8/9 (the compound) and adjacent to Area 14 exposed stratigraphic layers unique to the area. Examination indicated that some stratigraphic units might have represented agricultural fields. Backhoe trenches revealed that the buried horizons were located within slight depressions that extended laterally for 40–90 m. Such a depression would have trapped and retained more moisture than the surrounding sediments, thereby making it more suited to crop production in this arid setting.

SWCA researchers profiled, described, and sampled six horizons from the suspected fields (Figure 5.2) for physical and chemical analysis and also sampled four control horizons located away from the cultural activity. Profile descriptions followed procedures set forth in the Handbook on Soil Taxonomy (Soil Survey Staff 1975). Parameters such as color, texture, soil structure, and soil consistence were described in detail to accurately characterize the sediments.

Horizon 1 was the historic/modern disturbance zone, consisting of a plow zone and subsequent residential disturbance (Figure 5.2). Below this lay a flood deposit of massive sandy loam from the Salt River (Horizon 2) that had buried and preserved the prehistoric field. At 5.18–5.24 mbd, the physical characteristics of the sediments of the second horizon changed from a coarse-textured, light-colored sandy loam to a fine-textured, slightly reddish and dark-colored sandy clay loam. This was the agricultural zone, which represented the prehistoric ground surface (i.e., the agricultural fields). Two horizons were identified within the agricultural zone: Horizon 3, the upper portion, was characterized by strong clay enrichment; Horizon 4 exhibited weak clay enrichment. The physical characteristics representing the irrigated soils extended to approximately 5.36 mbd, although the lower boundary of this horizon was gradational and the textures became more loamy. Below this level, the soils were more representative of non-irrigation-affected sediments (Horizon 5). The described soil profile extended to approximately 5.68 mbd, where the soil became a sandy loam once again (Horizon 6).

The four samples from the natural stratigraphy lay within the same strata at the same depth as the four buried field horizons located during testing. Analysts expected that the soils of the natural strata would not contain the same physical and chemical signatures as the agricultural soils because they had not been irrigated (that is, field personnel had observed no evidence of irrigation). Thus, investigators could compare the control samples with potentially anthropogenic soils to arrive at an empirical identification of the agricultural soils.

Physical Characteristics

Irrigation-Altered Soils

Several characteristics of the geomorphic setting and the stratigraphy of the buried fields led to the assumption that the preserved sediments represented prehistoric agricultural fields. The changes in physical appearance brought about by adding irrigation waters to sediments are very similar to the changes produced by natural pedogenic processes (Birkeland 1984). Irrigation accelerates the natural processes of oxidation, increases the percentages of clay and silt, increases the rate of structural modification, and increases rates and amounts of bio- and pedoturbation (Table 5.1). Also, the addition of clay and silt and the increased availability of water through irrigation processes modifies soil properties such as structure, texture, color, type of lower boundary, amount of salts (including carbonates), presence of argillans, and type of surface horizons.

Table 5.1. Physical Characteristics of Irrigated Fields and Natural Substrate

Irrigated Fields	**Natural Substrate**
increased oxidation causing 5YR or 7.5YR color hues	little or no oxidation; 7.5YR or 10YR color hues
color values of fields are dark brown, 7.5YR 4/4 to 3/4	color values of nonirrigated soils are lighter, 7.5YR 5/4
strong angular blocky to prismatic structure	massive to weak angular blocky structure
high bio- and pedoturbation	low bio- and pedoturbation
mollic or umbric epipedon	no mollic or umbric epipedon
preserved primary depositional structures generally absent	primary depositional structures generally present
clay-coated peds and cracks	no clay-coated peds or cracks
high silt and clay content	gradual or no increase in fine materials
gradual or diffuse boundary	abrupt, gradual, or diffuse boundary
lateral change in soil properties	lateral change in geology (significant changes in texture)

Note: The properties of the natural substrate reflect the fine sands of the project area.

Irrigation water is capable of carrying large volumes of fine sediments (silts and clays). Therefore, the continuous addition of irrigation water to the soils will increase the relative amounts of these constituents, producing a marked effect on other soil properties and processes. The shrink/swell action of clays in the soil not only increases the amount of pedoturbation, destroying primary sedimentary features, but also causes alterations in the soil structure. Sandy soils tend to shrink and swell less than loamy or clayey soils and therefore typically retain both primary sedimentary features and a massive soil structure. The addition of clays and silts to sandy sediments by irrigation waters increases shrinking and swelling of the sediments, effectively increasing the rate of structural alteration of the soil from massive to angular blocky, parting to prismatic (Soil Survey Staff 1975).

Recurring additions of water also increase the rates of alternating wetting and drying episodes, thereby actively increasing the rates at which oxidation of the sediments occurs. This increase in oxidation affects the iron-bearing minerals in the sediments and oxidizes them to a redder hue, generally from a 10YR to a 7.5YR color ratio (Munsell Soil Color Chart 1988). Oxidation combined with an increase in the amount of organic matter will cause the soil to darken, thereby increasing the Munsell color value.

As discussed in Volume 1, Chapter 8, the beneficial effects of irrigation (increase in water, nutrients, organic matter, change in structure, etc.) eventually reach a threshold at which they become detrimental to agricultural pursuits. Too much water can lead to a rise in the water table that in turn can increase salinity and alkalinity. High clay concentrations can form hardpans that reduce the depth to which roots can grow and can retain water, thereby limiting the amount of water available to plants, and that decrease pore spaces in the soil, reducing the amount of oxygen available to the plant. Alkalinity and salinity adversely affect the availability of plant nutrients and can become toxic to plants (Thompson 1957).

Pueblo Salado Agricultural Fields

The physical properties of the soil in the irrigated fields at Pueblo Salado differed consistently from those of the nonirrigated sediments of the control samples. In all categories—color, texture, structure, and soil consistence—the pedogenic properties of the presumed fields resembled those of soils that have been irrigated (Table 5.1). The buried fields exhibited dark brown color, with lower color values than the nonirrigated soils (7.5YR 4/4 to 3/4 versus 7.5YR 5/4). This color difference is representative of the apparent increase in organic matter and clay in the irrigated soils. An apparent shift toward red that investigators observed in the field was not significant enough for laboratory analysts to read it as a change in the Munsell hue.

However, field personnel did perceive distinct textural changes in the soil. The surrounding sediments of unaltered substrate comprised sand, sandy loam, and silty loams, while the irrigation-altered sediments were loam and sandy clay loams. Such changes may be the result of turbid waters transported to the fields through canals (Huckleberry 1992).

Structurally, the irrigated sediments exhibited a stronger, angular blocky soil structure. The peds were easily seen in the profile, and when disturbed, fell out in nearly coherent blocks. This type of structure can be due to an increase in the amount of clay and the shrinking and swelling of the soil resulting from alternation between wetting and drying caused by irrigation. This shrinking and swelling causes the soil to break along consistent boundaries that become the boundaries of the individual peds. The structure of the non-irrigation-impacted soils, by contrast, was very weakly developed. When observed in profile, the sediments were either massive, with no visible structure, or were weak, angular blocky. When weak, angular blocky peds are disturbed, they easily fall apart into single grains. The weak structure strongly suggests a lack of shrink/swell action resulting from the absence of clay or repeated wetting and drying of the soil.

The consistence of the soil, which is a measure of the soil's ability to resist deformation under dry, moist, and wet conditions, also indicated an increase in the amount of clay and silt in the irrigated soils. The irrigation soils were generally slightly hard when dry, friable to firm when moist, and very sticky and very plastic when wet. The non-irrigation-impacted soils were soft, very friable, and nonsticky and slightly plastic when wet. The differences in consistence were due to the increase in the amount of clay in the irrigation-impacted soils.

PARTICLE-SIZE ANALYSIS AND SOIL CHEMISTRY

To either support or disprove their interpretation that the unusual geomorphic, stratigraphic, and physical properties observed in soils from Pueblo Salado represented prehistoric agricultural fields, SWCA analysts also had to rely on laboratory data. Initially, laboratory personnel undertook only particle size and nitrate-nitrogen analyses to determine the feasibility of detecting chemical signatures in the preserved soils. However, once these initial analyses indicated chemical and physical trends that supported the presence of a buried agricultural soil, laboratory staff analyzed nine other soil properties as well: pH, percent organic matter (%OM), salinity, computed percent sodium (%Na), and levels of calcium (Ca), potassium (K), magnesium (Mg), sodium (Na), and phosphorus (P) in parts per million (Table 5.2). These factors, along with nitrate-nitrogen concentration, characterize the chemical make-up of the soil horizons and enable analysts to identify a chemical signature for the buried agricultural deposits. Although characteristics of soil nutrients can also indicate fertility, nutrient availability depends not only on the concentrations of the various nutrients, but also on pH, organic matter, cation exchange capacity, texture, ion mobility and solubility, and interactions with other minerals (Janick et al. 1974). The concentration of nitrates, for example, does not by itself delimit prehistoric fertility and productivity but when used in conjunction with other data characterizes past chemical properties.

To carry out the laboratory analyses, project investigators took six soil samples from a vertical profile within Trench 3015 and analyzed the samples for particle size, nutrients, organic matter, pH, and salinity. In addition, they selected four control samples from unaltered natural sediments, at the same depth as the agricultural deposits, from a location farther south and outside of the identified agricultural field area. SWCA analysts determined particle size using the Bouyucous hydrometer method in the SWCA laboratory in Flagstaff and sent all other soil samples to IAS Laboratories in Phoenix for chemical analysis.

Particle-Size Analysis

The pattern derived from particle-size determinations shows a moderate proportion of clay, 10%, in the surface horizon (4.74 mbd) that decreases to 8% in the underlying horizon (Figure 5.3a). The buried agricultural horizon (5.2 mbd) shows a diagnostically higher concentration of 20% clay. Clay percents then decrease in a regular fashion with depth in the prehistoric fields, from 20% to 18% to 12% to 8% in the lowest horizon. This decrease in the clay percent represents a decrease in the direct effects of the irrigation waters applied to the prehistoric surface. The pattern of the control samples show no such dramatic increase in clay content (Figure 5.3b). The increase from 4% to 8% in subsurface horizons is within the range that would occur in natural floodplain deposits that remained unaltered by irrigation waters. In comparison with the control samples, the irrigation

Table 5.2. Analysis of Irrigated Fields and Control Area

Sample No.	MBD	PPM						pH	%OM	Salinity	%Constituents			%Na
		Ca	Mg	Na	K	N	P				sand	silt	clay	
Analysis of Irrigated Field Areas														
2381	4.74	5800.0	625.0	770.0	1330.0	142.7	13.0	8.0	4.2	8.5	60.0	30.0	10.0	8.2
2382	5.03	4300.0	445.0	500.0	270.0	94.2	4.3	8.1	2.0	5.5	60.0	32.0	8.0	7.7
2377*	5.21	6000.0	690.0	620.0	255.0	115.6	7.2	7.8	2.9	7.8	54.0	26.0	20.0	6.9
2379*	5.34	5500.0	580.0	370.0	215.0	75.6	2.3	8.1	2.2	4.8	50.0	32.0	18.0	4.7
2380	5.43	5600.0	550.0	240.0	165.0	50.0	2.7	8.4	2.1	2.1	40.0	48.0	12.0	3.1
2383	5.62	3900.0	365.0	140.0	120.0	18.0	3.8	8.6	1.2	1.1	60.0	32.0	8.0	2.6
Analysis of Control Area														
4596	5.21	4400.0	700.0	745.0	235.0	21.4	2.1	8.2	2.2	6.0	36.0	60.0	4.0	10.2
4597	5.39	5800.0	1200.0	930.0	285.0	24.3	2.1	7.9	3.2	10.0	52.0	40.0	8.0	9.2
4598	5.46	5900.0	830.0	780.0	250.0	20.7	4.2	8.0	2.4	8.8	54.0	40.0	6.0	8.4
4599	5.61	5400.0	710.0	750.0	220.0	18.6	3.9	8.1	2.2	7.5	54.0	42.0	4.0	8.9

Ca = calcium; Mg = magnesium; Na = sodium; K = potassium; N = nitrogen; P = phosphorus; %OM = percent organic matter; %NA = computed percent sodium
*agricultural zone

impacted sediments show an overall increase in the percent of clay, up to five times the concentration found in the natural sediments (20% versus 4%).

Chemical Analysis

Laboratory Methods

IAS analysts extracted four cation macronutrients (calcium, sodium, magnesium, and potassium) from the sediments with ammonium acetate and analyzed the nutrients using atomic absorption spectrophotometry. This technique measures nutrient availability but not the total amount of salts in the soil. Laboratory staff analyzed phosphorus, measured as phosphate, in a bicarbonate extraction solution using a calorimeter and spectrophotometer. They determined the level of nitrogen, measured as nitrate-nitrogen, using inductively coupled plasma spectroscopy (ICPS). The macronutrients all have important effects on the health and productivity of plants; deficiencies or too-high concentrations can cause poor crop production and lower yields (Thompson 1957).

IAS also measured the percentage of organic matter using the Walkley-Black wet oxidation method. Organic matter is important to crop production, soil moisture retention, aeration, soil structure, and cation exchange capacity and is generally the main source of microorganisms in the

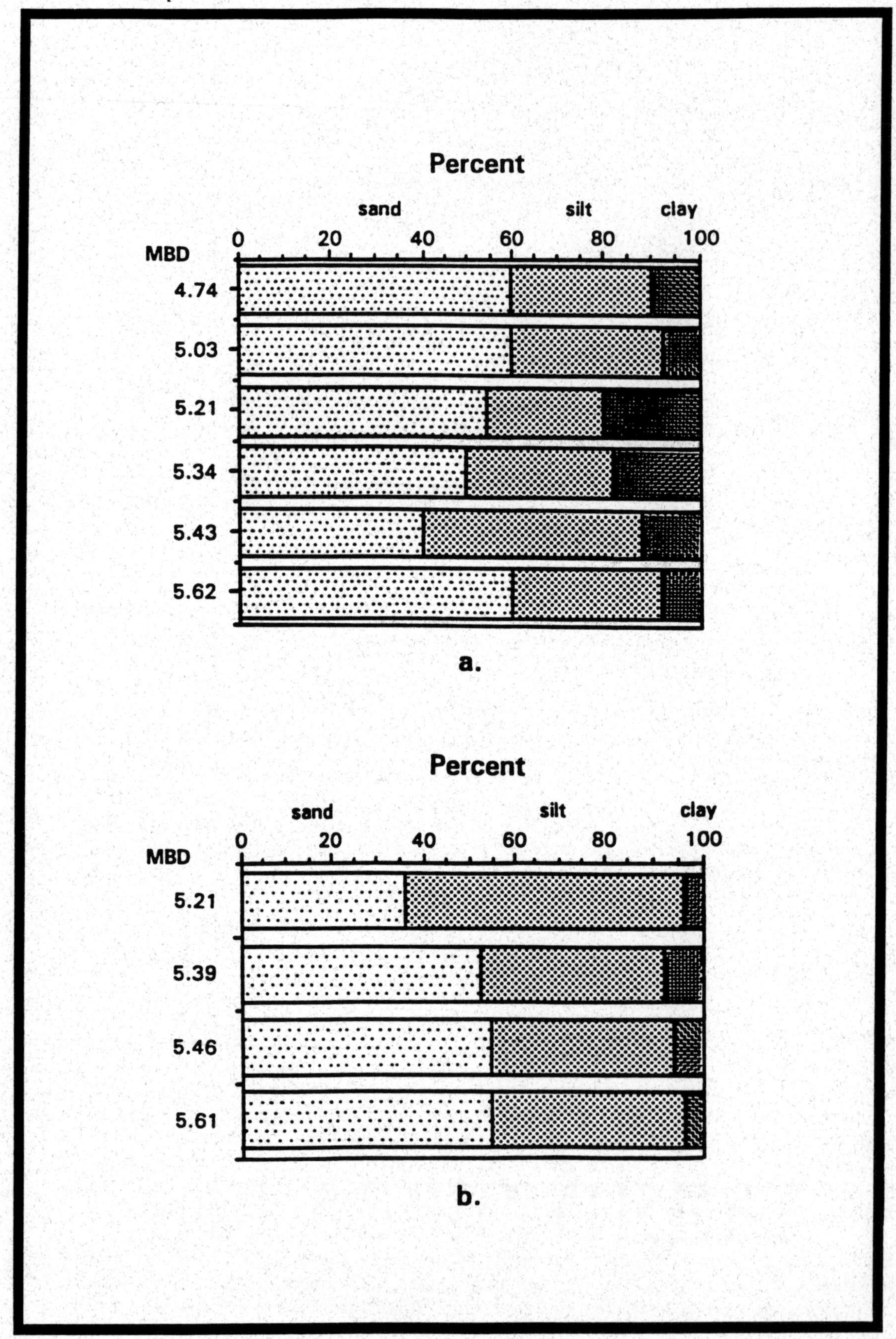

Figure 5.3. Textural analysis: (a) irrigated fields (note clay increase at 5.2 mbd); (b) control samples.

soil (Thompson 1957). The presence of increased amounts of organic matter indicates a horizon of plant growth and organic turnover.

The pH is an indication of the acidity or alkalinity of the soil as measured by the amount of hydrogen ions present. Values of pH are important to nutrient availability and mobility. In humid regions, acidic soil conditions usually prevail due to the high concentration of hydrogen ions. In arid areas, by contrast, the proportion of hydroxyl ions is usually higher, providing high pH measurements. Many plants can live within slightly acidic to slightly alkaline soil conditions. At very low (very acidic) or very high (very alkaline) pH levels, minerals either are restricted in their availability or are present in toxic concentrations (Janick et al. 1974).

Salinity is a measure of the electrolytic conductivity (EC) determined as the concentration of salts in a soil (Sposito 1989). In general, saline soils with EC measurements greater than 4.0 millimhos/cm (4.0 dS/m) exhibit greatly reduced plant growth. Saline soils are common in arid regions where the rate of evaporation exceeds the rate of precipitation, which results in salts remaining on or just below the ground surface. Saline soils have pH measurements greater than 8.5 (Thorne and Peterson 1954). As the pH (alkalinity) of the soil increases, salts accumulate in toxic levels, soil permeability decreases, and the soil structure closes and becomes difficult for roots to penetrate (Thorne and Peterson 1954).

General Characteristics of Irrigation-Altered Soils

Soils or sediments that have undergone irrigation will have been changed by the salts, suspended sediments, and infiltrating water in a predictable fashion (Howard and Huckleberry 1991; Huckleberry 1992; Thorne and Peterson 1954). Irrigated agricultural fields should have increased amounts of silts and clays deposited by the irrigation waters, increased levels of salts and salinity measurements as a result of high water tables and high rates of evapotranspiration, increased amounts of organic matter and nitrogen from surface soil horizons with high amounts of biotic material (Thompson 1957), and increased levels of phosphorus due to cultural activity (Eidt 1984).

The original chemical composition of the prehistoric soils has probably changed due to continued soil development, historic and modern irrigation practices, or both, so that the concentrations in Table 5.2 are not necessarily those of the prehistoric soils. Therefore, relative chemical composition and trends in concentrations of the nutrients from one horizon to another are important in identifying and characterizing the prehistoric soils. At Pueblo Salado, the chemical properties appear to have been sufficiently preserved to provide a diagnostic signature. However, SWCA investigators could not make a positive determination regarding the preservation of the signature because they had no area available for comparative study that had clearly been spared from historic disturbance.

Characteristics of Pueblo Salado Agricultural Fields

The data indicated some increases in the levels of chemical constituents and properties at the approximate depth of the buried agricultural fields, 5.18–5.36 mbd (Table 5.2). "Peaks" in seven chemical categories (%OM, P, N, salinity, Na, Mg, and Ca), plus clay content, all suggested that the buried fields retained a chemical signature from prehistoric irrigation and agriculture. After nearly a millennium of pedogenesis and a century of historic irrigation, the physical and chemical characteristics of these soils were remarkably preserved.

The diagrams illustrating the laboratory results of the chemical analyses of the agricultural fields reflect remarkably similar trends in their overall patterns (Figures 5.4 and 5.5). The data for each diagram can consistently be divided into three parts. The first part illustrates the relatively high readings for the present ground surface horizon at 4.74 mbd. Historic activities left high concentrations of salts (Ca, Mg, Na), OM, N, and P, which decreased with depth until the buried agricultural soil was encountered. Analysis of the pH showed inversely proportional values throughout. The second pattern is the peak in the amounts of salts, OM, N, and P (and low pH) in the buried agricultural horizon at 5.21 mbd. The third pattern is the consistent decrease in salts, OM, N, and P (and high pH) with increasing depth below the buried agricultural soil, reflecting the diminishing effect of irrigation waters on the underlying substrate.

At the depth of the agricultural horizon, the peaks in the levels of Mg and Na (Figure 5.4a), Ca (Figure 5.4b), and salinity (Figure 5.4c) most likely resulted from the retention of salts after irrigation waters underwent evapotranspiration. Analysts attributed high amounts of salts in the surface horizons to recent and historic activities. Absolute salinity levels were quite high (7.8 EC), possibly to the point that would have been detrimental to crop production. The addition of irrigation waters would have temporarily lowered these levels by leaching, but the high levels would have returned following drying of the soil.

The peaks in organic matter and nitrogen, although slight, represented a horizon of high biotic activity and plant growth suggestive of a weakly developed buried A horizon or plow zone (Figure 5.5a, b). The percent of organic matter increased to 2.9% in the agricultural horizon from 2.0% in the overlying sediments and decreased to 2.2% in the underlying sediments. By contrast, the nitrate concentration, measured in parts per million (ppm), was very high in the historic sediments (decreasing from 142.7 ppm in the surface horizon to 94.2 ppm in the subsurface historic horizon), representing modern vegetation and the addition of fertilizer from historic irrigation farming. The concentration in the field soils increased to 115.6 ppm from the 94.2 ppm in the horizon above. This increase, although slight, probably represents the addition of organic-rich clays from irrigation waters and organic matter from the fields. The concentration of nitrate-nitrogen then decreased again with depth below the irrigated soils of the prehistoric surface. Concentrations in the buried fields were more than five times higher than those in the control samples (115.6 ppm vs. 21.4 ppm), which increased very slightly with depth and then decreased at the bottom of the section.

The inverse relation between organic matter and pH may be due to the presence of organic acids that would reduce the pH in horizons containing high levels of organic matter (Figure 5.5a). Phosphorus, measured as available soil phosphate, shows peaks in both the buried agricultural horizon and the surface horizon (Figure 5.5b). The two phosphate peaks probably represent cultural activity (Eidt 1984).

Characteristics of the Control Samples

The physical properties of the control samples suggested that these sediments had not undergone alteration from prehistoric irrigation. Field investigators observed little or no oxidation and resultant reddening, only a slight increase in clay content (with silt loam to sandy loam texture), and poor soil structure. The dramatic contrast between the unaltered control samples and the altered field samples, however, ended with the differences in their physical properties. In nearly all analyzed chemical categories, with the exception of nitrate-nitrogen, absolute values were comparable to those of the field samples (Table 5.2). In addition to comparable values, the chemical constituents also exhibited similar trends (Figures 5.6 and 5.7), showing peaks at 5.4 mbd for the concentration of salts (Figure 5.6a, b, c) as well as percent OM and nitrate-nitrogen (Figure 5.7a, b). Values for

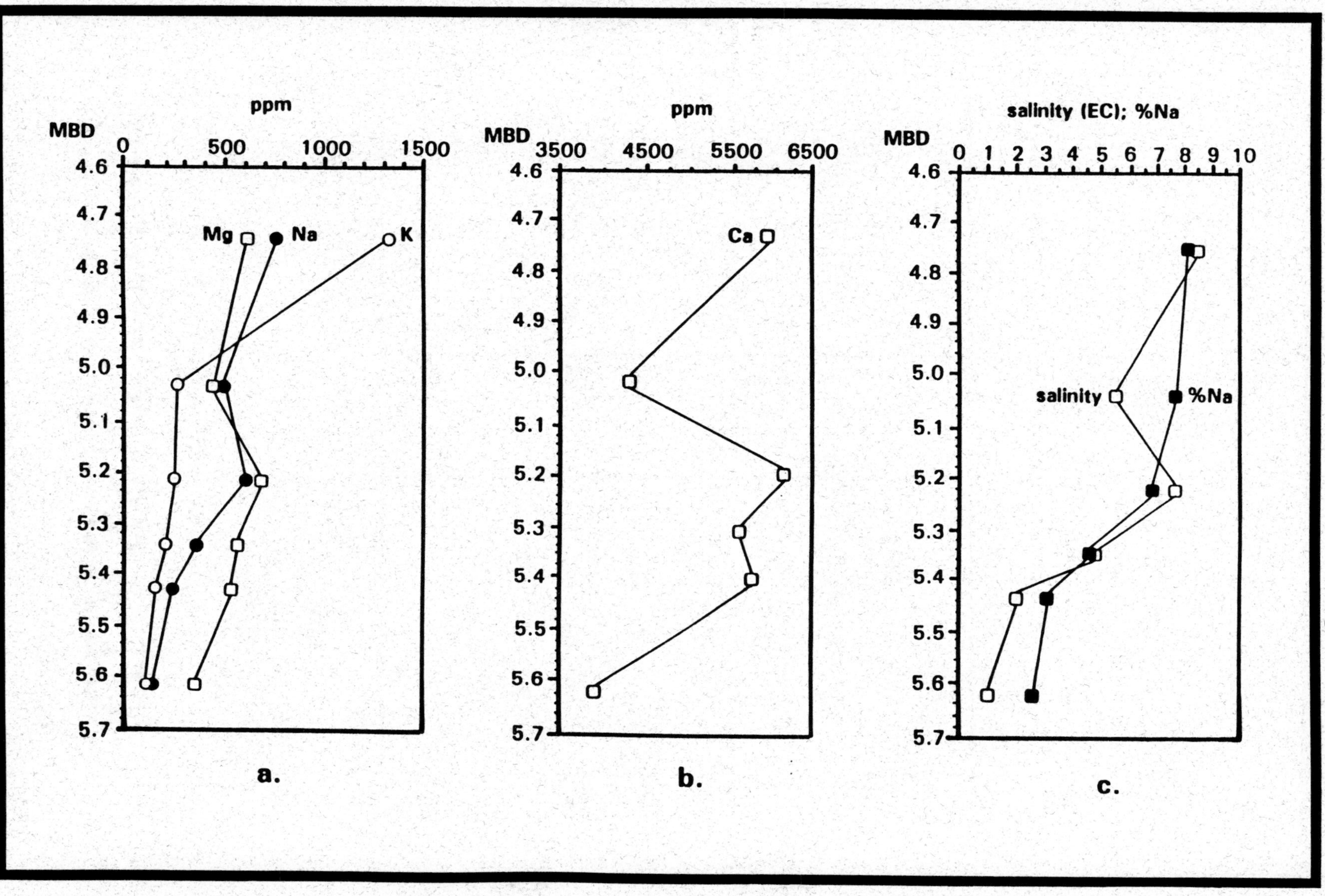

Figure 5.4. Chemical analysis of irrigated fields: (a) parts per million of magnesium, sodium, and potassium; (b) parts per million of calcium; (c) salinity and percent sodium. Note peaks at 5.2 mbd.

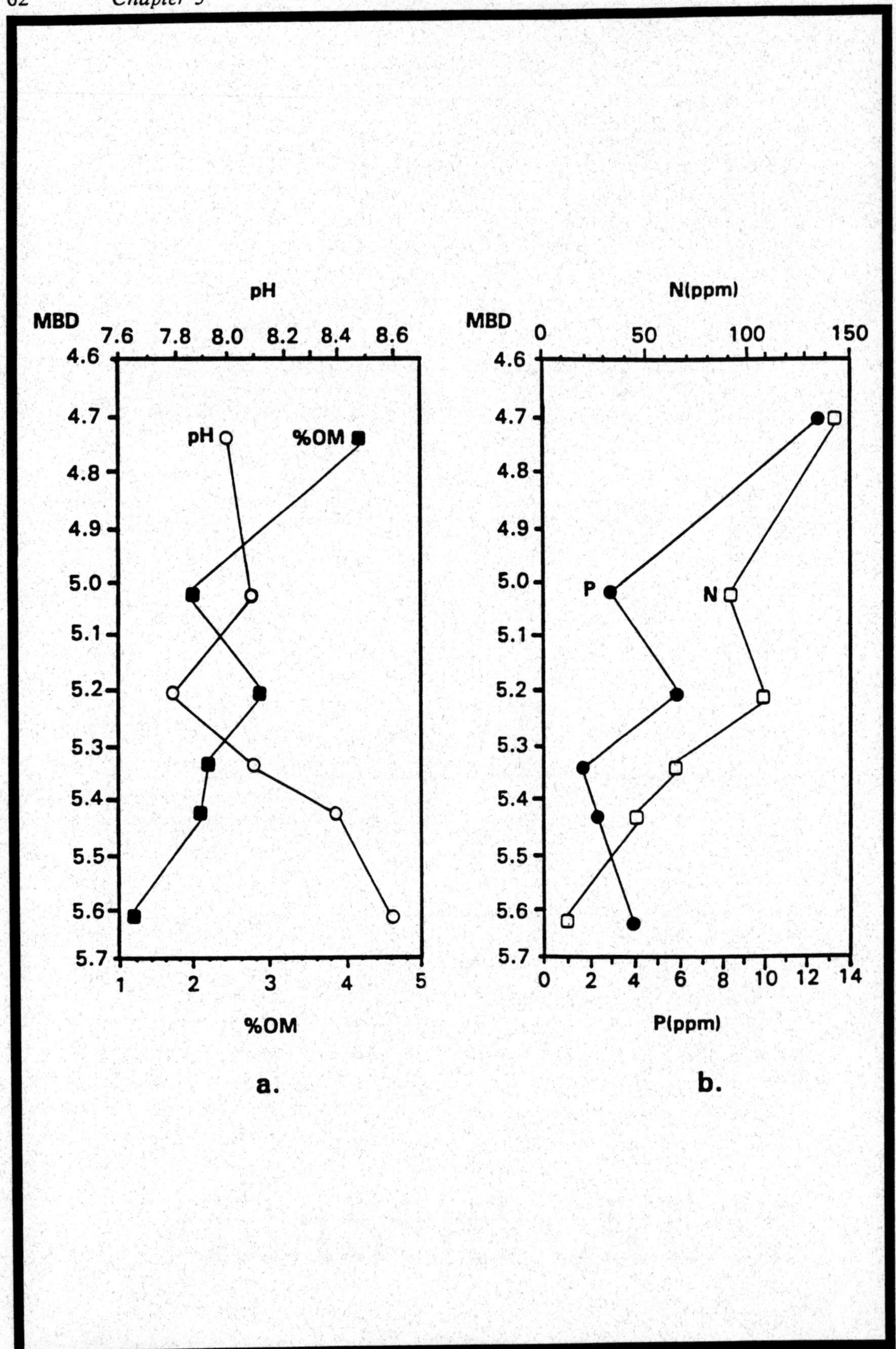

Figure 5.5. Chemical analysis of irrigated fields: (a) organic matter and pH analysis; (b) nitrate-nitrogen and phosphorus analysis.

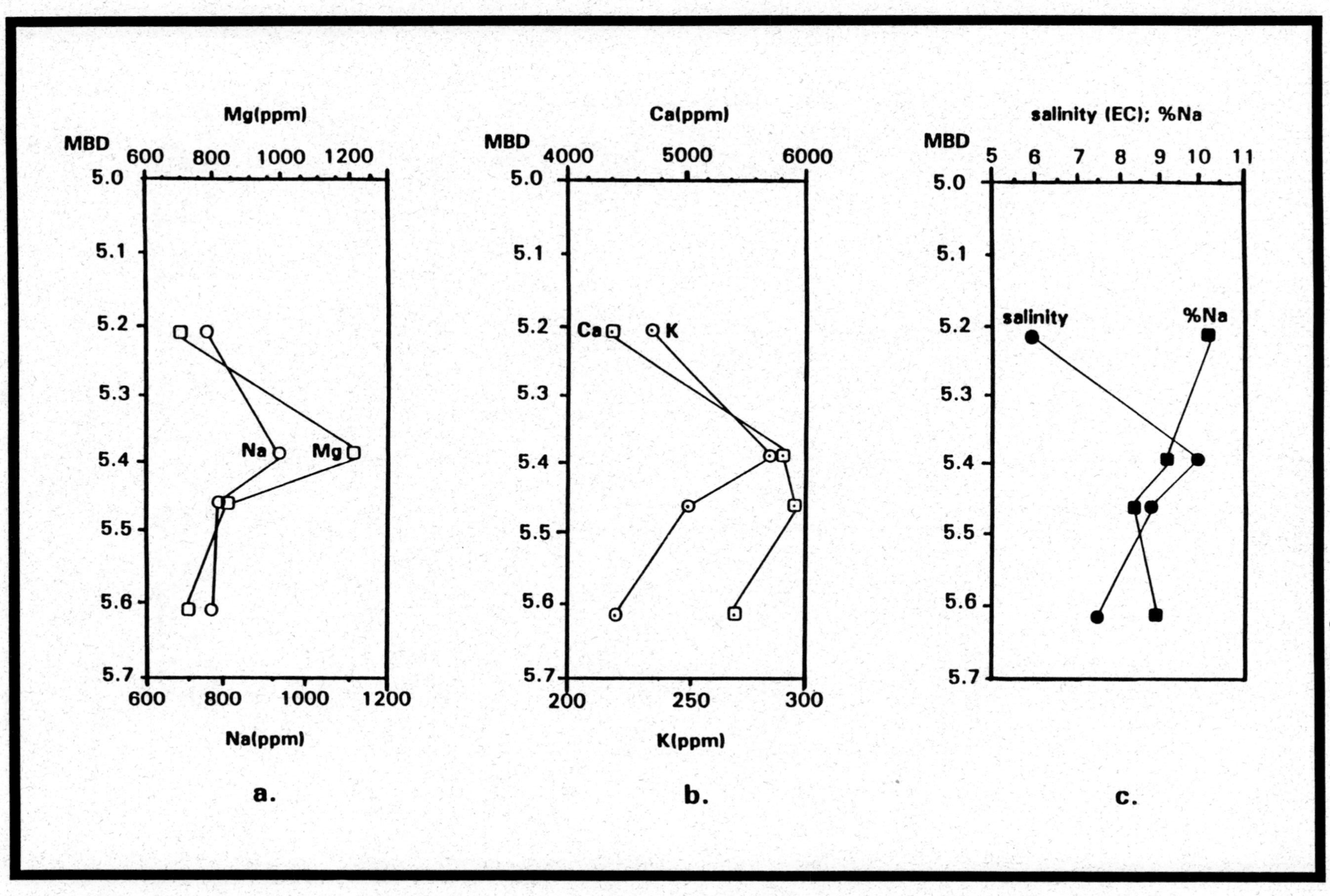

Figure 5.6. Chemical analysis of control samples: (a) magnesium and sodium; (b) calcium and potassium; (c) salinity and percent sodium.

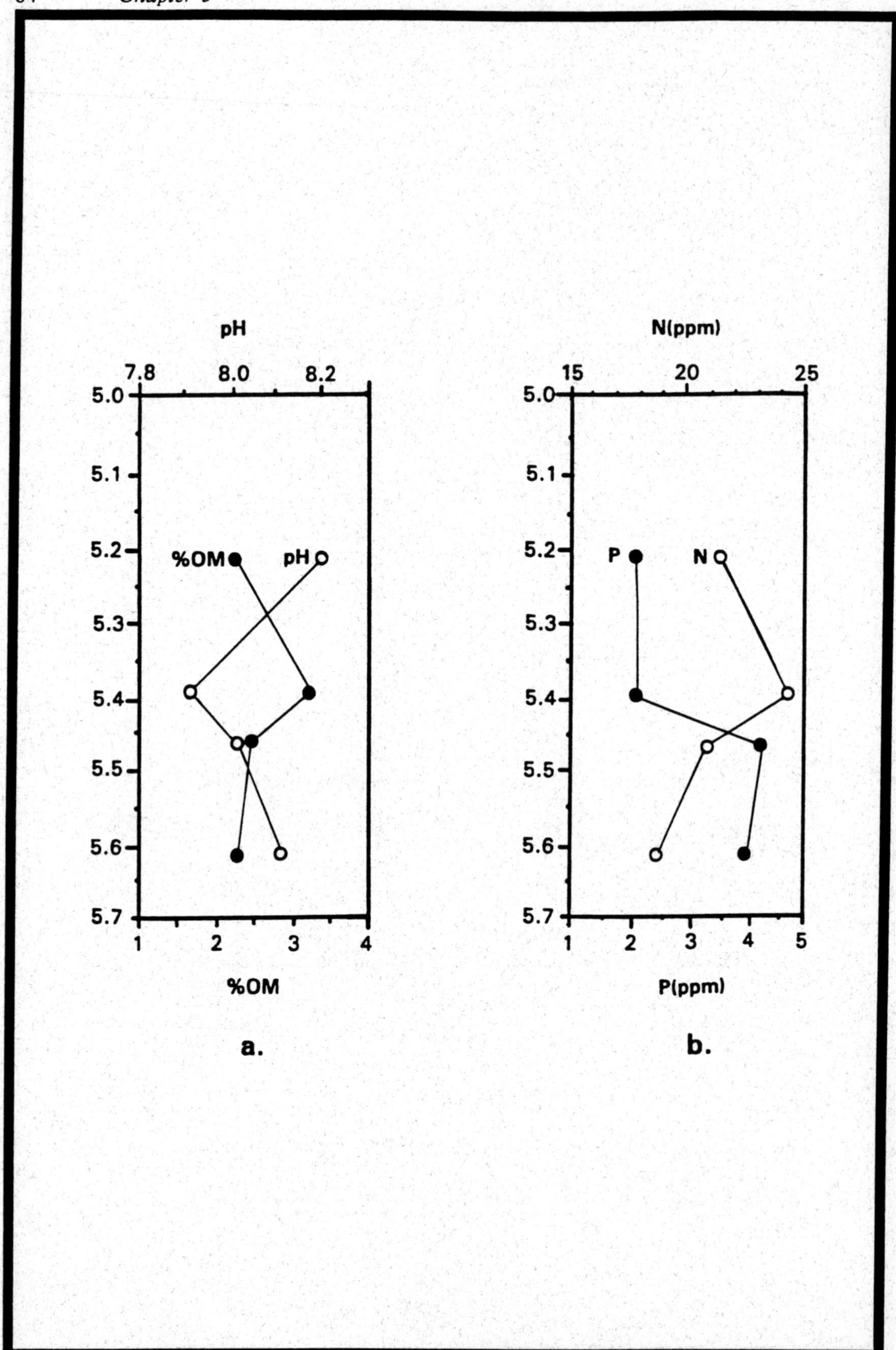

Figure 5.7. Chemical analysis of control samples: (a) percent organic matter and pH: (b) nitrate-nitrogen and phosphorus.

pH also indicated an inverse relationship to organic matter similar to that seen in the field samples (Figure 5.7b).

What the control sample results indicate is unclear, though several possibilities exist. First, the control samples may in fact have been from field sites; the chemical constituents may mirror those of the field samples because the Hohokam also farmed the control area, although perhaps less intensively than the identified field area. Second, an unfarmed prehistoric ground surface at 5.4 mbd may have contained high concentrations of organic matter and salts accumulating naturally within a depression. Third, the increase in chemical concentrations may be due to the slight increase in clay. Because of the important contribution of clay to soil chemistry, an increase in the percentage of clay dramatically increases the level of adsorbed cations (salts) that can be present within any soil horizon. The effect of clay on soil chemistry is suggested in the positive correlation between clay and organic matter (Figure 5.8a) and salinity (Figure 5.8b).

Soil Fertility

Soils of the Phoenix Basin are generally good for agriculture, with the obvious limiting factor being water. Once farmers solve the problem of water through the construction and use of canals, their food production yield can be quite high. Through time, however, with the addition of irrigation waters, two phenomena reduce the fertility of the sediments. The first includes the deposition and precipitation of salts from the irrigation waters after evapotranspiration. The second is a rise in the water table, which brings salts from the substrate closer to the surface and enhances the precipitation of salts upward in the soil profile through capillary action. The laboratory data presented above indicate that the amount of salts in the buried agricultural soil increased beyond that of the nonagricultural horizons. The pH values of the tested soils were also quite high, albeit somewhat lower in the agricultural soil.

Recommendations for such soils in modern agricultural practice would be to leach the fields with low-salt waters to remove the salts and to add sulfur to decrease the pH. Each time the prehistoric farmers added irrigation water to the soils, soluble salts would be put into solution and removed; each time the soils dried out, however, salts would reprecipitate. Although the Hohokam would not have known to add sulfur, addition of organic matter in the form of plant remains would have had the beneficial effects of adding nutrients, reducing pH, and increasing the water retention of the soil. Although analysts measured only a few of the macro- and micronutrients, nutrient levels appeared adequate, and the soils did not exhibit nutrient depletion.

Conclusions

To substantiate the existence of agricultural fields suggested by the geomorphic evidence, investigators analyzed trends in chemical and particle-size profiles by comparison to known trends in irrigation-affected soils. They then compared the presumed irrigation-affected soils to non-irrigation-altered soils as control samples. In general, well-managed irrigated soils in arid regions should be "richer" than nonirrigated soils. All things being equal, irrigated strata should have more silt, clay, and organic matter than nonirrigated strata. These additions give the soil a darker, "richer" appearance and alter the soil's physical and chemical properties as discussed earlier. In addition, the irrigation waters should deposit salts in the surface horizons, producing high salt concentrations. Analysts observed and measured these properties in the buried field sediments and the control samples.

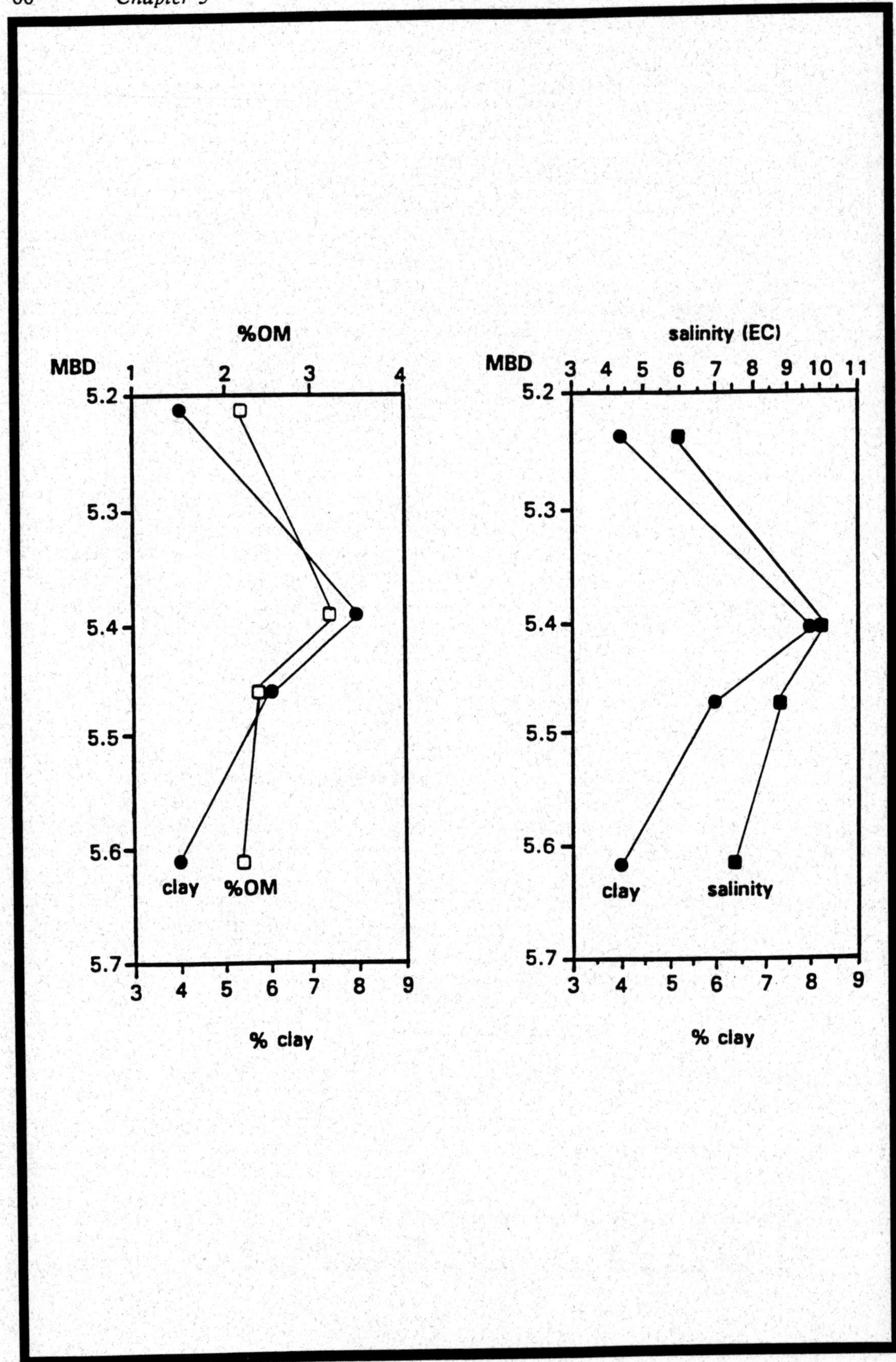

Figure 5.8. Control samples: (a) positive correlation between clay and organic matter: (b) positive correlation between clay and salinity.

The soils evidence supports the inference drawn from the geomorphic data and strongly suggests that the stratigraphy preserved at Pueblo Salado and discussed here represents prehistoric agricultural fields: (1) geomorphic setting within a low depression; (2) proximity to the prehistoric canal; (3) buried soil horizons that are darker and more oxidized, have better developed soil structure, and have higher levels of bioturbation than the control substrate; (4) high clay concentrations at the buried field level, which decrease with depth away from the irrigation horizon; (5) enrichment of the soils in calcium, magnesium, sodium, organic matter, phosphorus, and nitrates; and (6) high salinity readings. The physical and chemical properties of the soil all indicate the presence of an anomalous soil horizon located at about 5.2 mbd in the depression that investigators hypothesized was a buried field.

Problems with the interpretation of the soils as buried agricultural fields still remain. The comparison with control samples revealed that chemical trends can mirror textural trends and that (based on floodplain geomorphic processes) fine-grained sediments deposited in low-energy environments can produce chemical profiles similar to those produced by irrigation. In addition, stratigraphic units located within lower topographic areas may have different chemical and physical profiles than those on slightly elevated areas, such as the location of the Pueblo Salado compound.

Geomorphic conditions and analytically measured physical and chemical trends strongly suggested that prehistoric Hohokam farmers had left a weak signature in the soils. However, the geomorphic and soil evidence presented here remained equivocal on whether the properties of these sediments primarily resulted from natural or cultural processes. Without agricultural features, project investigators could not verify the buried fields of Pueblo Salado through physical and chemical analysis alone, and they therefore continued to rely on soil oxidation as supporting evidence.

FLOTATION ANALYSIS

SWCA personnel analyzed eight flotation samples for this study (Table 5.3). Investigators collected three control samples from adobe-lined pits (Features 8/9-52, 8/9-53, and 8/9-56) positioned stratigraphically below the sediments believed to represent irrigation soils and five from the irrigation soils. The agricultural field exhibited no evidence of in situ burning, and analysts therefore recovered few or no charred plant remains from the samples. Investigators expected the lack of charred plant remains, considering historic analogues and a recent review of the use of fire (Bohrer 1991) in Hohokam plant management.

The research design for the Phoenix Sky Harbor Center project (Volume 1:Chapter 10) specified a number of research objectives for the agricultural field study. Through the flotation study investigators asked questions such as what types of botanical remains were associated with fields and whether domesticates, economic native plants, and noneconomic plants were preserved. They also asked questions related to the types of plant communities present in the prehistoric fields and their association with cultivated plants. In particular, they sought evidence of economically important species that the Hohokam may have encouraged to grow within fields. They also hoped to develop methods for recognizing agricultural fields in other areas.

Table 5.3. Summary of Charred Plant Remains Recovered from the Possible Agricultural Field and from Control Samples

Feature or Bag Number	*Agave*	"Agavoid"	Cheno-am	*Descurainia*	*Leptochloa*-type	*Sphaeralcea*	*Sporobolus*-type	*Suaeda*
Adobe-Lined Pits Underlying the Possible Agricultural Field (control samples)								
Feature 8/9-52	—	—	—	—	—	—	—	—
Feature 8/9-53	—	—	—	—	—	—	—	—
Feature 8/9-56	1 fi	1 mf 3 rfi	1 sf	—	—	—	—	1 s
Possible Agricultural Field Samples								
Bag 8/9-2359	—	1 mf 1 rfi	2 sf	cf. 1 sf	—	—	—	—
Bag 8/9-2363	—	—	3 sf	—	—	—	—	—
Bag 8/9-2365	—	1 mf 1 rfi	—	—	—	—	cf. 1 gf	—
Bag 8/9-2368	—	—	1 sf cf. 1 sf	—	—	cf. 1 sf	—	—
Bag 8/9-2372	—	—	cf. 1 sf	—	1 g	—	—	—

cf. (compares favorably) = tentative identification; fi = fiber; g = grain; gf = grain fragment; mf = miscellaneous fragment with white styloid and/or raphide crystals; rfi = fiber with round cross section and white styloid and/or raphide crystals; s = seed; sf = seed fragment.

Processing and Analysis

SWCA personnel processed the flotation samples and sent them to Soil Systems, Inc. (SSI) for analysis, following procedures described in the section on flotation analysis for Pueblo Salado (Volume 3:Chapter 8), with two exceptions. First, because each sample had low light-fraction residue volumes after processing, analysts examined the material smaller than 0.25 mm during this study, even though they did not do so for the habitation features at Dutch Canal Ruin and Pueblo Salado. Furthermore, they also examined the heavy fractions without a lens, although these were not studied during the two previous analyses, because of the low light-fraction residue volumes. Material identified in the <0.25 mm and heavy fractions are noted in Appendix B to facilitate comparisons between this study and the flotation results from Dutch Canal Ruin and Pueblo Salado. Analysts observed no identifiable charred plant remains smaller than the 0.25-mm mesh in the light fraction (Appendix B).

Results

All samples contained few or no identifiable charred plant remains and no identifiable wood charcoal. The adobe-lined pits averaged 0.25 charred parts per liter (n=3; range=0–0.75; s.d.=0.43), and samples from the possible agricultural field averaged 0.35 charred parts per liter (n=5; range=0–0.75; s.d.=0.29). These figures do not include indeterminate, miscellaneous, or cf. (compares favorably) identifications. Samples from the main occupation of Pueblo Salado, by comparison, averaged 27.14 charred plant parts per liter (n=35; range=0–381.25; s.d.=64.21) (Volume 3:Chapter 8), a considerable difference that may be related to plant or food processing and the presence of fires or thermal features.

Discussion

Both the controls and the samples from the possible agricultural field contained low densities of small, charred seeds from economically important noncultivated plants (e.g., pigweed or goosefoot [Cheno-ams]), as well as low densities of "agavoid" remains that may represent the native cultivar agave. However, to conclude positively that Hohokam farmers had burned weeds in the immediate area, investigators would have had to recover higher charcoal densities, evidence for in situ field burning (e.g., ash or charcoal lenses overlying fire-reddened sediments), and higher densities of charred seeds from weedy taxa. If investigators indeed collected these samples from a prehistoric agricultural field, perhaps the farmers had weeded their fields by hand and then either left the weeds in the field to decompose or else disposed of them elsewhere. The charred plant assemblage from the possible agricultural field seems to represent low densities of carbonized plant remains that had either been windblown or washed into the sediments from cultural activities conducted at or near Pueblo Salado or at the margins of the fields.

Because the charred plant assemblage probably represented discard activities from Pueblo Salado and had been secondarily deposited in the agricultural fields, investigators could draw no conclusions regarding prehistoric plant communities in or near the possible agricultural fields. The Pueblo Saladoans could have collected any of the recovered plant remains at the site or at some distance away. Active fields and canals can provide a wide variety of economic resources along their margins (Chapter 7). Thus, some of the charred remains identified in the field samples could have resulted from intentional or accidental burning along fields and canals or the processing of economic resources adjacent to fields.

The lack of evidence for weed burning in the samples from the Pueblo Salado fields is consistent with a recent review of the use of fire in plant management by the Hohokam (Bohrer 1991:233) and with historic analogues. The Pima and Tohono O'odham historically weeded their cornfields two or three times each year with a hoe and then left weeds on the ground (Castetter and Bell 1942:173–174). They used no additives (e.g., trash) to fertilize the soil (Castetter and Bell 1942:172). River Pima agricultural fields are generally free of even edible weeds (Crosswhite 1981:64). Although Bohrer (1991:233) suggested that evidence indicates that the Hohokam managed large tracts of land through burning to encourage the growth of many weedy annuals, her argument deals specifically with noncultivated areas rather than with agricultural fields.

If the charred plant assemblage from the possible fields is typical of Hohokam agricultural fields, future researchers would probably have equally little success in using flotation analysis to distinguish them. Samples taken at almost any other nonfeature location near an archaeological site would likely appear similar, with the same likelihood that charred remains represent low densities of secondary refuse. This line of reasoning assumes, of course, that the area under discussion was indeed an irrigated field.

POLLEN AND PHYTOLITH EVIDENCE

Investigators submitted eight samples for microbotanical analysis of pollen and phytolith remains. They collected three control samples from adobe-lined pits (Features 8/9-52, 8/9-53, and

8/9-56) that were stratigraphically lower than the sediments thought to represent irrigation soils. They collected the other five from the irrigation-impacted soils, representing proveniences similar to those of the flotation samples discussed above.

Adobe-Lined Pits

The three control samples (Samples 2370, 2361, and 2356) came from adobe-lined pits overlain by what investigators believed to be agricultural field deposits and by later flood deposits that sealed both the fields and the pits. Field personnel sampled the pits to provide information on local vegetation prior to development of the irrigated fields. The pollen record from these features displayed moderate Cheno-am pollen frequencies, especially when compared with those recovered from structures and other features at this site (Volume 3:Figure 9.1). The quantities of Low-spine Compositae pollen were slightly elevated, and those of High-spine Compositae pollen relatively high. Gramineae (grass) pollen frequencies were low; those of *Larrea* (creosotebush) were slightly elevated. The levels of *Sphaeralcea* (globe mallow) pollen were also elevated, indicating that these weeds had grown in the vicinity. Analysts noted small quantities of *Typha* (cattail) pollen in Features 8/9-53 and 8/9-56, indicating that *Typha* pollen had been transported either by the wind or in the water that the Hohokam used in these mixing pits. *Zea mays* (corn, maize) pollen, noted in Feature 8/9-53, had probably been transported by wind from a nearby area.

The phytolith record from the adobe-lined pits exhibited few grass short cells. In contrast, Features 8/9-52 and 8/9-56 (Samples 2370 and 2356, respectively) exhibited moderately large quantities of sponge spicules and pillow-shaped phytoliths (Volume 3:Figure 9.2). The sponge spicules were probably derived from water from the Salt River that the Hohokam builders mixed with the adobe in these pits. Sample 2361 from Feature 8/9-53 was unusual in that it contained large quantities of starch grains derived from foods high in carbohydrates such as maize, beans, and cattail root. Analysts recorded three types of starch, all of them exhibiting centric hila. The individual granules ranged from mainly polygonal with distinct hila, forms commonly found in maize, to rounded or irregular shapes. The hila varied in shape from small and round to fissured or elongated slits. Maize starch may vary in form from polygonal to circular to irregular; maize hila may vary from small circular points to multiple fissures. Moreover, the size of individual maize granules can vary greatly over a range including all of the forms noted as starch granules in the phytolith samples from this project. Recovery of the large quantity of starch granules from this adobe-lined pit suggested that ground maize might have been added to the adobe mixture. If maize leaves or stalk fragments had been added as "temper" or bonding agents to prevent cracking as the adobe dried, they would have appeared in the phytolith record as increased panicoid short cells.

Agricultural Fields

Analysts examined for both pollen and phytoliths five samples (2360, 2364, 2366, 2367, and 2373) from the deposits identified through the presence of oxidized soil as possible agricultural fields. These samples exhibited moderate frequencies of Cheno-am pollen, occasionally accompanied by a few small aggregates. High-spine Compositae frequencies were generally elevated, with a few aggregates. The levels of *Eriogonum* (wild buckwheat) pollen were likewise elevated in all of the field samples compared to other samples from this site, suggesting that *Eriogonum* had grown in relative abundance in this area, possibly as a weed. *Larrea* pollen was common in these samples (Volume 3:Figure 9.1) in frequencies similar to those in the samples from the adobe-lined pits. Small frequencies of Gramineae pollen, also comparable to the adobe-lined pit samples, probably represented normal wind transport from natural vegetation. Analysts also noted Onagraceae (evening primrose family) pollen in four samples, probably indicating the presence of members of this family as weeds. *Cylindropuntia* (cholla cactus) pollen was present in the same four samples.

Recovery of *Larrea* pollen was consistent in all samples from the adobe-lined pits and agricultural fields but generally was more sporadic in samples from structure floors, the fill above floors, and hearths at Pueblo Salado. Recovery of *Larrea* pollen in samples from the adobe-lined pits and agricultural fields indicated that the Hohokam farmers had not cleared all native vegetation from areas near the settlement. The *Cylindropuntia* pollen suggested that cholla might have grown near the fields. Recovery of small frequencies of *Typha* pollen from two of the five agricultural field samples suggested that it might have been transported by wind or canal water from a riparian community. Consistent recovery of *Sphaeralcea* pollen, as well as recovery of slightly elevated frequencies in samples from the adobe-lined pits and agricultural fields, suggested that *Sphaeralcea* had been a common weed in this area. *Zea mays* pollen was absent from initial scans of the field deposit samples. After additional screening to concentrate large pollen grains, analysts screened the pollen-rich residue from the five samples through a 53-micron mesh and scanned it for evidence of *Zea mays*, which was present in two of the samples. *Zea mays* pollen was not entirely eliminated by the destruction of organics in the phytolith samples and was recovered from two of them; thus, a total of three of the five samples from the agricultural fields yielded *Zea mays* pollen.

With the exception of Sample 2366, the five samples exhibited relatively large quantities of sponge spicules, suggesting the introduction of irrigation water from the Salt River (Volume 3:Figure 9.2). The frequencies of pillow-shaped phytoliths were relatively high in four of these samples (2364, 2366, 2367, and 2373). Small quantities of *Opuntia* crystals recovered from Samples 2360 and 2367 indicated that either prickly pear or cholla had grown in the vicinity of the fields. Analysts recovered a large cross-shaped phytolith (20 microns on the long axis and 15 microns on the short axis), generally indicative of maize (Pearsall 1989; Piperno 1988), from Sample 2366. They also recovered small cross-shaped phytoliths that may be produced by Panicoid grasses, including maize, from Samples 2366 and 2367. Starches would not be expected to occur in agricultural field deposits, and analysts noted very few starch granules in the samples. Starches from cultigens, especially maize, would be present in an agricultural field context only under unusual circumstances, such as if kernels had been processed in the field; if processed kernels (e.g., in the form of ceremonial corn meal), had been introduced; or if kernels had failed to germinate, had been lost during harvesting, or had been taken by rodents and birds and had decayed in the field. The presence of maize starches, or starches resembling those of maize, might also have resulted from the growth and decay of grasses.

Analysts recorded *Zea mays* pollen in the scans of the large-sized pollen for Samples 2360 and 2366 (Volume 3:Figure 9.1) and in Samples 2360 and 2373 during phytolith analysis. Recovery of *Zea mays* pollen from three of the five samples suggested that the Pueblo Saladoans had indeed used this area as an agricultural field and had grown corn in this location. The presence in most of the agricultural phytolith samples of large quantities of sponge spicules was consistent with the premise of irrigation of agricultural fields; the spicules, which require moist or wet conditions, may have originated in the Salt River.

Evaluating the Pollen and Phytolith Evidence

In general, the pollen content from the samples from the three adobe-lined pits resembled that of the five samples from the irrigation soils. However, the samples from the irrigation soils exhibited higher frequencies and more regular occurrence of *Eriogonum* pollen. In addition, *Larrea* pollen occurred more regularly in the irrigation soil samples, although differences in frequencies were not significant between the two sample contexts. Pollen typically considered to represent agricultural weeds, including *Boerhaavia* (spiderling), *Sphaeralcea*, and *Kallstroemia* (carpetweed), was less abundant in samples from the irrigation soils than in samples from habitation features. *Zea mays* pollen was less abundant from the irrigation soil samples than from samples from habitation floors and hearths. *Typha* pollen was present in smaller quantities and in fewer samples from irrigation soils than from habitation features. Less *Typha* pollen also appeared in the irrigation soil samples than in the samples from the adobe-lined pits. *Zea mays* pollen occurred in the agricultural field samples, but in very small quantities. Analysts also noted it in the adobe-lined pits in small quantities. The recovery of maize pollen from Feature 8/9-53, one of the adobe-lined pits, might represent the addition of maize to the adobe, possibly as a temper agent or as a ceremonial item. The mere presence of *Zea mays* pollen may not be an effective method of identifying agricultural field deposits, because it is present in many types of samples at Hohokam sites.

The phytolith record might be more useful in distinguishing irrigation soils from other deposits. A comparison of the phytolith contents of samples from suspected agricultural fields with those from adobe-lined pits yielded a few areas of difference. Analysts recovered Panicoid phytoliths from all samples from the irrigation soils but not from all samples from the adobe-lined pits. *Zea mays* belongs to the Panicoid group of grasses and produces both typical Panicoid dumbbell-shaped phytoliths and cross-shaped phytoliths. In addition, analysts recovered both large and small crosses from the irrigation soils but found none in the samples from the adobe-lined pits. Sponge spicules were present in both types of samples (Volume 3:Figure 9.2), which would be expected if the sediments being examined were part of an irrigated agricultural field, a canal, an adobe-mixing pit, or any other area where water flowed or was introduced. Presumably sponge spicules would not be as abundant in samples from habitation proveniences, unless flood deposits were present.

Pollen or phytoliths may also assist in distinguishing different source areas for raw materials used in adobe manufacture, different periods (seasons) of building or remodeling, and use of organic matter to prevent shrinkage and cracking when the adobe dried. Analysis of pollen from historic adobe bricks (O'Rourke 1983:39–40) has identified the presence of vegetal matter used in the construction of the bricks. Although project investigators did not intend to use phytolith analysis to address this topic, the identification of grass and maize remains from the adobe-lined pits suggests it would be worthwhile to evaluate the possible use of these and other plants in prehistoric adobe construction. Further research is needed along these lines.

The pollen record from the irrigation soils did not exhibit specific evidence in the form of *Zea mays* or weedy pollen recovery that would conclusively separate these samples from either habitation features or adobe-lined pits, particularly on an individual basis. The examination of multiple samples from field areas, however, yielded a pattern of reduced frequencies of Cheno-am pollen and reduced quantities of Cheno-am aggregates, accompanied by increased quantities of pollen representing native plants such as *Eriogonum* and *Larrea* and a reduced variety of pollen representing economic plants, when compared with samples from living areas. The difference between the irrigation soil samples and the adobe-lined pit samples was considerably less.

In the pollen record from the irrigation soils at Pueblo Salado, botanical remains that may be associated with fields included a regular presence of *Larrea*, slightly increased *Eriogonum* frequencies, a regular presence of grasses, and, overall, a limited variety in the pollen record. These pollen samples did not exhibit a specific pollen assemblage that palynologists might identify as typical of irrigation agriculture. The phytolith record also did not display elements unique to agricultural fields, although the presence (usually in elevated frequencies) of sponge spicules appeared to be typical of irrigation soils, as it probably would be of canal deposits as well. This association should be examined during future research projects.

Evidence for domesticates, economic native plants, and noneconomic plants was preserved in both the pollen and phytolith records from the irrigation soil samples. However, rare occurrences of *Zea mays* pollen were the only indication of domesticates. Analysts noted pollen from economic native plants such as Cheno-ams, Gramineae, *Opuntia*, *Cylindropuntia*, and *Typha*. Non-economic plants represented in the pollen record included *Pinus* (pine), whose pollen had probably been transported from some distance on the wind; *Salix* (willow), possibly transported by wind or water from trees growing along the Salt River; Caryophyllaceae (pink family); *Tidestromia* (white-mat); various Compositae (sunflower family); *Ephedra* (Mormon tea); *Eriogonum*; *Euphorbia* (spurge); *Larrea*; *Boerhaavia*; and Onagraceae. The phytolith record contained evidence of a variety of grasses that included both cool-season and warm-season species, as well as short and tall grasses. The elevated levels of Buliform and Prickle-shaped phytoliths may be considered typical of agricultural field deposits, probably representing the increased water available to grasses growing in this area. Analysts also noted *Opuntia* crystals, though rarely, in the irrigation soils. A few starch granules noted in several of the irrigation soil samples probably represented the loss or decay of maize kernels in the field deposits.

The minimal variety of pollen from both weedy and native plants in the samples from the irrigation soils suggested that the Pueblo Salado farmers had probably intensively weeded their fields, encouraging few, if any, economically important plants. The absence of *Cucurbita* (squash family) and *Gossypium* (cotton) pollen, which were both present in samples from habitation features, is not unusual in these samples, as these pollen types are insect transported and normally very rare in the record. However, analysts found no direct evidence indicating that the Pueblo Saladoans had grown a domestic crop other than maize in these particular fields. The phytolith record exhibited limited evidence of the presence of maize in the form of a large cross-shaped phytolith and a few starch granules but no evidence of hook-shaped hairs from *Phaseolus* (beans) or from phytoliths typical of *Cucurbita* rind.

ARCHAEOLOGICAL EVIDENCE

Data recovery efforts concentrated on the southwesternmost area of oxidized soils (Figure 5.1), centered approximately on Trench 3015. Figure 5.1 shows the extent to which field personnel investigated the suspected fields and associated features. Backhoe trenching and horizontal stripping initially identified the oxidized soils and then exposed features for further investigation. These features were the focus during data recovery because the investigating team had initially discovered the oxidized soils and associated pit features in this trench. Hand excavations clarified the relationship of the oxidized soils and the pit features. Features 8/9-52, 53, 54, 55, 56, 141, and 145, all adobe-lined pits, predated the deposition of the flood sediments and development of the oxidized soils. Features 8/9-142 and 8/9-143, also adobe-lined pits, and Feature 8/9-144, a pit NFS (not further specified), postdated and intruded the oxidized soils. A layer of alluvium deposited during flooding by the Salt River protected these oxidized soils and the upper (intrusive) features from most historic and modern disturbances.

The few sherds recovered during hand excavations were not useful in establishing temporal associations for the individual features. Radiocarbon and archaeomagnetic samples were not available from these features for dating. However, the pits and oxidized soils could be dated by association. Investigators confidently assigned the earliest identified occupation at Pueblo Salado to the Soho phase (A.D. 1100–1300). The compound, approximately 35 m to the southeast, appears to have been occupied during the Civano phase (A.D. 1300–1350). Abandonment of the site occurred during the Polvorón phase (post–A.D. 1350). The development of the oxidized appearance of the soils could have occurred only after the use of the adobe-lined pits but prior to site abandonment, within a maximum span of 400 years. Therefore, the archaeological evidence supported the hypothesis that the oxidized soils had developed in a short period of time, probably as the result of repeated applications of irrigation water.

SUMMARY AND CONCLUSIONS

The SWCA investigating team employed a multidisciplinary approach to define the cause of the oxidized soils at Pueblo Salado, making use of geomorphological, soil chemistry, macro- and microbotanical, and archaeological data. Archaeologists have not previously defined the morphological attributes of buried fields in the Phoenix Basin. Other researchers have used soil sampling and botanical analyses to identify possible locations where the Hohokam conducted agricultural activities and have used surface remains such as rock alignments and piles to identify field locations. Within the Phoenix Sky Harbor Center, however, investigators had sufficient evidence to attribute the anomalous oxidized soil horizons to prehistoric agricultural activity rather than to natural formation processes.

Although analysts did not discover conclusive evidence, such as in Central America where excavators have found cultivation rows (rills) and casts of maize stalks (Zier 1983), each disciplinary study generated information to support the association of the oxidized soils with irrigation agriculture. Similar soil deposits may exist only in areas that have been subjected to catastrophic events such as floods or volcanic eruptions that would have buried and preserved them. In the Phoenix Basin and elsewhere, the preservation of features such as agricultural fields also depends on postabandonment processes. Historic farming and erosion may be the most devastating activities affecting the preservation of agricultural features (Huckleberry 1992). However, the floodplain of the Salt River appears to exhibit high potential for the discovery of other buried fields, due primarily to thick accumulations of alluvium.

Processes relating to the oxidized soils resulted in increased concentrations of nutrients, increased salinity, accumulations of finer soil constituents, and the occurrence of economic plant remains. The alteration in soil color appeared to have come about through repeated wetting and drying episodes. Analysts did not determine, however, if the extent of the oxidized soils accurately represented the actual field locations. Given the natural slope of the area, irrigation water may have accumulated in higher quantities at the "bottom" or lower end of the fields, resulting in excessive moisture levels in those locations; oxidation is more likely when soils are saturated rather than wetted just sufficiently for crop maturation. The level of oxidation of the irrigation-impacted soils differed from that of the oxidized soils commonly observed at the base of canals. Moreover, investigators did not observe iron and manganese oxides in the irrigation-impacted soils, suggesting that their saturation levels never reached those in canals. However, the red hue of the soils did support the notion that oxygen had been introduced into these soils through infiltration of water.

In conclusion, the oxidized soils at Pueblo Salado probably resulted from agricultural activities, namely, the application of irrigation water. These soils contained elevated nitrate levels and an increase in finer sediments, perhaps the result of fine-grained particles being dropped from suspension when the Pueblo Salado farmers applied irrigation water to their fields. The botanical remains included both maize and wild economic plants that archaeologists often find in association with agricultural fields. However, the botanical remains could also have come from plants that grew in the general area, perhaps within the compound, and could have been deposited within the oxidized soil horizon and pit features through natural or cultural processes. Ideally, this analytical approach should have included samples from a similar environmental and physical area that was not used historically or prehistorically for agriculture. Despite the problems inherent in this study, the geomorphological and archaeological data provided strong support for the contention that the oxidized soils represented prehistoric agricultural fields and thus that the oxidation was more likely the result of repeated irrigation than of natural conditions.

CHAPTER 6

THE PHOENIX SKY HARBOR CENTER CANALS

David H. Greenwald
Kirk C. Anderson

Canal studies have been a part of Hohokam research since the earliest historic records of Hohokam remains. In 1874, General James E. Rusling published the first map of the Phoenix Basin that illustrated alignments of prehistoric irrigation canals (Rusling 1874). Adolf Bandelier was the first to produce a written record that mentioned the prehistoric canals of southern Arizona during his visit to the Phoenix Basin in 1883 (Bandelier 1884). Several researchers have recounted the history of pre-A.D. 1900 archaeological investigations (Howard 1991b:2.2–2.8; Turney 1929; Wilcox 1987:9–19), including Emil W. Haury's (1945) report on contributions made by Frank H. Cushing (1890, 1892) for the Hemenway Southwestern Archaeological Expedition. Much of what archaeologists know about some sites in the Phoenix Basin they have gathered from early maps produced by members of the Hemenway Expedition, including those by Frederick W. Hodge and Charles F. Garlick (Matthews, Wortman, and Billings 1893) and James Goodwin (1887). Herbert R. Patrick (1903), a local surveyor, independently began mapping the canals and sites of the Phoenix Basin as early as 1878. Surface indications of archaeological sites and canals were obvious features that spread across the natural landscape and were readily recognized. In fact, in their attempt to secure federal aid for water storage (Turney 1929:7), settlers cited the existence of prehistoric canals as evidence that they could farm the Phoenix Basin through canal irrigation (Chapter 1).

Turney began compiling information on Phoenix Basin canals by using data gathered previously by Hodge and Garlick, Goodwin, and Patrick. He viewed the often-complex network of canals as integrated systems (Figure 1.2). Following Turney's death, the Smithsonian Institution undertook the first attempt to record Hohokam canal systems on a regional basis through the organized efforts of Odd S. Halseth (1932) and Neil M. Judd (1931). The Smithsonian intended to conduct aerial photography along the Salt and Gila rivers, recording prehistoric canals. Although this project was never completed, the data collected impressed later researchers with the extent and complexity of the Hohokam canal system.

Researchers involved in Hohokam canal studies have taken two directions. Early on, Hohokam scholars focused principally on the spatial aspects of the canals: mapping them, identifying their extent, and defining the characteristics of the areas they served, such as soil condition, slope or grade, and topography. In 1937, Haury (1965) made one of the first attempts to study canal morphology and function. After investigations at Snaketown had defined the chronological sequence of the Hohokam, researchers studied temporal ordering of the canals as well. In the late 1950s, Woodbury (1960, 1961b) conducted studies at Pueblo Grande that included descriptions of canal morphology. His interests led him to further studies at Snaketown in 1964 with Haury (1976). Frank Midvale (1965, 1968, 1970, 1974), however, continued as Turney's protégé in mapping the areal extent of canal networks throughout the Salt and Gila valleys, efforts that have proven invaluable to other researchers.

Since 1970, the number of canal studies has increased many times over the amount of work done prior to the era of cultural resource management projects. Highway projects initiated some of the earliest canal studies (Herskovitz 1974; Masse 1976) and have continued to be the primary impetus of many later studies (Ackerly and Henderson 1989; Ackerly, Howard, and McGuire 1987; Greenwald and Ciolek-Torrello 1988a; Howard and Huckleberry 1991; Huckleberry 1988, 1989;

Masse 1987) in the Phoenix metropolitan area. Academic studies have made major contributions as well, discerning system detail from aerial photographs (Nicholas 1981), searching for clues to social organization by examining interrelated canals (Howard 1987; Howard and Huckleberry 1991; Nicholas and Neitzel 1984), and studying canal morphology and mechanics (Howard 1990; Howard and Huckleberry 1991; Lombard 1988; Nials and Gregory 1989). The Phoenix Sky Harbor Center Project, the most recent of these studies, provided an unusual opportunity within metropolitan Phoenix to examine canals in an area that exceeds one square mile, with canal segments up to one mile in length available for study. In tandem with projects sponsored by the Arizona Department of Transportation and the Federal Highways Administration, archaeological research on the largest prehistoric canal systems in North America has yielded information on the complex and changing structure of Hohokam sociopolitical organization. The canal systems in southern Arizona, especially those in the Phoenix Basin, exceed those of Mesoamerica in terms of size and technology (Doolittle 1990). Despite the rise of today's Phoenix over the Hohokam ruins, archaeologists have managed to glean important information about the canals and their significance in the prehistory of the Phoenix Basin.

This chapter examines the irrigation capacity of the project area and estimates the amount of arable land at Pueblo Salado and Dutch Canal Ruin, including areas beyond the project boundaries. All canal features (primarily prehistoric, but historic canals as well) and other components of the canal system are described along with interpretations of their function. The chapter concludes with a discussion of irrigation methods based on the various canal-related features found during the investigations.

PALEOHYDRAULIC RETRODICTIONS

Paleovelocity and Paleodischarge

Geomorphologists and archaeologists can use paleovelocity and paleodischarge retrodictions to characterize past canal usage based on extant canal dimensions. Of course, numerous problems exist when an investigator is attempting to reconstruct hydraulic conditions that existed hundreds of years ago. Nevertheless, paleovelocities of water flow within canals provide information with which investigators can identify conditions of erosion or deposition, sizes of clasts (rock fragments) transported by the flow, and whether human intervention might have been necessary for the maintenance of efficient irrigation systems. The most efficient system requires little labor investment to achieve optimum water conditions for a particular crop (Howard and Huckleberry 1991). In general, open-channel irrigation systems must have velocities below one meter per second (m/s) to maintain a balance between erosion and deposition (Israelsen and Hansen 1962). The sandy loam texture of the substrate at the Phoenix Sky Harbor Center would have required velocities less than 1.0 m/s; however, below 0.8 m/s, deposition of silts would have occurred and slope would have decreased, and infilling with sediments may then have become a problem.

After calculating paleovelocities, analysts apply the continuity equation to retrodict discharges. They then use the derived discharges to estimate the amount of land that the canals could have irrigated. Finally, they estimate crop yields and potential food production levels from the amount of land under irrigation. In this chapter, analysts also compare the number of hectares (ha) that farmers of the project area most likely had enough water to irrigate (based on the discharge retrodictions) with the amount of land that potentially could have been placed under irrigation. This comparison, however, lacks information regarding topographic anomalies that may have restricted the use of some areas for fields; the ratio of available water to available arable lands therefore may be conservative.

To calculate paleovelocities and paleodischarges for the relict canals at Dutch Canal Ruin and Pueblo Salado, project analysts used the Manning equation. This equation incorporates slope, channel bottom roughness, and hydraulic radius to determine flow velocities in open-channel systems. The hydraulic radius is simply the cross-sectional area divided by the wetted perimeter according to the following equations:

$$R = A/P$$

$$A = 2/3w*d$$

$$P = 2/3w+2d$$

where R is the hydraulic radius, A is the cross-sectional area, w is the width, d is the depth, and P is the wetted perimeter. Analysts employed a one-third freeboard calculation to evaluate the presumed hydraulic conditions during canal use at two-thirds full (Lombard 1988; Masse 1976). The Manning equation used for metric calculations is

$$Vm = R^{2/3} * s^{1/2}/n$$

where Vm is the Manning velocity, s is the slope, and n is the Manning roughness coefficient.

The value for s is 0.0024, which is the regional gradient (Greenwald 1988). Although project investigators measured canal gradients and considered these measurements when describing sediments and canal morphologies (see below), the average for all of the measured canal gradients approximated the regional gradient. In addition, field investigators measured the gradient from the bottom of the canal channel system and not the bottom of the individual canals used for paleovelocity and paleodischarge retrodictions.

Manning's n is a coefficient of roughness, which attempts to incorporate an empirically derived value for frictional resistance to flow generated at the contact between the flowing water and the channel bottom. The value of n is 0.03, which is for straight, relatively smooth alluvial channels with some vegetative growth (Chow 1959). Smooth, unvegetated channels have low n values, whereas vegetated channels have high n values, indicating the presence of rough bottoms impeding flow and decreasing velocity.

As mentioned previously, analysts calculated discharge from the continuity equation, which is the product of flow velocity and the cross-sectional area:

$$Q = AV$$

where Q is discharge (m^3/s), A is the cross-sectional area (m^2), and V is the calculated Manning velocity (m/s).

Paleodischarge and the Amount of Irrigable Land

Analysts for this project used the calculation provided by Haury (1976:144) and used by Lombard (1988) to determine the amount of land that could have been irrigated by the canals given the retrodicted discharges. This calculation indicates that Hohokam farmers would have needed 0.028 m^3/s of water to irrigate a minimum of 16 ha and a maximum of 28 ha. The following section presents paleodischarge figures and the potential number of hectares that those discharges could have irrigated (Table 6.1). However, the potential area that could have been irrigated by each canal may have been greater than the derived calculations, since the discharge values are based on canal segments located more than 3 km below their heads. Canal dimensions decrease with an increase in distance from the head; this could affect the accuracy of the calculations, which err on the conservative side. Refer to Figure 6.1 for location of project canals.

Canal Salado

Retrodicted discharges for the north branch of Canal Salado within the Phoenix Sky Harbor Center ranged from 0.47 to 1.10 m^3/s, with an average discharge of 0.79 m^3/s. This average discharge would have provided enough water to irrigate between 449 and 785 ha. The amount of land occupied by Pueblo Salado is approximately 148 ha, and the tract of land between the intake along the Salt River and the terminus of Canal Salado at Turney's Gully is approximately 385 ha. Thus, the discharge provided by Canal Salado apparently would have been sufficient to irrigate the areas contained by the "alluvial island" on which Pueblo Salado was situated.

The North and South Main Canals and Alignment 8002 at Dutch Canal Ruin

The retrodicted discharges for the North Main Canal ranged from 0.41 m^3/s to 1.83 m^3/s, with an average discharge of 0.95 m^3/s, enough water to irrigate between 543 and 950 ha. The retrodicted discharges for the South Main Canal alignment had a broader range (from 0.10 to 2.49 m^3/s), with an average discharge of 0.98 m^3/s, enough water for between 558 and 976 ha. Alignment 8002, which connected the North Main Canal to the South Main Canal, had a range of discharges from 0.47 m^3/s to 1.30 m^3/s, with an average discharge of 0.97 m^3/s. This average discharge could have provided enough water to irrigate between 554 and 969 ha. These canals thus all exhibited similar discharge capacities. It is important to note that the canals that comprise the North and South Main Canal alignments were not contemporaneous but represent re-engineered channels that made use of earlier canal segments.

Moreover, the above discharge values compare favorably to Lombard's (1988) calculations for the North and South Main Canal alignments in the Squaw Peak Parkway. By examining topographic maps, project analysts estimated the amount of irrigable land within Dutch Canal Ruin to be 128 ha. The tract of land between the Park of Four Waters along the Salt River and the possible terminus of these canals covers approximately 448 ha. This expanse includes land to the east and west of Dutch Canal Ruin (as currently defined) where investigators documented the main canal alignments. According to the discharge retrodictions presented above, the combined discharges from both the North and the South Main Canals would have been sufficient to supply water to the fields of Dutch Canal Ruin as well as fields farther to the east or west. Perhaps the westernmost extent of land serviced by the South Main Canal lay immediately east of Casa Chica, where other archaeologists recently documented Canal Colinas, which may have serviced the area west of modern-day 7th Street (Schroeder 1991). Similarly, the North Main Canal may have served the area as far west as 7th Avenue, as illustrated by Turney (1929; see Figure 1.2), and south to Canal Colinas.

Table 6.1. Velocity and Discharge Retrodictions for Prehistoric Canal Alignments

Feature	Trench	W(m)	D(m)	W/D	P(m)	A(m^2)	R	V(m/s)	Q(m^3/s)	Ha(min)	Ha(max)
Canal Salado											
3113	3051	2.30	0.80	2.88	3.13	1.23	0.39	0.89	1.10	628	1100
3113	3075	2.40	0.60	4.00	2.80	0.96	0.34	0.82	0.79	450	788
3113	3070	1.50	0.65	2.31	2.30	0.65	0.28	0.72	0.47	268	468
Average		**2.07**	**0.68**	**3.06**	**2.74**	**0.95**	**0.34**	**0.81**	**0.79**	**449**	**785**
North Main Canal											
8001[3]	2040	1.60	0.55	2.91	2.17	0.59	0.27	0.70	0.41	235	411
8001[1]	1147	2.70	1.00	2.70	3.80	1.81	0.48	1.01	1.83	1048	1834
8001[3]	1145	2.30	0.65	3.54	2.83	1.00	0.35	0.83	0.83	475	832
8001[2]	1144	2.90	0.55	5.27	3.03	1.07	0.35	0.83	0.89	506	885
8001[1]	1144	2.30	0.90	2.56	3.33	1.39	0.42	0.93	1.28	734	1285
8001[2]	1120	1.90	0.50	3.80	2.27	0.64	0.28	0.71	0.45	259	453
Average		**2.28**	**0.69**	**3.46**	**2.91**	**1.08**	**0.36**	**0.84**	**0.95**	**543**	**950**
Alignment 8002											
8002a[2]	1309	3.10	0.65	4.77	3.37	1.35	0.40	0.90	1.22	697	1220
8002b[3]	1309	2.50	0.40	6.25	2.47	0.67	1.27	0.70	0.47	266	466
8002a[2]	1308	2.10	1.00	2.10	3.40	1.41	0.41	0.92	1.30	742	1298
8002b[3]	1308	2.10	0.75	2.80	2.90	1.06	0.36	0.85	0.89	511	893
Average		**2.45**	**0.70**	**3.98**	**3.04**	**1.12**	**0.61**	**0.84**	**0.97**	**554**	**969**
South Main Canal											
8003	2098	2.50	0.70	3.57	3.07	1.17	0.38	0.88	1.03	586	1026
8003	2093*	1.00	0.30	3.33	1.27	0.20	0.16	0.49	0.10	56	98
8003	1177	1.60	0.70	2.29	2.47	0.75	0.30	0.75	0.56	322	563
8003	1173	2.20	0.60	3.67	2.67	0.88	0.33	0.80	0.70	402	704
8003	1074	3.40	1.00	3.40	4.27	2.28	0.53	1.09	2.49	1425	2493
8003	1082	2.10	0.80	2.63	3.00	1.13	0.38	0.86	0.97	556	973
Average		**2.13**	**0.68**	**3.15**	**2.79**	**1.07**	**0.35**	**0.81**	**0.98**	**558**	**976**
Northern Canal Alignments											
8501	2022	4.15	0.90	4.61	4.57	2.50	0.55	1.11	2.79	1593	2787
8501	2022	2.80	1.00	2.80	3.87	1.88	0.49	1.03	1.93	1101	1926
8537	2093	4.10	1.80	2.28	6.33	4.94	0.78	1.41	6.98	3989	6982

Note: Depths calculated after excavation.
W = width
D = depth
P = wetted perimeter
A = area
R = hydraulic radius
V = Manning velocity (for Manning equation, n=0.03, s=0.0024)
Q = discharge
Ha(min) = minimum hectares; Ha(max) = maximum hectares

[1]North Main Canal 1
[2]North Main Canal 2
[3]North Main Canal 3
*data do not reflect complete information

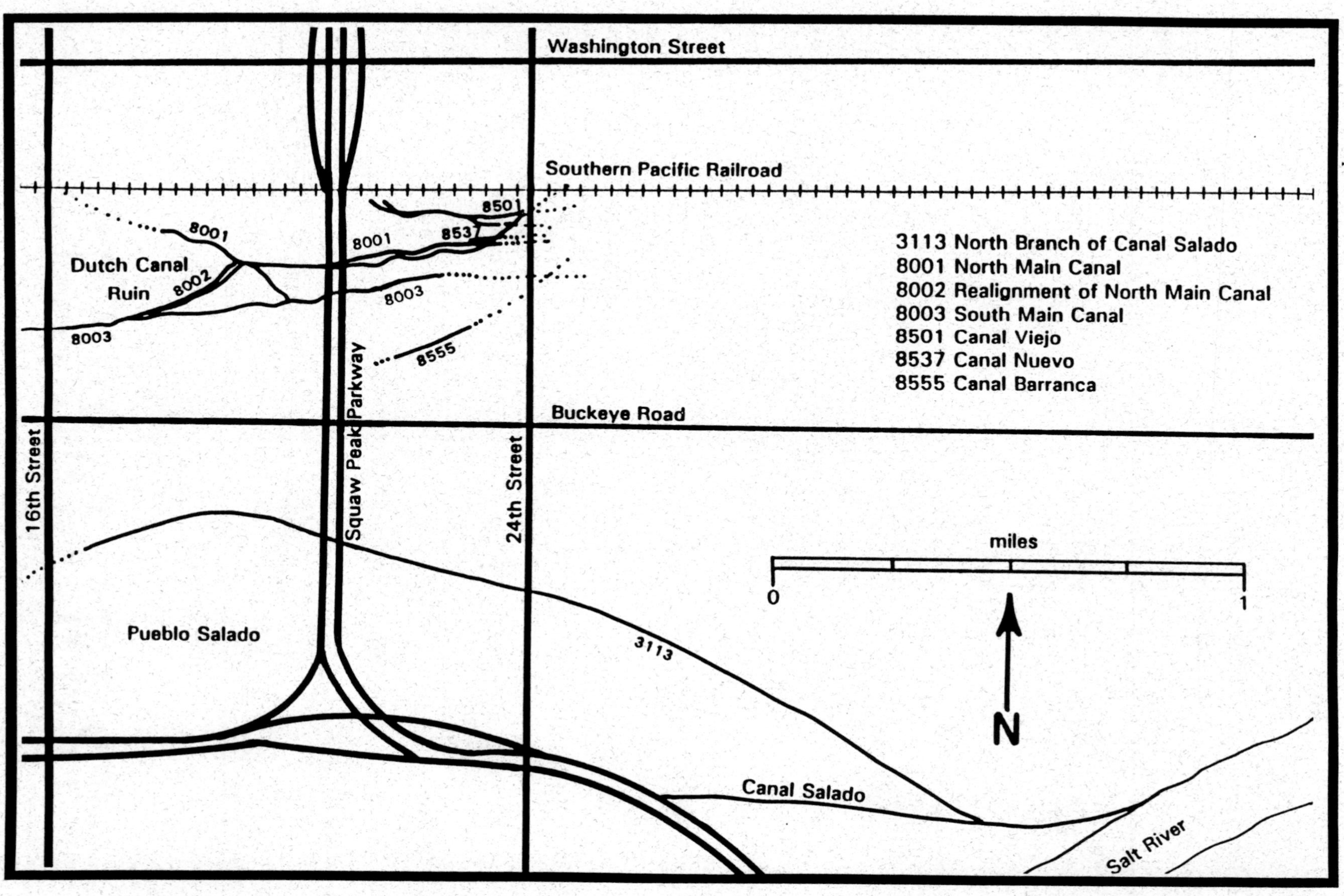

Figure 6.1. Canal alignments in the Phoenix Sky Harbor Center.

Canal Viejo and Canal Nuevo at Dutch Canal Ruin

Alignment 8501 (Canal Viejo), located in the northern portion of Dutch Canal Ruin, had estimated discharges ranging from 1.93 m^3/s to 2.79 m^3/s. Such discharges would have been capable of irrigating between 1101 and 2787 ha. This canal exhibited a northwesterly trend and may have supplied water to field areas in that direction. Alignment 8537 (Canal Nuevo), also located in the northern portion of Dutch Canal Ruin, had the highest discharge of all the canals in this study: 6.98 m^3/s, enough water to irrigate between 3989 and 6982 ha.

Each of these two northern canal alignments potentially could have carried enough water to irrigate far more arable land than is available in the Dutch Canal Ruin area. These noncontemporaneous canals, with higher velocities, discharges, and hydraulic heads, may have carried water to fields farther west, generally bypassing the field areas already irrigated from the more southerly canals at Dutch Canal Ruin.

Paleohydraulics and Erosion

Scholars in various fields have made numerous attempts to quantify erosion, deposition, and transportation velocities for various grain sizes and textural classes of sediments. Hjulstrom (1938), for example, empirically derived such velocities for clay, sand, silt, and gravel sizes. The limitations of this method include the fact that alluvial channels are a mixture of various grain sizes. In response, Fortier and Scobey (1926) derived limiting velocities for mixed textures (Table 6.2). Huckleberry (1991) provided an excellent review of paleohydraulics and sediment transport as it relates to canals in the Phoenix Basin.

Table 6.2. Limiting Velocities for Essentially Straight Canals after Aging

Material	Velocity (m/s) (water with colloidal silts)
Fine sand	0.80
Sandy loam	0.80
Silt loam	0.91
Alluvial silts (noncolloidal)	1.07
Ordinary firm loam	1.07
Alluvial silts (colloidal)	1.52
Fine gravel	1.52

Note: Modified from Fortier and Scobey 1926

The textures of the substrate traversed by the canals of the project area ranged from sandy loam to noncolloidal alluvial silts. Erosion of these materials occurs at limiting velocities ranging from 0.80 m/s to 1.07 m/s (Table 6.2). Indeed, except for the northern canal alignments, which fell far outside the velocity discharge and irrigable acreage averages of all the other alignments, the average velocities for the main canals ranged from 0.81 m/s to 0.85 m/s (Table 6.1), and even the smallest channels apparently lay at a critical break-even point between erosion and stability.

Velocities for the northern canal alignments, by contrast, were much higher, ranging from 1.03 to 1.41 m/s, and probably had caused problems with erosion.

Problems Associated with Relict Canal Retrodictions

Retrodicting paleohydraulics based on width and depth measurements is a convenient means of analyzing past canal characteristics. However, numerous problems exist in retrodicting velocity and discharge measurements from relict canals. Due to erosion, natural channel migration, and human re-engineering, channels are poorly preserved, and excavators often have difficulty in identifying a single channel. Canals exposed in backhoe trenches may be up to 20 m wide, with numerous channels and channel remnants only a few meters wide.

An individual channel may be only partly preserved if historic or recent irrigation has masked or destroyed its upper portions. When determining canal width and depth, investigators must estimate the upper measurements at the bottom of the plow zone, even though they may not be able to discern the channel dimensions. Since accurate retrodictions depend on accurate width and depth measurements, problems arise when complete dimensions are unavailable. Measurement of canals following excavation often yields dimensions different from those recorded during testing. In addition, even if dimensions are clearly observable and accurately recorded, analysts cannot know the prehistoric water depth when the canals were in use, making the resulting calculations, again, only estimates (Lombard 1988).

The exposure of the channels in backhoe trenches is important and must be at right angles to the canal alignment. Exposures at obtuse angles indicate wider channel dimensions than actually exist, yielding errors in width measurements and thus errors in velocity and discharge retrodictions.

Evidently the canals at Dutch Canal Ruin and Pueblo Salado progressed through fairly predictable stages. Initially, channels would have had parabolic shapes and wide, sandy bottoms, suggesting high flow velocities in an erosional system. Later, each canal would have deposited silt and clay instead of sand because with decreased water velocity the canal could no longer transport sand into the system. With few exceptions, canals in the Phoenix Sky Harbor Center illustrated this progressive infilling with an increase in time and use. If left to the natural processes at work in alluvial channels, this progressive infilling would have persisted uninterrupted. However, due to human intervention in altering slope and water volume (hence velocity and discharge), variation in the sediment texture was not uncommon.

As a result of channel migration and erosional processes, the latest channel will be most completely preserved and is the one measured. This channel often appears to be the smallest channel within an alignment, and velocity calculations derived from this channel can therefore be misleading. The most recent channel often records the dying energy of a canal, not the most vibrant, active period. It is a mere remnant of the former maintained canal, when maximum flow and minimum erosion and sedimentation would have been the goal. The characteristic clays that represent the later periods of flow regimes for most canals result from a decrease in canal maintenance leading to sluggish flow and deposition. Although slow flows are desirable to decrease erosion, sedimentation of clays to the point of channel filling is not.

Discharge retrodictions based on channel measurements are further complicated by the fact that a high volume of water is lost to seepage and evapotranspiration. Although paleohydrologists have derived figures for determining the amount of water lost through these processes, project analysts made no effort to improve on Haury's (1976) estimate, which takes these problems into consideration. In addition, the calculation for determining the number of hectares irrigated does not consider the type of crop, season of growth, preexisting soil moisture conditions, or stage of plant growth. Such complications are beyond the scope of this work but may be important for refining paleohydraulic retrodictions in future studies.

PROJECT CANALS AND IRRIGATION SYSTEMS

Prehistoric Canals

Within the project area excavators identified seven prehistoric canal alignments; some alignments comprised multiple channels, illustrating the complexity of even the simplest canal networks. Each canal alignment is defined below in terms of its function, morphology, discharge and gradient, and temporal associations (see also Figure 6.1). Two historic ditches known to have extended through the Phoenix Sky Harbor Center project area also are described.

Alignment 8001

Investigators traced Alignment 8001, the North Main Canal at Dutch Canal Ruin, for approximately 1160 m, nearly the entire width of the project area. They identified three separate main canals within this alignment: North Main Canal 1, the earliest; North Main Canal 2, the second in the sequence; and North Main Canal 3, the youngest. North Main Canal 2 contained two channels that diverged at Area 6 (Volume 2:Chapter 7). The earlier channel followed the same alignment as North Main Canal 1, while the later channel paralleled North Main Canal 3 to the southwest. The change in orientation of these three canals was temporally separate. Once the orientation changed from a northwest trajectory to a southwest trajectory, the northwest alignment was never re-established. For this reason, the following discussion focuses on North Main Canal 1 but also includes the initial alignment of North Main Canal 2.

Morphologically, North Main Canals 1 and 2 (Figure 6.2) exhibited scoured bottoms and coarse basal materials, suggesting an initially erosive history. Often a shallow V-shaped groove extended from the base of the canals into the culturally sterile deposits below the canals, exemplifying erosion associated with early use of this alignment. Through time and as the canals began to fill, flow episodes were controlled and water velocity became less erosive. In general, the canals seemed to have maintained parabolic shapes, although field investigators obtained few complete profile exposures of North Main Canal 1 because North Main Canal 2 often lay directly within the earlier channel. Numerous inner channels were present in both of these canals, and the complexity of the use history of these canals was further indicated by the microstratigraphic units represented by laminated bedding planes.

Analysts calculated the discharge of North Main Canal 1 at between 1.28 and 1.83 m^3/s, with a velocity of 0.93–1.01 m/s, using selected trench profiles of the various canals (Table 6.1). Similarly, they calculated the discharge of North Main Canal 2 at 0.45–0.89 m^3/s, with a velocity of 0.71–0.83 m/s. The gradient was 2.14 m/km for North Main Canal 1 within the project area and 2.60 m/km for North Main Canal 2, steep for canals and likely to result in erosive conditions. These gradients, however, closely mirrored the natural gradient of the area, calculated at 2.40 m/km (Greenwald 1988:85) and 1.90 m/km (Graf 1983), and their orientations closely followed the natural slope of the geologic floodplain. The numerous cut and fill episodes exhibited in this alignment probably resulted from natural channel migration, cleaning episodes, or attempts at realigning these canals. Because canal breaches occurred (Greenwald 1988) and were probably labor intensive to

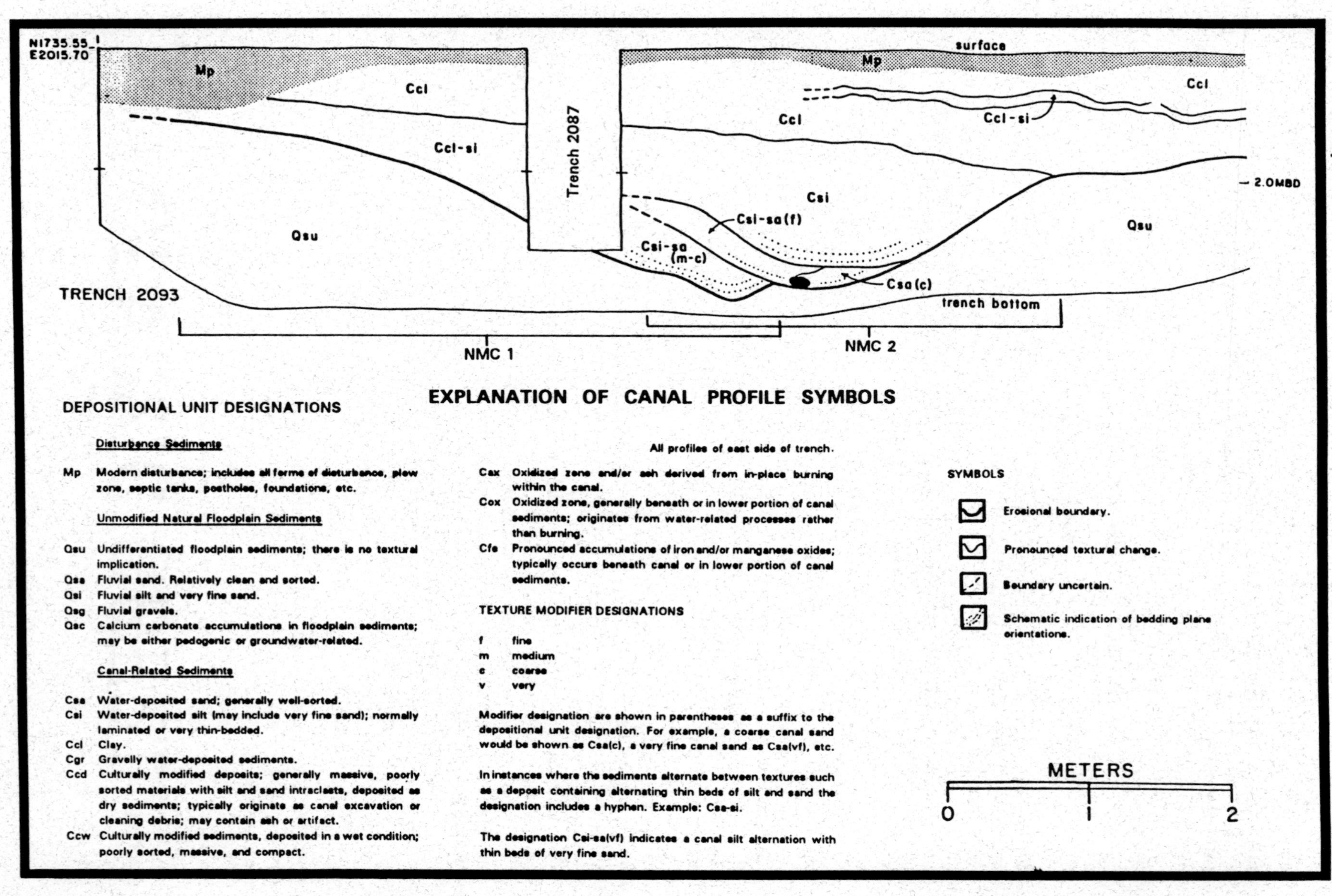

Figure 6.2. Profile of Alignment 8001, showing North Main Canals 1 and 2. Note the partial superpositioning of North Main Canal 2 on North Main Canal 1. This figure provides the complete key to canal profiles in the following figures.

repair, and because of the steep gradient, the farmers of Dutch Canal Ruin probably avoided achieving the maximum capacity of these canals to prevent erosional problems. Discharge and velocity did fluctuate, evidenced by radical changes in the texture of sediments and scouring rather than filling events.

Alignment 8001 appeared to be associated with the late Pioneer and early Colonial periods. Through archaeomagnetic dating and stratigraphic associations, analysts determined that the dates for North Main Canals 1 and 2 ranged between A.D. 650 and 730 (Volume 2:Chapter 19). Project investigators believed this alignment was Canal Patricio, as the canals compared closely with the alignment illustrated by Midvale (1966; Turney 1924, 1929).

Alignment 8002

This alignment corresponded with two channels of the North Main Canal: North Main Canal 3 and that section of North Main Canal 2 that extended to the southwest of the bifurcation in the western portion of Dutch Canal Ruin and North Main Canal 3 (Volume 1:Figure 8.1). Investigators traced the later channel, North Main Canal 2, only a short distance east of the bifurcation (Volume 2:Chapter 7) through horizontal exposure; they traced the earlier channel, North Main Canal 3, the entire width of the project area.

Following the testing phase, project analysts defined Alignment 8002 west of the bifurcation as a crosscut canal. This assignment seemed appropriate, as both channels had carried water from one alignment to another: specifically, from the North Main Canal alignment to the South Main Canal alignment (8003). When the canal builders diverted water to the southwest, however, they no longer used the northern alignment. As their water needs changed, they apparently chose to revitalize an earlier canal alignment (the South Main Canal) rather than create a system that allowed water to be diverted in two directions at one time or alternately. Therefore, Alignment 8002 was simply a continuation of a new alignment and had not functioned as a crosscut canal according to strict definition.

Investigators gained little knowledge about the morphology of the southern branch of North Main Canal 2 west of the bifurcation, although it was apparently stratigraphically shallower than its northern counterpart. Since they did not identify it until data recovery efforts in Area 6 of Dutch Canal Ruin (Volume 2:Chapter 7), they restricted to that area their attempts to isolate it in trench profiles. By contrast, a substantial amount of information exists for North Main Canal 3 (Volume 1:Chapter 8; Greenwald and Ciolek-Torrello 1988a; Lombard 1988) both east and west of the bifurcation. Morphologically, North Main Canal 2 was parabolic in shape, although at least two channel episodes were present (Figure 6.3). North Main Canal 3 exhibited a broad parabolic shape, which was directly related to channel migration (Figure 6.3). This shape was a consistent feature of North Main Canal 3 (Greenwald 1988) from the eastern end of the project area to the point at which it entered the South Main Canal (8003) alignment. Neither of these channels exhibited the basal erosional groove present in North Main Canal 1 and the earlier alignment of North Main Canal 2, suggesting that by the time these canals were put into service, the canal builders had learned to control problems relating to erosion. As with other project canals, numerous inner channels were present. Coarse materials were uncommon, as were fine laminates.

Discharge and velocity rates for the southern branch of North Main Canal 2 downstream of the bifurcation were 1.22–1.30 m^3/s and 0.90–0.92 m/s, respectively (Table 6.1), while those for North Main Canal 3 near the center of the project area (in the Squaw Peak Parkway) were 1.25 m^3/s and 0.87 m/s, respectively (Volume 1:Table 8.1). Gradient calculations are not available for North Main

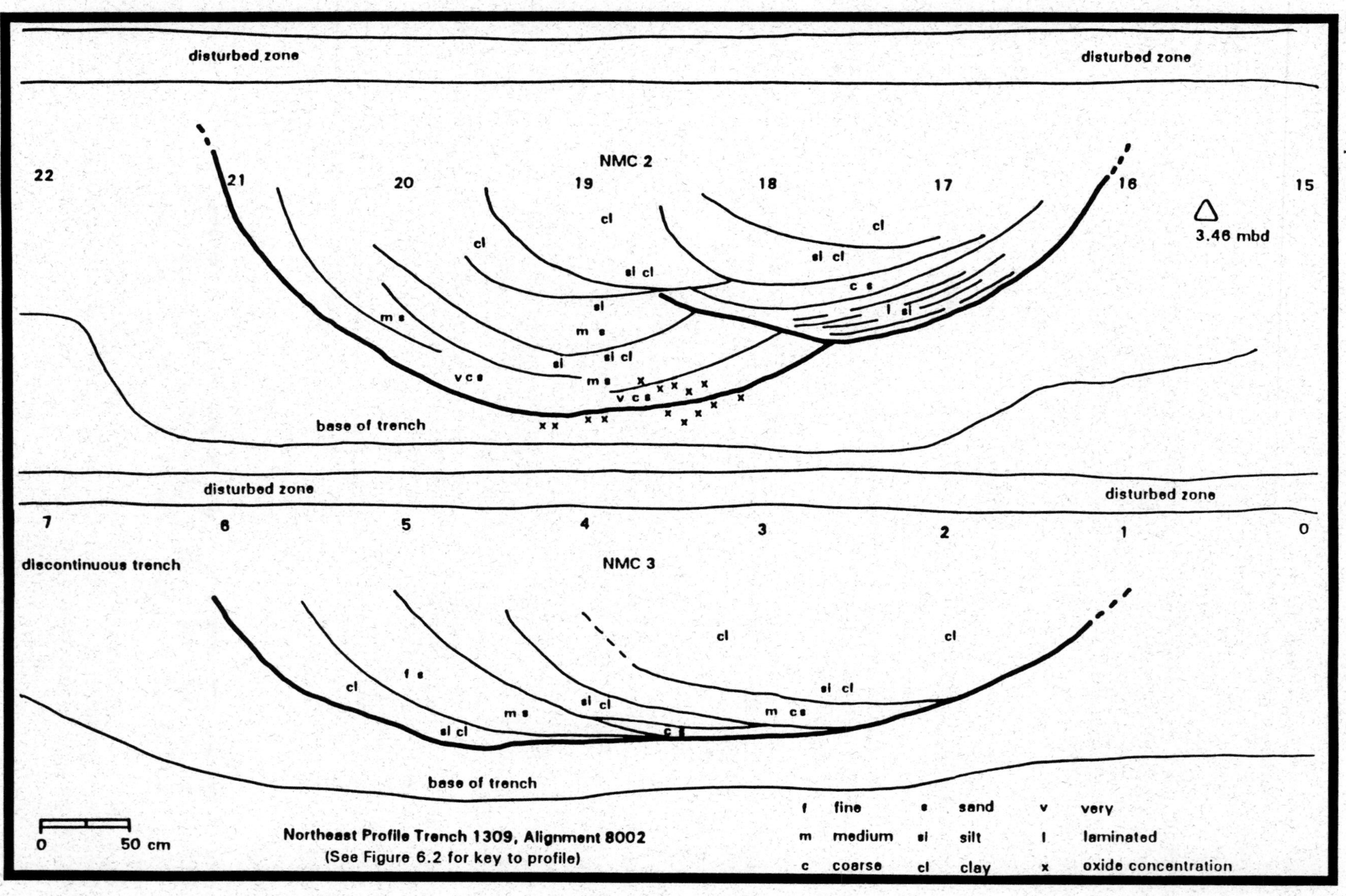

Figure 6.3. Profile of Alignment 8002, showing North Main Canals 2 (above) and 3 (below).

Canal 2 because of its late discovery. North Main Canal 3, however, maintained a gradient of 1.39–1.43 m/km across the project area. Given the natural gradient of the area, 2.40 m/km (Greenwald 1988:85), the canal builders would have needed to employ inner-channel devices such as dams to retard erosion and maintain the trajectory of this canal.

Project investigators used various means to make temporal assignments for these canals. North Main Canal 2 clearly postdated its northern counterpart based on stratigraphic evidence, which allowed archaeologists to date it to post-A.D. 730. It had been abandoned prior to the construction of North Main Canal 3, which investigators dated through archaeomagnetism (supported by stratigraphic evidence) to A.D. 830–930 (Volume 2:Chapter 19). These two channels probably represent the lower branch of Canal Patricio illustrated by Turney (1924, 1929) and named by Midvale (1966).

Alignment 8003

Alignment 8003 included the South Main Canal and, in the extreme western portion of the project area, North Main Canal 3 and the southern branch of North Main Canal 2. The canal builders had abandoned the South Main Canal prior to the realignment of the North Main Canal. Stratigraphic evidence clearly indicated that Alignment 8002 had been constructed within the South Main Canal channel (Figure 6.4). Field personnel traced Alignment 8003 for 1300 m, from the western project boundary to approximately E1800. Project investigators used information gathered from previous investigations in the Squaw Peak Parkway (Greenwald and Ciolek-Torrello 1988a) to define the function, morphology, and temporal associations of this canal alignment.

The South Main Canal exhibited considerable evidence of both basal and lateral channel erosion, and the fill sequence observed in various profile exposures indicated that it had carried a coarse bed load. Channel migration, due to erosion, was common, and builders had probably made numerous attempts to clean and rechannel the canal. Most inner channels exhibited a wide parabolic shape; however, investigators could not determine to what extent erosion had affected the widths of the channels. As with the North Main Canal alignment, the South Main Canal followed the natural slope of the geologic floodplain. Due to this orientation, the gradient would likely have been steep, accounting for the extensive erosion observed from one profile exposure to the next. Only in the extreme western portion of the Phoenix Sky Harbor Center did the extent of erosion appear to be somewhat less. This may have been a result of lower volumes or velocities of discharge in this area or may have been related to the topographic high on which the canal was located. A change in the natural gradient may have resulted in less bank erosion and channel migration as well as a lower gradient. In the farthest west trench in which the South Main Canal was exposed, the canal still exhibited a broad parabolic shape; however, the number of channels was much reduced compared to nearly every other exposure in the project area.

Discharge and velocity rates for Alignment 8003 ranged between 0.56 m^3/s and 2.49 m^3/s and 0.75 m/s and 1.09 m/s, respectively, exclusive of calculations based on Trench 2093 (Table 6.1). The wide range in the volume calculations was due principally to the difference between areas of extensive lateral erosion in the eastern portion of the project area and the less eroded profile exposure in the western portion. Overall, the South Main Canal had a gradient of 1.77 m/km. Although other canals at Dutch Canal Ruin had higher gradients, the South Main Canal exhibited substantially more erosion than those canals, except in the extreme western portion of the project area. The increased erosion may be related to discharge levels that approached the maximum capacity of the canal, while the western extent of the canal appeared to have been built on a topographic rise that reduced the natural gradient, thereby affecting the canal's gradient as well.

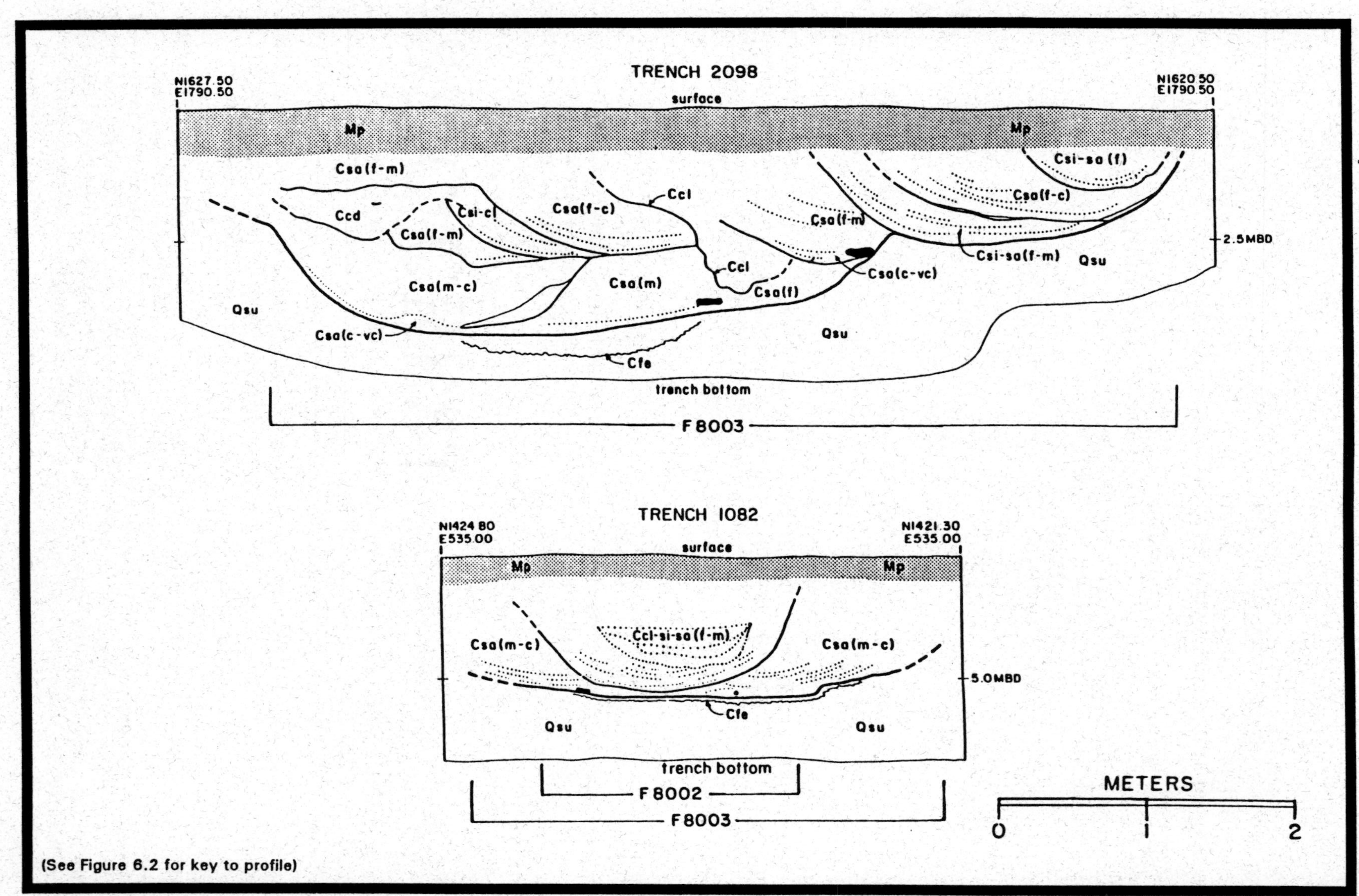

Figure 6.4. Profiles of the South Main Canal (Alignment 8003) in the eastern (above) and western (below) portions of the project area. Note the width of the South Main Canal in Trench 2098 compared to that in Trench 1082. Also note the inner channel in Trench 1082, which is the merged channels of North Main Canals 2 and 3 (Alignment 8002).

According to the fill sequence of the various flow episodes, the coarse materials and bedding planes supported the premise of high-velocity discharge.

Analysts determined that the South Main Canal, dated through archaeomagnetism to A.D. 600 to pre-A.D. 700, was the earliest at Dutch Canal Ruin. Two other lines of evidence supported this early assignment: stratigraphic data and diagnostic ceramics. The stratigraphic data include the later use of the South Main Canal by the lower branch of the North Main Canal alignment and the occurrence of Alignment 8005, a small channel that originated in the South Main Canal and extended to the northwest where the North Main Canal alignment was superposed on it (Figure 4.1). Although field personnel recovered few diagnostic sherds from the canal, they did find incised varieties, probably Snaketown and Gila Butte Red-on-buff types, in the fill (Greenwald and Ciolek-Torrello 1988a). Two trench exposures also indicated that the canal had been abandoned prior to the establishment of nearby field house loci, from which residents had dumped trash into the channel of the canal. These field house loci dated no earlier than the Gila Butte phase.

Alignment 8501

Alignment 8501, Canal Viejo, entered the Phoenix Sky Harbor Center at the eastern project area boundary (24th Street), followed a westerly course, and exited the project area at the intersection of the Squaw Peak Parkway and the Southern Pacific Railroad tracks (Volume 1:Figure 8.1). This was the earlier of two canals that followed this general trajectory and constituted the northernmost canal alignments in the project area. However, the two canals were temporally separated by at least 400 years.

Canal Viejo exhibited considerable erosion and rechanneling (Figure 6.5). Lateral channel migration was generally to the south, following the natural slope of the project area. Coarse sands common in the various channels indicated high-velocity discharge; these sands were most often present in the lower levels of the individual channels. Finer-textured sands were present at higher elevations, often interfingered with silts and clays in the upper levels. Basal erosion was present but not as common as in the North Main Canal alignment. A distinct clay lens located in the upper canal fill was present in most canal profiles, which enabled field investigators to trace Canal Viejo from east to west. This clay lens appeared to represent standing or very slow-moving water, possibly resulting from the last use of the canal or from postabandonment deposition.

Discharge and velocity reconstructions for this canal ranged from 1.93 m^3/s to $2.79^3/s$ and 1.03 m/s to 1.11 m/s, respectively (Table 6.1), based on two channels within Trench 2022. The degree of erosion expressed in the profile of this trench plus the slightly oblique angle of the trench to the canal may have affected volume and velocity reconstructions, resulting in exaggerated rates. The gradient for this canal, calculated over a distance of 300 m, was 2.57 m/km, one of the steepest within the project area. The steep gradient, although not manifested in the extent of basal channel erosion, was indicated through channel migration. The coarse-grained materials common in the lower portions of the channels were probably directly related to the steepness of the gradient.

Analysts did not obtain absolute dates for this canal and thus could not conclusively determine its period of use. Investigations in Area 5 at Dutch Canal Ruin (Volume 2:Chapter 6), however, determined that Canal Viejo temporally preceded North Main Canal 3, which was dated to A.D. 830–930. Therefore, Alignment 8501 was probably associated with the late Pioneer or early Colonial periods.

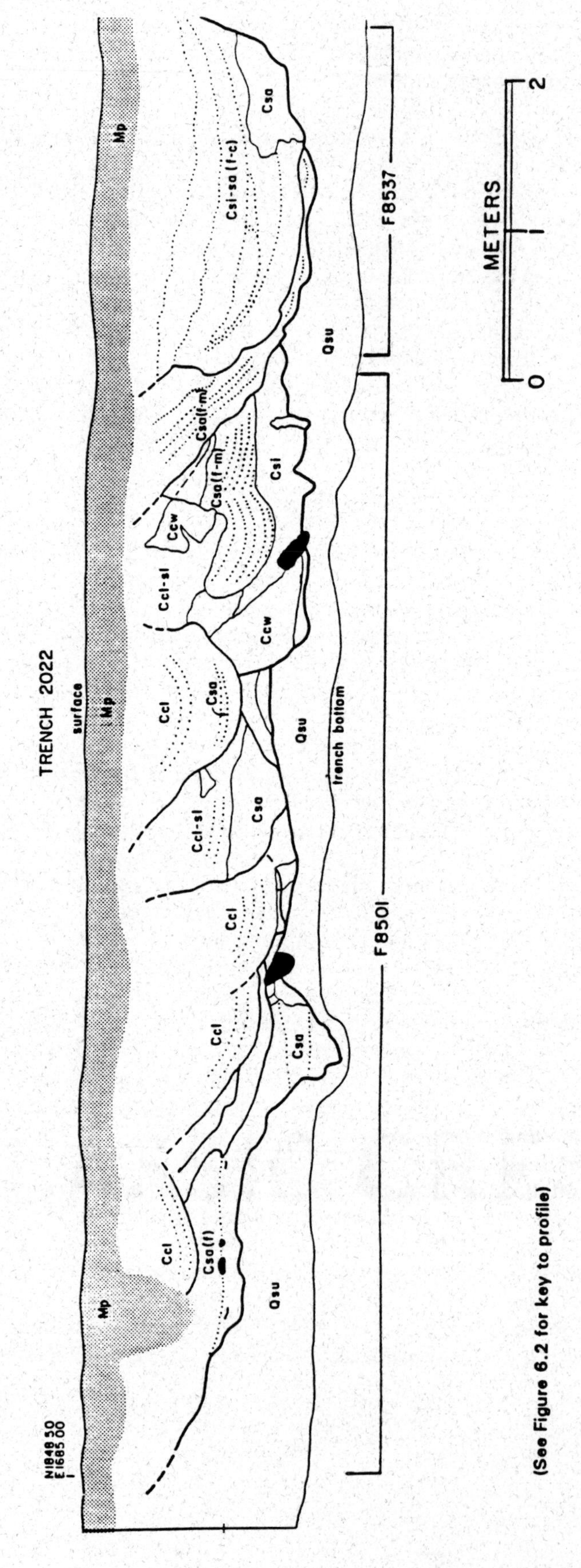

Figure 6.5. Profile of Canal Nuevo and Canal Viejo (Alignments 8501 and 8537). Note the morphological differences between these two canals.

Alignment 8537

This alignment, Canal Nuevo, followed a trajectory similar to that of the earlier Canal Viejo, entering the project area a short distance south of Alignment 8501, then merging with the earlier canal until a short distance before it exited the project area in the vicinity of the Squaw Peak Parkway and the Southern Pacific Railroad tracks (Volume 1:Figure 8.1). This canal exhibited a consistent parabolic shape and contained sediments composed of fine-grained materials with limited amounts of organic matter (Figure 6.5). Investigators identified no more than three rechanneling episodes in this canal, indicating either a short use history or little requirement for maintenance. Although bedding planes were present, the sediments suggested a non-erosive, low-velocity discharge.

In comparison with other project canals at Dutch Canal Ruin, Canal Nuevo shared many morphological similarities with North Main Canal 3, a Colonial period canal. Basal and lateral channel erosion did not occur in Canal Nuevo and was considerably reduced in North Main Canal 3 when compared to earlier canals in the Phoenix Sky Harbor Center. The fine-textured sediments in this canal suggested that water volume and velocity had been well regulated, although analysts could not determine whether flow had been intermittent or continuous. In general, Canal Nuevo was larger than North Main Canal 3 and exhibited few flow episodes, although this could be a result of increased irrigation technology and strategies for water regulation.

Discharge and velocity rates for Alignment 8537 were 6.98 m^3/s and 1.41 m/s, respectively (Table 6.1), the highest of any of the project canals; the gradient was 0.67, the lowest for any of the project canals. Although volume was high, velocity was controlled by the low gradient, resulting in a non-erosive flow history. The fill sequence and low organic content of the sediments indicated that the canal had functioned either very efficiently or for only a short period of time. Accordingly, standing water had not accumulated in this canal, which resulted in an organically rich environment.

Attempts to date this canal through archaeomagnetic means resulted in a wide alpha 95 with the date plot off the currently defined Southwest Magnetic Curve. Stratigraphic evidence indicated, however, that Canal Nuevo postdated Canal Viejo and probably postdated North Main Canal 3. Excavations in Area 7 of Dutch Canal Ruin (Volume 2:Chapter 8) produced two Roosevelt Red Ware Gila Polychrome sherds from the lower fill of the canal. The presence of these sherds supported the placement of this canal in the Classic period. Such a placement was further supported by the canal's morphology and sedimentology, which compared closely with other Classic period canals on the second terrace rather than with the pre-Classic period canals on the first terrace (Huckleberry 1988:119–146).

Alignment 8555

Alignment 8555, Canal Barranca, was an anomaly at Dutch Canal Ruin. It was the southernmost canal at the site and was oriented from northeast to southwest. Project investigators derived its name from the westernmost portion of the canal extending into one of the braided channels of the Salt River, referenced as Turney's Gully (Figure 3.1). The section of the canal within the Phoenix Sky Harbor Center appeared to be the terminal portion, and it may have been intentionally constructed to drain into Turney's Gully.

Morphologically, Canal Barranca exhibited considerable lateral erosion, resulting in exaggerated profile exposures. Basal erosion was not as apparent, perhaps because the base of the canal was restricted by the shallow gravel and cobble substrate common to this portion of Dutch Canal Ruin (Figure 6.6). Alignment 8555 was exposed in only three trench profiles, including a construction trench for a storm drain. Sediments became coarser with an increase in distance to the west, and the number of channels appeared to increase as well. These characteristics, which investigators interpreted to be related to a decrease in canal maintenance with an increase in proximity to the canal terminus, compare closely with those observed on the western extent of Alignment 3113 at Pueblo Salado (discussed below).

The low gradient (0.19 m/km) calculated for Canal Barranca should be expected near the terminus of a canal where depth is not a particular concern and only a gradient sufficient for drainage would be needed. The coarse-grained materials found within the canal may have accumulated through two processes. First, discharge velocity may have been radically reduced immediately upstream of the preserved canal section by a change in gradient that allowed rapid accumulation of coarse-grained materials there. Second, lateral (and to a lesser extent basal) erosion due to unregulated water movement may have resulted in the accumulation of coarse materials. Water velocities may have been intentionally increased at times to facilitate the removal of canal sediments, resulting in fine- and medium-grained materials being flushed downstream while coarse-grained materials accumulated in areas where velocity could not be maintained to transport them farther down the canal. Furthermore, the orientation of Canal Barranca in the project area and the association of its terminus with Turney's Gully suspiciously resembles a drainage ditch, designed to remove excess water from fields or aid in reducing groundwater levels that would waterlog fields, resulting in elevated levels of alkali and salts. Alternatively, this canal may have been intentionally designed to debouch its discharge into Turney's Gully during cleaning episodes. It may be possible to explore these alternatives during future investigations as part of the Phoenix Sky Harbor International Airport Master Improvements Project in the vicinity of the west end of the north runway (Greenwald and Zyniecki 1993).

Field personnel did not recover datable samples or ceramics from Canal Barranca, and therefore analysts made no age determination. The canal was somewhat isolated from the others at Dutch Canal Ruin, preventing determination of stratigraphic relationships. However, future investigations in the Phoenix Sky Harbor International Airport (Greenwald and Zyniecki 1993) may provide the opportunity to establish temporal associations by defining the stratigraphic relationship between Canal Barranca and other dated canals and perhaps by obtaining absolute dates.

Alignment 3113

Alignment 3113, the north branch of Canal Salado, was located in the extreme northern portion of Pueblo Salado within the Phoenix Sky Harbor Center. Field investigators traced this canal within the project area from the Squaw Peak Parkway to 16th Street (Figure 3.1). Recent investigations within the Phoenix Sky Harbor International Airport (Greenwald and Zyniecki 1993; Huckleberry 1993) identified a portion of this canal, and data gathered as a result of that project are included here for comparison.

As exposed in the various profiles, the north branch of Canal Salado exhibited considerable morphological variation (Figure 6.7). The canal appeared to have been more stable near 18th Street than farther downstream. Lateral channel movement, although visible in all profiles, was more apparent in the westernmost trench profiles, in the area considered the canal terminus. Many of the channels had well-formed parabolic shapes, indicating that the channels themselves had experienced little or no erosion, with channel migration having resulted from human intervention such as cleaning episodes or channel modification. The sediment history of the north branch of Canal

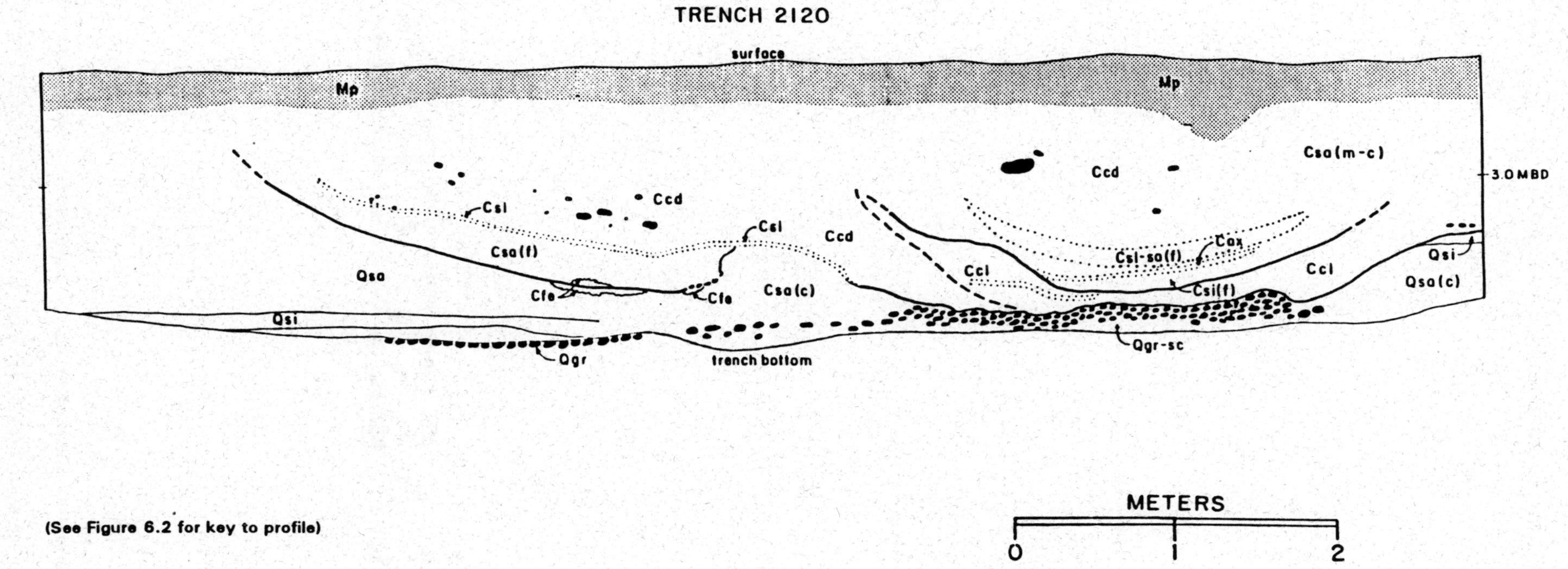

(See Figure 6.2 for key to profile)

Figure 6.6. Profile of Canal Barranca (Alignment 8555) showing lateral channel movement and cobble substrate.

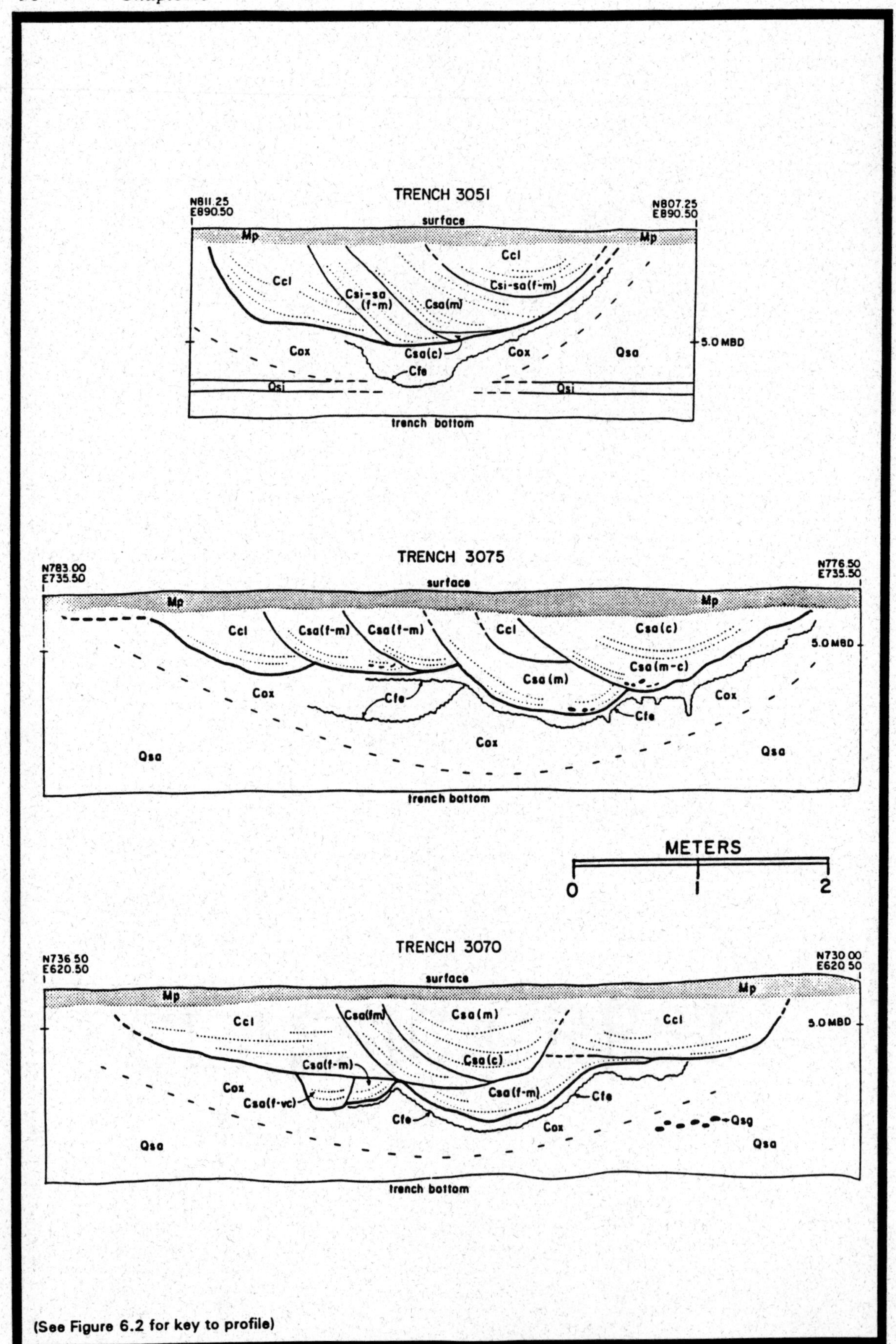

Figure 6.7. Profile of the north branch of Canal Salado (Alignment 3113), showing variations in channel morphology.

Salado was somewhat reversed from sediments of the canals at Dutch Canal Ruin, with the exception of Canal Barranca. The sequence of fine to coarse sediments from the earliest to the latest flow episodes suggested that such sediments were related to the proximity of the canal terminus or to the steep gradient of the canal in the western half of the project area.

Using measurements from three profile exposures, analysts calculated discharge estimates for the north branch of Canal Salado between 18th and 16th streets ranging from 0.47 m^3/s to 1.10 m^3/s, generally decreasing from east to west, and velocity estimates from 0.72 m/s to 0.89 m/s, also decreasing from east to west (Table 6.1). Gradient reconstructions based on that portion of the canal, a distance of 400 m, yielded a very steep gradient of 3.2 m/km; the natural gradient in this same area was calculated at 3.7 m/km, indicating that the canal grade was closely related to the natural slope of the area. In the area of Phoenix Sky Harbor International Airport, the north branch of Canal Salado maintained a gradient of 0.1 m/km over a distance of 460 m. Over that same distance, the natural gradient was level (Greenwald and Zyniecki 1993; Huckleberry 1993). As illustrated by these examples, canal gradient and natural gradient appear to be closely related on the geologic floodplain because of the linear nature and topographic constraints of this setting.

Sedimentation in the north branch of Canal Salado indicated that, in this particular canal, a correlation existed between sediment texture and gradient. Fine-textured sediments characterized the fill sequence of the canal in the Phoenix Sky Harbor International Airport where the gradient was low. Coarse-textured sediments characterized the fill sequence near the terminal portion of the canal where the gradient was steep. The coarse-grained materials present in the lower reaches of the canal had not been introduced from the river; instead, they must have originated from erosion of the base and sides of the canal through increased water velocity in those portions of the canal where the gradient increased and produced turbid, erosive waters.

Analysts did not determine the temporal placement of the north branch of Canal Salado during the current project. However, because of its spatial association with Pueblo Salado, the Hohokam had probably built and used this canal during the Classic or post-Classic periods. Recent investigations at Phoenix Sky Harbor International Airport produced radiocarbon intervals that supported a late Classic, post-Classic, or Protohistoric period affiliation (Greenwald and Zyniecki 1993:Appendix A).

Historic Canals

Dutch Ditch

Settlers constructed the historic Dutch Ditch in late 1868 or early 1869, following the construction of the Swilling Ditch. Figure 6.8 illustrates the location of the Dutch Ditch according to five sources consulted by Rodgers and Greenwald (1988b): Maricopa County Immigration Union (MCIU) 1887; U.S. Geological Survey (USGS) 1912; U.S. Geological Survey, Reclamation Service (USGSRS) 1902–1903; Salt River Valley Water Users' Association (SRVWUA) 1921; Turney 1924. Each source identifies a different alignment. Project researchers did not determine whether these differences represent actual changes in location through time or errors in map plotting. Documentation (Volume 1:Chapter 4) does show at least one realignment in 1899, although this alignment had probably been located to the north of the northern project boundary, as descriptions of this undertaking make reference to the ditch paralleling the railroad tracks.

Field personnel located the alignment indicated on the 1912 USGS map within the southwest quarter of Section 10 (the northwest quarter of the project area). In cross section, this alignment

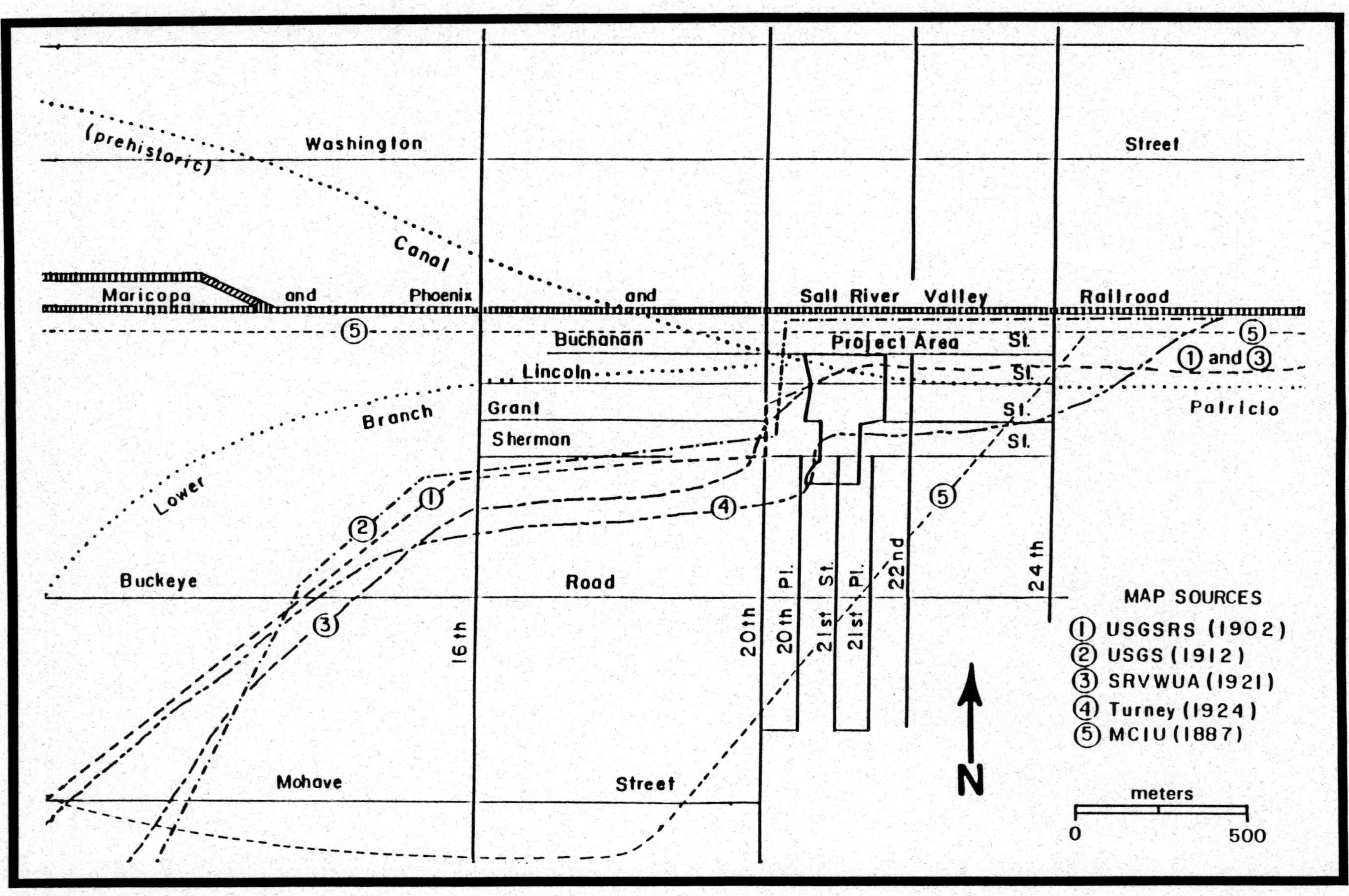

Figure 6.8. Various mapped locations of the Dutch Ditch (from Rodgers and Greenwald 1988b).

consisted of a broad parabolic shape having at times a nearly flat bottom. The fill of this ditch was very sandy, varying from fine to medium, often with a bleached-white appearance. Historic and modern trash was common in the fill, much of it intermixed with the flow deposits. Oxides were common in the sandy fill, but these appeared to have originated from decomposing metal cans rather than through the natural processes observed in the prehistoric canals. Above the flow deposits investigators observed more historic and modern trash, apparently deposited after this alignment had been abandoned in favor of the underground concrete pipe that paralleled the south side of Grant Street. After abandonment, the ditch was backfilled and leveled. Its orientation could be observed during the fieldwork and was easily traced due to its slight elevation.

Researchers did not determine the periods of use of this alignment. Its earliest indication is on the Phoenix, Arizona, 1912 USGS topographic map. It is illustrated again on the 1936 version of William H. Becker's Phoenix subdivision map (Volume 1:Figure 4.8). It is not illustrated on the Phoenix, Arizona, 1952 USGS topographic map, and by 1952 the ditch may have been replaced by the underground pipe.

Unnamed Ditch through Pueblo Salado

On a 1934 aerial photograph (Soil Conservation Service 1934), an alignment of what appears to be an irrigation ditch runs through the project area from 18th Street to 24th Street (Volume 1:Figure 4.15). This alignment has not been documented in the field or through archival research but appears to follow a route similar to that of the north branch of Canal Salado (following similar contours), although the course of the historic ditch appears to be more deliberate and angular, sometimes paralleling streets or forming boundaries of fields. Prior to intersecting Turney's Gully, the ditch appears to angle to the southwest, crossing 16th Street near Mohave Street and proceeding in a southwesterly direction to 14th Street. An early historic ditch, the Wilson Canal, may have paralleled the north side of the Salt River and entered the Phoenix Sky Harbor Center at the approximate location of the ditch shown in the 1934 aerial photograph (see also Greenwald and Zyniecki 1993:Figure 2.2). The ditch shown in the photograph may be a later alignment of the historic Wilson Canal. It may be possible to confirm this speculation during future investigations associated with the Phoenix Sky Harbor International Airport.

Canal Systems and Project Canals

Geographically, the Phoenix Sky Harbor Center lies within Canal System 2 as defined by Turney (1929). Canal System 2 (Figure 1.2) is spatially extensive, originating immediately downstream of the Papago Buttes on the north side of the Salt River. System growth reached its maximum areal extent during the Classic period, with the construction of canals that extended as far north as present-day Camelback Road and perhaps as far west as Tolleson, Arizona, a distance of approximately 19 miles from the head of the system. Although many of the canals originated along the Salt River between Pueblo Grande and the Papago Buttes, some canals—for instance, Canal Salado, Canal Colinas, Canal Riverside, and Canal Rio (Figure 3.2; Midvale 1966), which Turney (1929) labeled Canal Ten, Canal Five, Canal Four, and Canal Six, respectively—headed at various points along the river. As has been demonstrated by this project and recent investigations by the City of Phoenix in the Pioneer and Military Memorial Park (Schroeder 1991), the components of Canal System 2 represent a wide temporal range. Schroeder's investigations have revealed that Canal Colinas may have been built and in use during the sixth century A.D. (Kelly Schroeder, personal communication 1992), whereas the north branch of Canal Salado may have been used from the Soho phase through the Polvorón phase (Greenwald and Zyniecki 1993). The range represented by these extremes supports the premise that the Hohokam used canal irrigation north of the Salt River for a period of 1000 to 1100 years.

Chapter 3 of this volume discusses the degree to which all canals and their associated settlements belonged to or participated in the sociopolitical sphere identified as Canal System 2. Pueblo Salado displayed characteristics of an autonomous settlement, having an independent canal system and exhibiting greater interaction with groups or exploitation of resources south of the Salt River than within the area of Canal System 2. The pre-Classic period occupation at Dutch Canal Ruin, by contrast, appeared to have been directly tied to Canal System 2, with residents participating in that system. The canals were probably extensions of those that headed on the Salt River upstream of the Park of Four Waters, and the field house and farmstead settlements had probably been occupied by residents of larger sites within Canal System 2, such as La Ciudad, Pueblo Grande, and Pueblo Patricio. Associations of the Classic period and post-Classic period occupations at Dutch Canal Ruin were more difficult to assess in terms of sociopolitical interaction. The post-Classic period occupation of Dutch Canal Ruin shared similarities with Pueblo Salado and may have been an extension of that site. The orientation of Canal Nuevo, a Classic period canal at Dutch Canal Ruin, provided little insight into where it may have originated along the Salt River. Its gradient and orientation within the project area left open the possibility that it headed on the Salt at some distance downstream of the majority of canals in Canal System 2.

In summary, it is possible that, although both Dutch Canal Ruin and Pueblo Salado lay within the geographic boundaries of Canal System 2, residents of the sites had not participated in the sociopolitical sphere at all times. In fact, Pueblo Saladoans may have had little interaction with the Hohokam of Canal System 2, especially during the later occupation of the site. However, the investigating team found little reason to question the pre-Classic period interaction of Dutch Canal Ruin with Canal System 2, as residents of the canal system had probably used the general site area as an agricultural zone (Volume 2:Chapter 19; Chapter 4, this volume).

CANAL COMPONENTS

Field investigators observed various types of canals and canal features in the project area. The diversity of feature types was somewhat limited compared to sites on the lower bajada, and this lack of diversity may be related to the specific functions of the features and to poorer preservation of features within the unconsolidated sediments of the geologic floodplain.

Main Canals

Main canals are "first order canals, functioning as the main water transport system" (Howard 1990:233) of irrigation networks. They carry water from the intake or source to the vicinity in which it is used. Distribution canals, second-order canals, generally deliver water from the main canal to its intended locations of use, such as fields or domestic areas. Lateral canals then distribute the water to the fields (Masse 1991:212–213). Investigators identified no distribution canals within the Phoenix Sky Harbor Center. In fact, at Dutch Canal Ruin, laterals extended directly from the mains (Greenwald 1988), possibly because the fields lay immediately adjacent to the canals.

Laterals

A lateral canal, also referred to as a field lateral, is a third-order canal, used to transport water from a distribution or main canal into the fields (Howard 1990:233). Because the volume of water transported by lateral canals is generally much smaller than in mains or distribution canals, laterals usually are limited in area and often in length as well, as they are designed to serve small areas (Masse 1991). Masse (1991:213–214) defined two types: the lateral network canal and the lateral isolate. Lateral network canals are interconnected, forming a "weblike effect" or grid pattern. The lateral isolate, as the term implies, functions separately or independently from other laterals. The lateral isolate was the only type identified at Dutch Canal Ruin (Greenwald 1988) and also may have been present at Pueblo Salado.

Crosscut Canals

A crosscut canal transfers water from one main canal to another. In a sense, the crosscut would provide an optional source of water for a canal that served a distinctly separate location. The Hohokam may have used crosscuts when insufficient water was available, augmenting water requirements at another source. The use of crosscuts may have been related to, but not limited to, increased water needs or the erosion of a canal head or section of a canal that could not be repaired as quickly as a crosscut channel could be constructed. At Dutch Canal Ruin, Alignment 8002 appeared to have redirected water from one area to another by connecting two canals (North Main Canal and South Main Canal) that had served different geographic areas at different times (Figure 4.1). Alignment 8002 then followed the previously abandoned alignment (8003) beyond the project area.

Headgates

Headgates, not to be confused with diversion weirs at the heads of canals, are structures used to control the water elevation in canals, which in turn controls water velocity. Headgates function in two ways. First, within main canals, headgates elevate water levels to divert water into other channels. Second, at the heads of smaller-order channels and field turnouts they control flow into and the volume of water in smaller-order channels (Masse 1991:Figure 9.2). Headgates are sometimes called tapons (Howard 1990:234; Nials and Gregory 1989:53). They may have been made from a variety of organic materials, resulting in poor preservation.

Field Turnouts

Field turnouts are generally associated with lateral canals, although Masse (1987:191) identified two turnouts associated with a "primary distribution canal" at AZ U:9:69(ASM). Turnouts generally exhibit a parabolic opening through the canal bank. In tracing the turnouts at AZ U:9:68(ASM), Masse found that they became shallower and wider with an increase in distance from the canal, allowing water to be distributed in floodlike fashion directly into fields. At Pueblo Salado, a turnout in a small distribution or lateral canal incorporated gravels as reinforcement on the canal banks (see below). This turnout appeared to be nothing more than a simple break in the canal bank, allowing water to be dispersed over the landscape immediately adjacent to the canal.

Gravel Ridges

Two gravel ridges at Pueblo Salado, Features 8/9-115 and 8/9-116, appeared to have functioned in direct association with a field turnout and a headgate. They were located on each side of the canal (Feature 8/9-118), partially forming the upper banks, and apparently had prevented bank erosion. The gravel ridge on the south side of the canal (Feature 8/9-116) contained a distinct gap, thought to represent a turnout (see above). Both gravel ridges lay immediately upstream of a plunge pool (Feature 8/9-135), possibly created by the presence of a headgate, although field personnel did not find the headgate.

Plunge Pools

Feature 8/9-135 was an eroded area within the course of Feature 8/9-118, a small canal. This eroded area had distinct boundaries, cutting below the basal level of the canal, and exhibited characteristics that could only have been created by a sudden change in water velocity such as occurs when water spills over a headgate or dam. Such features are commonly referred to as plunge pools, for the pool or eroded basin formed by the force of the falling water. A plunge pool may also have been present in Alignment 8001 in the Squaw Peak Parkway east of Backhoe Trench 21 (Greenwald 1988). Investigations in that location revealed a turnout that fed a large (16.0 m north-south by 14.0 m east-west by 1.1 m in depth) sediment basin. The plunge pool, however, was not investigated, and investigators for this project can only speculate about its nature based on the presence of the turnout and a radical change in the base of North Main Canals 1 and 2 between Backhoe Trenches 20 and 21.

Baffles

Archaeologists have observed various types of devices in prehistoric canals that appear to have restricted the velocity of water or reduced the erosional effect of high discharge rates. Interpreted as baffles, these devices include cobbles and boulders, brush and pole structures, or a combination of posts and rocks (Ackerly, Kisselburg, and Martynec 1989). Structures of posts or brush and poles identified as baffles differ from structures identified as headgates or tapons in their orientations. Headgate structures are generally constructed perpendicular to the canal, whereas baffles are constructed parallel to the canal. When rocks are used, they are generally scattered along the length of the canal. The baffle identified at Dutch Canal Ruin in Alignment 8501 was of pole and rock construction (Volume 2:Chapter 8). The poles, located in the center of the channel, were apparently aligned with the canal. The rock, intermixed with canal sediments, was piled in a linear fashion along the south bank of the canal. This rock feature was located on the outside radius of a curve in the canal and may have been built to reduce bank erosion. This particular location would have been susceptible to lateral cutting, leading eventually to a breach in the canal bank, if some form of reinforcement had not been used.

IRRIGATION METHODS

Although the Hohokam may have used two basic irrigation methods, flood irrigation and rill irrigation, they most likely preferred some form of flood irrigation. During the late historic period, farmers generally adopted rill irrigation along with mechanized agriculture. The Hohokam probably planted in hills or in a random fashion rather than in rows, in which case flood irrigation would have been less labor intensive. Prehistoric farmers may have used two types of flood irrigation in the Phoenix Sky Harbor Center: wild flooding and border-ditch flooding.

Wild Flooding

This form of irrigation allows water to be randomly spread across the field, fed from the canal by a turnout. Farmers often determine the spacing of the turnout after observing the extent of the random coverage; the more the water fans across the surface, the wider the spacing of the turnouts can be. When enough water has been applied to the field through a given turnout, the farmer closes the turnout to stop the discharge. If existing turnouts do not adequately cover the field, irrigators can dig new turnouts in the canal banks. The authors have used this method, which requires very little labor beyond the effort required within the field to divert water from low areas to high areas by building low dikes or by digging shallow ditches. As Phelan and Criddle (1955:259) pointed out, this method of irrigation is inefficient because water spreads unevenly across the field surface. Invariably, some areas receive excess amounts of water, while others receive too little. When wild flooding is used, irrigators must retard the velocity of water being delivered from the canal to the field to prevent extensive erosion and downcutting along the canal bank. Wild flooding is most effective when used in conjunction with lateral canals.

Border-Ditch Flooding

The Hohokam also may have used border-ditch flooding. Aerial photographs of the Phoenix Basin indicate that Hohokam farmers had constructed lateral networks of irrigation ditches, an irrigation system that was very common during the Classic period (Masse 1991:214). Lateral networks include a variety of irrigation methods in addition to the border-ditch method, including basins, contour ditches, border strips, and bench borders (Nials and Gregory 1989:42). However, given the physiographic restrictions of the geologic floodplain, where the area available was long and narrow, especially as at Pueblo Salado, border ditches may have been more efficient than other lateral networks. Border-ditch flooding uses parallel ditches that follow the natural slope of the land. Water is diverted from either side of the ditches to water the fields. Border-ditch flooding is similar to wild flooding except the parallel ditches allow water to be applied to fields with greater control. In comparison to other network systems, border-ditch flooding is inefficient (Phelan and Criddle 1955:259), but modern farmers still use it in the Phoenix Basin for crops such as wheat, barley, and alfalfa. As with wild flooding, differential wetting can occur due to microtopographic variability within fields. Low areas tend to collect larger volumes of water and become saline more rapidly than surrounding areas. Low areas also tend to become overgrown by weedy annuals encouraged for their economic potential.

CONCLUSIONS

The retrodicted velocities and discharges for the project canals indicated strong similarities between the average velocities and discharges of four of the main canals associated with Dutch Canal Ruin and Pueblo Salado: Alignments 8001, 8002, 8003, and 3113 (the North Main Canals), the crosscut, the South Main Canal, and the north branch of Canal Salado. Average velocities fell between 0.81 m/s and 0.84 m/s and discharges between 0.79 m^3/s and 0.98 m^3/s (Table 6.1). Average velocities for North Main Canal (0.84 m/s), South Main Canal (0.81 m/s), and the crosscut (0.84 m/s) were nearly equal and suggested that the last phase of use of each of these canals, incorporating the best-preserved and most recent channels, exhibited consistent flow regimes across the project area.

The uniformity of these retrodicted values may suggest that these canals, although near the erosion limit of the geomorphic setting, reached and maintained a fairly stable state. This detail suggests that for at least the late stages of canal use, overall velocities were kept to the approximate erosion velocity, thereby maintaining a balance between erosion and sedimentation. This near-equilibrium would have allowed for minimum maintenance and adequate water supply from the irrigation system to the fields. Although individual channels within a single canal alignment sometimes varied with respect to calculated hydraulics and sediment textures, it appears significant that the average velocities and discharges of the three alignments were nearly equal. The mechanisms for this near-equilibrium state are unclear and may be either cultural or natural. Cultural regulation of water intake from the headgates and maintenance of proper channel dimensions would have led to stable conditions. Stability also may have been due to a canal system that was achieving equilibrium between the geomorphic conditions of the natural gradient and the substrate textures and maintaining a fairly continuous, predictable amount of water entering the system from the headgates.

Although discharges seemed low, they were sufficient to irrigate the amount of land near the canals at both Dutch Canal Ruin and Pueblo Salado. The similarity of the discharge figures from one canal alignment to another indicated that overall the water supply to the fields had been fairly consistent. Although the calculations for determining the amount of hectares that can be irrigated by a given flow neglect several complicating factors involving seepage and vegetation, the amount of water provided by these canals appeared to have been sufficient for the amount of available land thought to have been under irrigation. However, soil porosity on the geologic floodplain appears in general to be greater than on the adjacent bajada slopes, and thus a greater amount of water per hectare may have been needed to irrigate crops adequately on the floodplain.

Canal morphology appeared to show a tendency toward technological advancement over time. In other words, the earliest canals (North Main Canals 1 and 2, South Main Canal, and Canal Viejo) exhibited the steepest gradients, the greatest amounts of erosion, the coarsest sediments, and the most instability; the later canals, beginning with North Main Canal 3 (a Colonial period canal) and including the Classic period canals (Canal Nuevo and Canal Salado), exhibited less infilling, channel migration, and downcutting along with much-reduced canal grades. Although gradients of canals on the floodplain were directly affected by the natural gradient, through time the Hohokam engineered and regulated their canals to reduce the effects of this association. They may have closely monitored discharge and velocity rates and may have used other canal features such as drop structures that project investigators did not discover.

Throughout the volumes for this project, the investigating team has questioned the association of the project canals to Canal System 2. Investigators have little doubt that the pre-Classic period canals at Dutch Canal Ruin were components of that system. Similarly, the Classic period canal, Canal Nuevo, probably was associated with the complex of canals in that system. Canal Salado, by contrast, may represent a separate and much smaller system, consisting of an intake located at the upstream confluence of Turney's Gully and the Salt River. Due to the geomorphic setting of Pueblo Salado, this site may have maintained greater autonomy than village sites within the more complex network of canals in the larger systems.

Field personnel found fewer canal-related features in the Phoenix Sky Harbor Center than they had anticipated, perhaps because the canals and their associated features had been built in unconsolidated sediments, resulting in poor preservation. In comparison to other investigated canals

(Ackerly and Henderson 1989), very little rock was present, which, as in the case of the baffle found in Canal Viejo, would have aided in detecting internal features. The remnants of many features may have been eroded away by later cleaning episodes or scouring events. The amount of rechanneling that occurred in the project canals also would have been detrimental to the preservation of interior features. Results of the canal investigations were thus somewhat disappointing in relation to many of the functional questions.

During the testing phase, field crews excavated trenches in a north-south orientation because previous excavations indicated that canals extended from east to west through the project area (Greenwald and Ciolek-Torrello 1988a). Although investigators placed judgmental east-west trenches along the south side of the South Main Canal, they discovered no lateral canals in this area. The only information regarding lateral canals at Dutch Canal Ruin was recovered from investigations in the Squaw Peak Parkway (Greenwald and Ciolek-Torrello 1988a).

To summarize, canal studies within the Phoenix Sky Harbor Center have contributed to the growing data base regarding prehistoric irrigation strategies. The width of the project area enabled archaeologists to examine canal segments over considerable lengths, demonstrating that canal characteristics were fairly uniform over long distances. They also tracked functional changes and the abandonment and reuse of canal segments. Although the social aspects are not clearly understood, these changes were probably related to system upgrade and the need to supply water to previously unfarmed areas or fallow fields that required new sources of irrigation water.

CHAPTER 7

OPTIMIZING THE LANDSCAPE

Dawn M. Greenwald
David H. Greenwald

The Phoenix Sky Harbor Center project area lay within two different environmental zones, the riparian and the desertscrub. Human activities modified the habitat zones within the Phoenix Sky Harbor Center, and natural processes acting on these modifications created microenvironmental zones. This chapter reviews the types of environmental zones available to the prehistoric inhabitants and evaluates their potential importance for subsistence and other economic opportunities. Both natural features and modifications resulting from human impact on the natural landscape may have provided various types of resources. Prehistoric use of these resources is examined through the project botanical record, ethnographic analogy, and previous environmental studies.

ENVIRONMENTAL ZONES

Natural Zones

Prehistorically, the floodplain and terraces of the Salt River contained a wide variety of plant and animal species. Desertification and reduction in this habitat (Crosswhite 1981:67; Hastings and Turner 1965; Rea 1983) have decreased species diversity and changed the type of flora and fauna that characterize the landscape. Although Sonoran Desert biota have been evolving over millennia (Lowe 1959; Turner 1959), human influence over the past 100 years has changed the amount of surface streamflow, groundwater, and erosion and has depleted native vegetation along the river systems, affecting their ecology (Carothers 1977; Rea 1983). According to Rea (1983),

> On nearly all these desert streams the delicate fabric of the ecological community has been destroyed or warped beyond recognition and restoration. Emergent vegetation has all but disappeared. The riparian forest is largely gone or replaced by feral saltcedar [sic]. Other weedy species proliferate. The water table of the floodplains, formerly but a few feet below the surface, now averages hundreds of feet underground [Rea 1983:3].

Traces from the archaeological record in the form of plant and animal remains and early historic accounts by missionaries and explorers have documented the fact that the habitat occupied by prehistoric and early historic people was quite different from what it is today. Historic records thoroughly describe the habitat along the Gila River, which is environmentally similar enough to the Salt River for comparative purposes. Riparian woodland, including extensive mesquite bosques, occurred on the floodplain of the Gila. Riparian habitats are biotic communities present along streams or rivers that vary from the communities in the immediately surrounding habitats (Brown, Lowe, and Hausler 1977). Riparian vegetation in the Sonoran Desert consists of broad-leafed trees that can survive extreme temperatures and water fluctuations, weedy annuals and perennials, saltbush (*Atriplex* sp.), and woody species such as bursage (*Ambrosia* sp.). Trees associated with this type of environment include both native species (cottonwoods [*Populus* sp.], mesquite [*Prosopis* sp.], willow [*Salix* sp.], Arizona sycamore [*Platanus wrightii*], Arizona walnut [*Juglans major*], and the velvet ash [*Fraxinus velutina*]) and non-native taxa (salt cedar [*Tamarix* sp.] and Russian olive [*Eleagnus angustifolia*]) (Rodiek 1981:Table 1). Outside of the riparian zone is xeric vegetation that, in the Phoenix Sky Harbor Center, is composed of low desertscrub such as saltbush and creosotebush-bursage communities. The interface between these two major zones is an ecotone, a transition area that accommodates species common to both.

Riparian Habitat

The riparian zone or habitat is characterized by surface water and mesic vegetation. The habitat along water courses provides a diversity of species not found elsewhere within the surrounding arid environment. Uniquely adapted trees, such as the cottonwood, willow, and mesquite, attract breeding birds, and the multistory canopy, in turn, provides a large variety of microhabitats for small mammals. The combination of riparian vegetation and surface water encourages the presence of some reptile and amphibian species that do not live elsewhere within the desert (Hubbard 1977:17). Fish are also attracted to waters close to land because of the large amount of organic matter, the temperature, and overbank shelter.

The Salt River and its adjacent riparian communities provided a wide variety of resources in prehistoric times. Water was the most important resource in the desert environment, necessary for the support of all life. Twice a year the flow in the river would have risen, once carrying the spring snowmelt and again during the late summer storms, allowing water to run through the irrigation canal systems (Rea 1991:3). The irrigation water not only provided moisture for crops and enhanced the productivity of natural vegetation but also supplied extended riverine habitats for desert fauna. Nutrients contained in the silts of irrigation waters fertilized the soils in the fields (Chapter 5; Brown 1869), especially after flood episodes. In fact, Rea (1979) related that the Akimel O'odham ("River People"; formerly Pima) would wait to irrigate their fields until flooding introduced the correct amount of silt into the river. Overbank deposits also added nutrients to fields, especially those on the floodplain. Depending on the location of the field relative to the valley floor, flood deposition may have been a better source of nutrients because thick deposits could accumulate in a short period of time, rejuvenating fields and flushing salt accumulations from the soils (Chapter 5).

Riverine harvest also played a role in prehistoric and historic diet; because of inconclusive and conflicting evidence, archaeologists have not yet determined the extent of that role. Rea (1991:6) suggested that fishing produced a major source of protein for the Akimel O'odham, although an ethnographic study from nearly a century earlier (Russell 1975:83) made few references to fish, perhaps due to long years of drought. Additionally, Russell (1975:83) recorded only one term used by the Akimel O'odham for fish (*vatop*) and noted that early observers alluded to, at most, three different species: *Gila robusta*, *Gila elegans*, and *Xyrauchen cypho*. Nevertheless, older Akimel O'odham described eight species of fish to Rea (1991:6). Evermann and Rutter (1895) collected ten different species, including those listed above, mostly along the Gila River.

Excavations at the site of Pueblo Grande identified a large number of fish remains (Steven R. James, personal communication 1992): bonytail chub (*Gila elegans*), roundtail chub (*Gila robusta*), Colorado squawfish (*Ptychocheilus lucius*), humpback sucker (*Xyrauchen texanus*), Gila coarse-scaled sucker (*Catostomus insignis*), flannelmouth sucker (*Catostomus latipinnis*), and Gila mountain sucker (*Catostomus clarki*). Miller (1955) identified fish remains from the village of Snaketown as Colorado salmon or squawfish (*Ptychocheilus lucius*) and humpback sucker (*Xyrauchen texanus*). These types were also identified at the site of Quiburi, located farther south along the San Pedro River. Evidence from Snaketown regarding the average weight of harvested fish was conflicting. Haury (1976:115) indicated in two separate analyses that the Hohokam had harvested both large and small fish. He assumed the large fish had been procured from the river and that the fish that were too small for open stream fishing had probably been taken from the irrigation canals. Given the apparent lack of fishing devices in Hohokam assemblages, researchers have little evidence to support either possibility. Stone plummets resembling some prehistoric stone sinkers (Tuohy 1968) may have been fishing sinkers, but their lack of context and limited numbers (Chapter 12) make this interpretation very tentative. Although the fish undoubtedly came from the river, they may have found their way into canals where the Hohokam trapped them and caught them by hand. Crosswhite (1981:65) observed fish stranded by irrigation in Piman fields that could be taken in this way. Although the exploitation of riverine resources is debated, one of the best sources for information about riparian fauna is Hohokam Red-on-buff ceramics, which illustrate fish, aquatic birds, reptiles, and amphibians (e.g., Gladwin et al. 1965; Haury 1976). Archaeologists have found few designs that represent actual exploitation of these resources by the Hohokam, but their frequent depiction indicates they were of some importance.

Riparian vegetation provided a variety of food items and construction materials. Fruits, seeds, cattail (*Typha* sp.), mesquite pods and beans, and wild sorrel or dock (*Rumex* sp.) were all available. Mesquite was one of the staples in the Hohokam diet (Gasser 1981–1982:226) and may have become increasingly important during periods of drought. Crosswhite (1981:65) noted that riverine Akimel O'odham built their houses from riparian plants, such as cottonwood, mesquite, willow, and cattail. Willow and acacia were available as materials for basket construction, and cattails and reeds were woven into mats.

Desertscrub

Desertscrub vegetation is less diverse and dense than riparian vegetation and is "low in stature and simple in composition" (Hastings and Turner 1965:187). Creosotebush is dominant, with saltbush, bursage, and other herbaceous and woody species, as well as some hedgehog (*Echinocereus* sp.) and prickly pear cacti (*Opuntia* sp.), grasses (Gramineae), and weedy species. The low stature accommodates various medium to small mammals, including lagomorphs (rabbits and hares) and rodents. Although this environment promotes limited hunting and gathering, areas dissected by washes include a wider variety of floral and faunal types, such as mesquite. Historically, the Akimel O'odham used desertscrub vegetation for medicine, for fuel, and to supplement their diets exploiting food resources in a planned manner (Goodyear 1975). Such activities, however, being of an ephemeral nature, leave few clues to specific behavior and are often difficult to recognize and interpret in the archaeological record. Turner (1974) illustrated the native vegetation communities for the Salt River Valley, showing that most of the Phoenix Sky Harbor Center supported a desert saltbush community north of the Salt River riparian zone. Along the northern margins of the project area, the creosotebush community begins to encroach on the desert saltbush community. Turner's reconstruction of native vegetation implies that the project area provided even less diversity than the "typical" desertscrub community, which is reflected in the results of the botanical analyses for Dutch Canal Ruin (Volume 2:Chapters 15 and 16).

Bajada

The desert bajada was another environmental zone exploited by the prehistoric inhabitants of the project area. Forming the lower slope of the mountains and extending to the interface of the first and second terraces along the north side of the Salt River, these areas are well suited to a variety of cacti and other plants, such as agave, that require higher elevations and well-drained soils. Creosotebush (*Larrea tridentata*) and bursage communities constitute the lower bajada areas where soils are suitable for agriculture. Saguaro (*Carnegiea gigantea*) and cholla (*Cylindropuntia* sp.), common to the upper bajada, were heavily exploited prehistorically and historically (Goodyear

1975) for their fruits; the saguaro also was used for construction. In the bajada zone archaeologists have found rock pile features associated with dry farming techniques. These features may have been used in growing agave and possibly other cultigens (Fish, Fish, and Madsen 1985; Fish et al. 1985).

Modified Zones

Human disturbance of the landscape often promotes changes to the habitat that may be either beneficial or detrimental to natural resources and to those who rely on them. Prior to Euroamerican influence, what effects did the prehistoric and historic inhabitants of the Salt and Gila river valleys have on the natural environment, and did these inhabitants maintain the natural habitats so that they not only provided food and materials for human consumption, but also regenerated successfully without damage to the ecological balance? Researchers have noted that habitats were modified so that some of their desirable characteristics extended beyond their natural boundaries or were artificially induced outside of their natural limits (Fish and Nabhan 1991:44). The historic Akimel O'odham created at least three modified environments: irrigation canals, agricultural fields, and secondary gardens (Crosswhite 1981:64–65; Gasser 1981–1982:222). Aspects of each modified zone represent an ecotone, where the biota of the river and desert environments meet. These areas provide greater plant and animal diversity than the surrounding desertscrub habitat because of their increased moisture and the variety of biota that they consequently support and attract. As the canals carry water from the river, they also transport seeds, organisms, and nutrients that will be transplanted into a new environment. Wild plant seeds, exposed by human disturbance, will germinate with the addition of moisture.

Irrigation Canals

The effect of irrigation canals on the natural landscape can be compared to the ribbon of vegetation surrounding washes or arroyos. In-use canals maintained an artificial water source that supported various economic annuals and perennials. Abandoned canals functioned more like washes or arroyos, as they accumulated moisture from precipitation and runoff; as temporary receptacles during periods of moisture, these water-collection features enhanced the growth of vegetation, and thus the density of fauna, along their lengths. Roadways today create the same sort of effect; rainfall is effectively washed off of the road in sheets to be collected in the roadside drainage, concentrating moisture and encouraging roadside vegetation to thrive. Irrigation canals and intermittent washes have the same moisture-retaining quality and often have riparian growth along their margins. Many canals were purposely designed to follow natural contours rather than to crosscut them as do arroyos and washes. As a result, they became barriers retarding natural runoff, creating denser vegetation than in the natural drainages and "creating biotic conditions dissimilar to locally prevailing natural ones" (Fish and Nabhan 1991:45).

Canal margins and irrigation ditches promote vegetation growth that is, in turn, attractive to a number of animal species; "plants such as *Anemopsis*, *Spergularia*, and *Heliotropium* thrive in irrigation ditches. Flowing water and increased humidity attract certain insects, and in turn attract birds (e.g., phoebes)" (Nabhan et al. 1982:141). The Tohono O'odham ("Desert People") gather the aforementioned plants for medicinal purposes. Irrigation ditches and canals also are prime habitat for cattail and reeds, grasses, and, later in the cycle, mesquite. Therefore, canals can be important sources of a variety of economic resources.

Agricultural Fields and Secondary Gardens

Agricultural fields represent considerable modification to the landscape. Whether a field is plowed or disturbed by digging sticks, the turned soil exposes wild seeds to new conditions, such as sunlight and a thinner layer of soil, that enhance the germination process, and it also exposes a new food supply to the bird population (Nabhan et al. 1982:141). Residential locations are primary loci of human disturbance that greatly affect the native flora and fauna, promoting the growth of weeds and attracting rodents. Agricultural field margins, field borders, and fence lines are also disturbance areas that have been used historically to promote the growth of edible weeds as a "second garden" (Crosswhite 1981:64). According to Crosswhite (1981:64), the Akimel O'odham allowed irrigation water to drain off their fields to produce this second garden, and because they encouraged weeds to grow in this particular area, they kept the agricultural fields relatively free of weeds. The Hohokam may have domesticated amaranths and other weedy taxa by the Classic period (Gasser 1981–1982:221) and encouraged these types of plants, which are often present around fallow, abandoned, or untended fields. Indeed, weedy taxa often make up a substantial portion of the archaeobotanical remains from agricultural sites, sometimes in frequencies that suggest that they were a primary resource, or diet staple.

The enhanced environment of agricultural fields and secondary gardens also attracted small game, providing hunting opportunities (Linares 1976; Rea 1979). Animals took advantage of the edible plants in the fields and gardens and proliferated, creating a local supply of faunal resources, primarily small rodents, lagomorphs, and birds (Rea 1991:6).

EVIDENCE FROM THE PHOENIX SKY HARBOR CENTER PROJECT AREA

Pueblo Salado

Pueblo Salado contained evidence of a higher incidence of riparian resources such as fish, cottonwood/willow-type charcoal, and cattail than did Dutch Canal Ruin or other documented Classic period sites in the Salt River Valley (Table 7.1). Macrobotanical remains indicated that structural and fuel wood was mostly limited to cottonwood/willow-type and mesquite (i.e., plants found in a riparian setting, although mesquite is not restricted to such habitats). When analysts compared the wood charcoal assemblage from Pueblo Salado to that of other sites in the Phoenix area, they found it to be unusually abundant in cottonwood/willow-type. Because the cottonwood/willow-type was a poor heat producer, the Pueblo Saladoans had probably used these woods either because their source, along the Salt River, would have been convenient or because other sources had become exhausted (Volume 3:Chapter 8). The riparian species reach maturity faster than hardwoods and may simply have been more readily available. Certainly, long-term occupation of a given area would have impacted, possibly even depleted, the preferred hardwood species.

Cattail (*Typha*) indicates marshy habitats, and its pollen, although sometimes transported by wind or water from a riparian environment, provides evidence of close proximity to such a community. Pollen remains at Pueblo Salado indicated a heavy reliance on this plant through time, and it was one of the more common economic types recovered (Volume 3:Chapter 9). From the context in which some cattail pollen was found, analysts inferred that it had been processed for consumption during the entire duration of occupation at the site.

Table 7.1. Resources of Environmental Zones Exploited at Dutch Canal Ruin and Pueblo Salado

Site	Environmental Zone				
	Mountain	Bajada	Desertscrub	Riparian	Modified[1]
Dutch Canal Ruin	bighorn sheep	cholla saguaro agave?	cheno-am creosotebush bursage	mesquite cattail	little barley grass maize possible agricultural weeds
Pueblo Salado		saguaro agave?	saltbush hedgehog cactus creosotebush grasses	freshwater fish cottonwood/ willow cattail mesquite common reed	little barley grass maize domesticated bean cotton possible agricultural weeds

[1]Includes irrigation canals, agricultural fields, and secondary gardens
?=possible source

Fish remains were the second-largest faunal taxonomic category recovered from all contexts at Pueblo Salado, after small mammal remains. They seemed to have been an integral part of the prehistoric diet at the site, supplied by either the riparian habitat along the river or the enhanced environment produced by the canal system. Recovered from the site were humpback sucker (*Xyrauchen texanus*) and cf. *Cyprinidae* (carp and minnow) remains, both potentially important food categories.

Residents of the site also exploited the desertscrub biome, as indicated by the carbonized remains of grass grains, saltbush, hedgehog cactus, and other species. Bajada resources such as saguaro also were present. None of these categories, however, occurred in high frequencies, and they may have constituted a very low proportion of the subsistence base. Pueblo Saladoans utilized a variety of small and medium-sized mammals, similar to types recovered from Dutch Canal Ruin. These faunal resources may have been obtained by casual hunting, with residents taking advantage of wildlife encountered while working in the fields or exploiting nearby resource zones.

Residents of Pueblo Salado made use of the diversity of local resources and also modified their environment to increase their subsistence yield. They intensively practiced agriculture, if indications from the flotation record are correct. Although irrigation agriculture appeared to be an important activity at the site for its duration, agricultural activity increased during the Civano phase of the Classic period. In addition, Pueblo Saladoans used fewer cacti in comparison to residents of other Classic period sites, perhaps because these products grew at some distance from the site. Agave also was utilized to a lesser extent than at other sites of the same time period, which may have been because the floodplain was not suited (geomorphically) to agave growth (Volume 3:Chapter 8). The pollen and phytolith records agreed with the flotation data in indicating that irrigation agriculture was an important subsistence practice. Pollen remains from agricultural field deposits also suggested that fields had been kept relatively clean of weeds and other native plants, so that if weedy plants had been encouraged prehistorically (as their presence in the pollen record suggested), they probably had been restricted either to secondary gardens or to other disturbed areas such as field borders. The diversity of the faunal remains from Pueblo Salado, including the many small and medium-sized mammals represented in the assemblage, suggested that residents had exploited local breeding populations when these animals were available in and near the site.

Dutch Canal Ruin

Occupants of Dutch Canal Ruin also exploited a variety of habitats. Situated prehistorically within a desert saltbush community, Dutch Canal Ruin also was close to a creosotebush-bursage community just north of the project area (Turner 1974) and to a riparian habitat to the south. The site was on the geomorphic floodplain, just below the interface with the lower bajada. This interface appears to correspond closely with the vegetation pattern mapped by Turner (1974), in which the floodplain supported a saltbush community and the lower bajada supported a creosotebush community. Although an upper bajada environment did not occur within a 5-km catchment radius, it was available just beyond that limit. Pollen studies at the site found evidence of resources from all three habitats (Volume 2:Chapter 16). Desertscrub land was represented by high frequencies of Cheno-ams (>18%) and by the presence of creosotebush and bursage (Low-spine Compositae). The relatively low frequencies of creosotebush and bursage indicated that they had been available but not abundant in the immediate site vicinity (Volume 2:Chapter 16), concurring with Turner's (1974) data. Mesquite and cattail provided evidence for the use of a riparian or water-enhanced habitat, and remains of cholla and saguaro cacti were evidence of probable bajada exploitation. In addition, evidence of bighorn sheep suggested that the residents of the site had also visited the mountain zone to procure resources, although they could have acquired upland resources through exchange as well.

Agriculture at the site was well represented by maize in both the pollen and macrobotanical records. Although the pollen study identified agricultural weeds, the flotation results found little evidence of charred weedy species. Perhaps weedy taxa were not important components of the diet during the pre-Classic period at this site due to the obvious emphasis on maize cultivation. The diet was, however, supplemented by mesquite, agave, and cacti during the pre-Classic period and by agave, cotton, cattail, and little barley grass during the Classic period. The most common faunal taxa were medium-sized mammals, followed in frequency during the Classic period by large mammals. The archaeological record, therefore, did not seem to support the premise of either secondary gardening or opportunistic hunting at Dutch Canal Ruin; larger frequencies of charred weedy species, implying greater consumption of edible weeds, and a higher incidence of smaller mammals, suggesting exploitation of local resources, would be expected under those conditions. Instead, occupants of Dutch Canal Ruin apparently focused their subsistence activities on maize and agave. One exception may be Area 5 at Dutch Canal Ruin (Volume 2:Chapter 6), possibly a late pre-Classic period use of the site that postdated the abandonment of the canals. Charred maize remains were associated with a small structure constructed in the postabandonment deposits of two canals. Although maize could have been introduced at this location, the Hohokam may have attempted to use the abandoned canals or field drainage areas (secondary garden settings). This strategy would have been dependent on the retention of moisture by the abandoned canals, increased moisture levels due to irrigation runoff from upslope fields, natural runoff, or some combination of these water enhancement factors.

OPTIMIZATION OF THE LANDSCAPE

The proximity of Pueblo Salado to the Salt River clearly enhanced the diversity and influenced the intensity of use of resources exploited by site occupants. For residents of the site, riparian resources supplanted certain bajada or scrubland resources used by populations occupying the river terraces. Pueblo Saladoans exploited plants and animals found farther afield, such as saguaro or bighorn sheep, less than residents of Dutch Canal Ruin did. In addition, local diversity was greater at Pueblo Salado than at Dutch Canal Ruin, where floral and faunal resources were less dense due to the environment's inability to sustain as large a population. According to Stein's (1979:80–81) analysis of sites along the Salt-Gila Aqueduct (see also Fish and Nabhan 1991), natural resources in nonriverine resource zones were also available in riverine resource zones, but the reverse was not true.

Exploitation of modified microenvironments was well represented by evidence for irrigation agriculture. The site settlement pattern and composition, the canal system, and analyses of maize remains indicated that occupation of Dutch Canal Ruin had primarily been based on agricultural production. At Pueblo Salado, additional evidence included oxidized soils, phytoliths, chemical and nutrient data, and particle-size data (Chapter 5). Moreover, compared with neighboring sites, Pueblo Salado contained more maize, which was ubiquitous in sample contexts.

Evidence for the use of modified microenvironments other than for agriculture was not as abundant. Carbonized weedy plant species were low in frequency, providing little evidence of opportunistic gathering. However, the absence of charred weeds does not mean that secondary gardens did not exist or that residents of a site did not exploit field and canal borders for these plants. Weedy species would probably have been picked while they were still young and succulent and would have been eaten either raw or cooked in water (Bye 1981:109). The archaeological evidence for such a strategy would be scarce, even if field personnel attempted to locate agricultural field margins or to take samples from canal sediments and field margins. A dramatic increase in disturbance plants would agree with other evidence of prehistoric modification, but it would not indicate that the inhabitants had exploited the subsistence potential of these areas. The only processing strategy that would produce evidence of wild greens exploitation would be seed parching, which was documented at the project sites, although not extensively.

It is unlikely that land-use patterns were so intensive that residents of the project sites farmed all arable lands at one time. In fact, they probably allowed lands to lie fallow because of nutrient loss or elevated saline levels. In periods when less irrigation water was available, they may have farmed only prime land. Previous disturbance processes and land modification may have resulted (in the short term) in secondary gardens, which residents may have extensively exploited for economic resources. They may also have exploited other microenvironments, such as in-use and abandoned canals, washes and arroyos, braided stream channels, field borders, and field and canal drainage areas, as supplemental resource zones.

If Hohokam sites located on canal systems were focal centers supplying goods and services for a surrounding population, then the distribution of some of these sites may have been based, if not on marketing principles, at least on specialized resource exploitation. Archaeologists may be able to determine spatial arrangement by interpreting the economic potential of catchment areas. Residents of areas that were marginal for agriculture may have opted to supplement their subsistence base with other staples or to exploit resources available within their catchment area as an exchange item to obtain surplus produce or other economically important resources from other locales. Pueblo Salado, in its situation on the floodplain, could have offered other members of the Hohokam regional system access to riparian products not normally available to them or available only at an excessive energy investment. For instance, temper analysis and all ceramic type categories indicated that Pueblo Saladoans obtained as much as 33% of their plainware and 49% of their redware ceramics through exchange with southern and northern production areas (Chapter 3). Although investigators did not discover the reciprocal exchange item, it was probably a product of the highly diverse riparian habitat.

Another environmental factor guiding settlement location may have been periods of stress, when the Hohokam may have attempted to maximize resource diversity. Settlement of Pueblo Salado occurred during that period when levels of effective moisture were at or somewhat below normal (Graybill 1989). Streamflow reconstructions indicate that this was also a time of reduced likelihood of Salt River flooding. Therefore, the settlement of Pueblo Salado may have been a response to decreased water availability elsewhere in the Salt River Valley, as residents relocated to an area that offered greater diversity of resources and a more reliable water source for domestic and agricultural use. In the process, Pueblo Salado, located on its own canal system, would have become more self-sufficient than sites that were located in multiple-village canal systems (Chapter 3).

CONCLUSIONS

Two natural zones, riparian and desertscrub, occur within the Phoenix Sky Harbor Center. The riparian zone offers a wider variety of both plant and animal resources than the desertscrub zone, and Pueblo Saladoans appeared to have made extensive use of the riparian zone while practicing intensive agriculture. Residents of Dutch Canal Ruin during the pre-Classic period also relied on maize agriculture and, to a lesser extent, on wild resources. During the Classic period at Dutch Canal Ruin, agave also became an important agricultural resource. The botanical assemblages from both sites exhibited limited use of the bajada zones.

Use of the physical landscape by Hohokam groups resulted in modifications that may have provided additional resource exploitation areas. Such areas included in-use and abandoned canals, agricultural fields and their margins, and secondary gardens or field drainage areas. Although archaeologists cannot conclusively demonstrate that these areas provided supplemental resources, such artificially enhanced microenvironmental areas often provide a wealth of economic resources. Field borders are often convenient to the day-to-day activities related to farming, and wild resources can quickly be gathered without site residents having to travel to a more distant source. Elevated moisture levels and disturbance activities related to agriculture result in greater quantities and varieties of economically important plants when they are allowed to grow.

Pre-Classic period occupation at Dutch Canal Ruin consisted of field house locales used in conjunction with agricultural activities. Project investigators gathered few data that support extensive wild plant and animal resource exploitation by residents of the site during this time. Later occupants of the site relied on a wider variety of resources, with the dominant resources being maize and agave. The botanical assemblage from Pueblo Salado exhibited evidence of maize agriculture as well as a wide variety of natural resources that had been procured from the Salt River floodplain with species common to the riparian habitat. Extensive exploitation of the floodplain and riparian zone produced resources used for construction, fuel, and food and perhaps products for the local exchange network.

CHAPTER 8

THE HOHOKAM POST-CLASSIC POLVORÓN PHASE

Mark L. Chenault

Excavations at the site of El Polvorón (Sires 1984), in the Queen Creek drainage, revealed a small prehistoric settlement dating late in the Hohokam chronological sequence. Inhabitants of the site lived in pit houses, practiced agriculture, and had a material culture in the Hohokam tradition. As a result of those findings, Sires (1984) proposed a new phase, the Polvorón, dated A.D. 1350–1450, for the Hohokam sequence. This addition places the end of the Civano phase at A.D. 1350 rather than at the apparent Hohokam abandonment of the region around A.D. 1450, as traditionally held. Sires considered the Polvorón phase to be a post-Classic period manifestation of Hohokam culture. The Hohokam chronology used for this project (Figure 8.1) is based on Dean 1991, with the addition of the Polvorón phase.

This chapter examines the validity of the Polvorón phase concept, drawing on the results of the archaeological investigations in the City of Phoenix Sky Harbor Center. Although the concept of the Polvorón phase seems to be gaining acceptance among practitioners in the field, some Hohokam archaeologists have been reluctant to accept it. Doyel (1991a:263), for example, suggested that the phase was hastily formulated and that it should not include such adobe-walled Civano phase assemblages as those from Escalante Ruin. Andresen (1985) also originally considered the pit houses in Compound F at Casa Grande to be simply a variation of Civano phase architecture, but he later (1985:627) indicated that the Casa Grande pit structures might date to the Polvorón phase.

Therefore, in the interest of more carefully formulating and defining the Polvorón phase, this chapter examines the evidence for its existence in light of investigations at Dutch Canal Ruin, Area 8, and Pueblo Salado, Area 8/9, along with information from the excavations at El Polvorón and at other late Hohokam sites in the Phoenix Basin and Hohokam region.

CHARACTERISTICS OF THE POLVORÓN PHASE

The concept of the Polvorón phase is very much like that of the Bachi phase proposed by the Gladwins (Gladwin and Gladwin 1935) to cover the time period between A.D. 1400 and 1700 (Doyel 1991a; Sires 1984). Concerning the Bachi phase, Sires (1984:323) stated: "For a number of reasons, primarily the lack of archaeological data specifically from this time and the fact that it was interpreted as resulting from the exodus of the Salado from the Phoenix Basin, this phase has not gained wide acceptance." In a recent paper, however, Doyel (1991a:263) recommended retaining the Bachi phase for denoting the period between the Polvorón phase and the Spanish entrada (A.D. 1450–1700).

Sires (1984) characterized the Polvorón phase as consisting of a pattern of dispersed small settlements, possibly located along small canal segments. According to Sires, this pattern can occur either through the continued occupation of rooms within Civano phase compounds or through construction and occupation of jacal pit structures forming house clusters. Material culture of the Polvorón phase includes ceramics dominated by plainware and redware types, with decorated types consisting almost entirely of Roosevelt Red Ware (Salado Polychromes). Intrusive ceramics found in Polvorón phase assemblages include Hopi orangeware and yellowware (e.g., Jeddito Black-on-yellow and Awatovi Black-on-yellow). Lithic assemblages do not appear to differ appreciably from

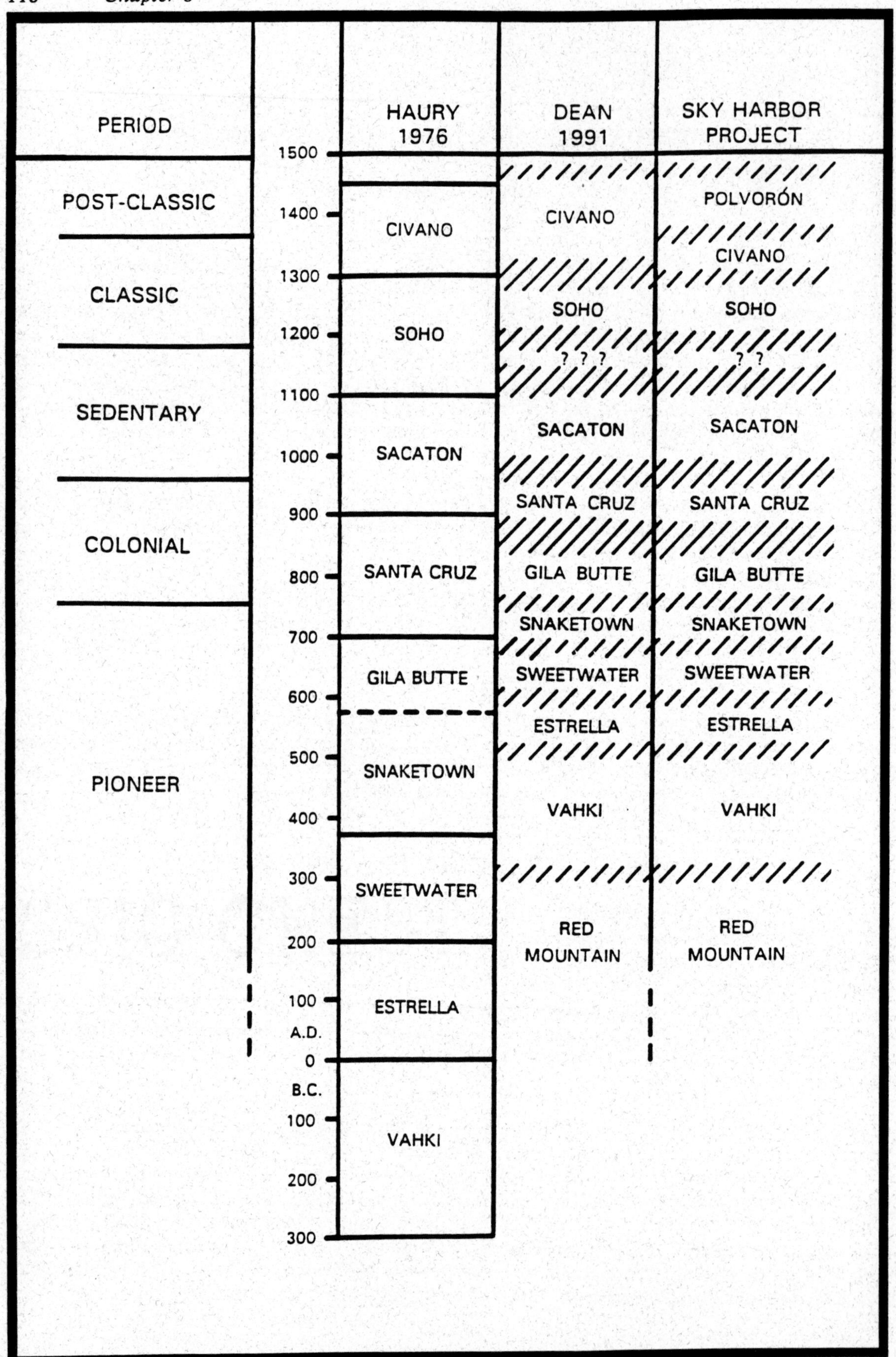

Figure 8.1. The Hohokam chronology used for the Phoenix Sky Harbor Center Project, compared to those devised by Dean (1991) and Haury (1976).

those in the Civano phase, with the exception of an increase in the use of obsidian, but shell artifacts exhibit a decrease both in number and in types of shell. Treatment of human remains appears to have involved both inhumation and cremation. Subsistence practices do not seem to have changed drastically, with most of the traditional wild and domestic foods continuing to be exploited (Sires 1984:324–325). Other archaeologists have found new data that indicate continued use of canal irrigation at this time, although D. H. Greenwald (personal communication 1992) stated that he believes that the Hohokam used small-scale systems on the floodplain during the Polvorón phase.

Area of Distribution

Investigators have found remains attributable to the Polvorón phase primarily in the Salt-Gila Basin (Crown and Sires 1984; Sires 1984), also known as the Phoenix Basin (Doyel 1991b) (Figure 8.2). These manifestations typically occur as small concentrations of habitation features in a dispersed settlement plan. Excavated sites at which archaeologists have identified Polvorón components include El Polvorón (Sires 1984); Pueblo Salado (Volume 3:Chapter 2); Dutch Canal Ruin, Area 8 (Volume 2:Chapter 9); Las Colinas (Hammack and Sullivan 1981); Brady Wash, Locus E (Greenwald and Ciolek-Torrello 1988b); Casa Buena (Howard 1988b); Pueblo Grande (Bostwick and Downum 1994; Downum and Bostwick 1994; Foster 1994); and Casa Grande, Compound F (Andresen 1985; Hayden 1930).

Temporal Range

Sires (1984) placed the beginning of the Polvorón phase at A.D. 1350 and the end at 1450, based primarily on archaeomagnetic dates obtained for samples from the site of El Polvorón. Dates from other suspected Polvorón phase components have confirmed this range. Radiocarbon and archaeomagnetic dates obtained for the suspected Polvorón phase components in Area 8 of Dutch Canal Ruin and Area 8/9 at Pueblo Salado also fell within that time span but suggested that the Polvorón phase may have extended to as late as A.D. 1500.

Architecture

Polvorón phase architecture consists primarily of rectangular pit houses with jacal superstructures (Sires 1984). These structures are true pit houses, in which the sides of the pit form the lower walls (Figure 8.3), as opposed to the "houses-in-pits" common to the Hohokam pre-Classic period. Often, Polvorón phase components are present within earlier Classic period settlements (Figure 8.4; Sires 1984). It also appears that collapsed adobe-walled rooms were sometimes rebuilt with jacal superstructures, utilizing the existing wall remnants for support (Figure 8.5; Gregory 1988).

Polvorón phase structures at Dutch Canal Ruin (n=2) and Pueblo Salado (n=6) conformed to the plan of those at El Polvorón. They were true pit houses and were primarily rectangular with rounded corners, although one structure (Feature 8-2 in Area 8 of Dutch Canal Ruin) was more D-shaped than rectangular (Figure 8.6). The builders of a pit house apparently constructed it by digging a shallow pit (approximately 0.40 m in depth in one example) and applying a lining of adobe to the sides of the pit and sometimes the floor. The superstructure above the sides of the pit was constructed of jacal. Figure 8.7 shows the plan and profile of a Polvorón phase structure from Pueblo Salado.

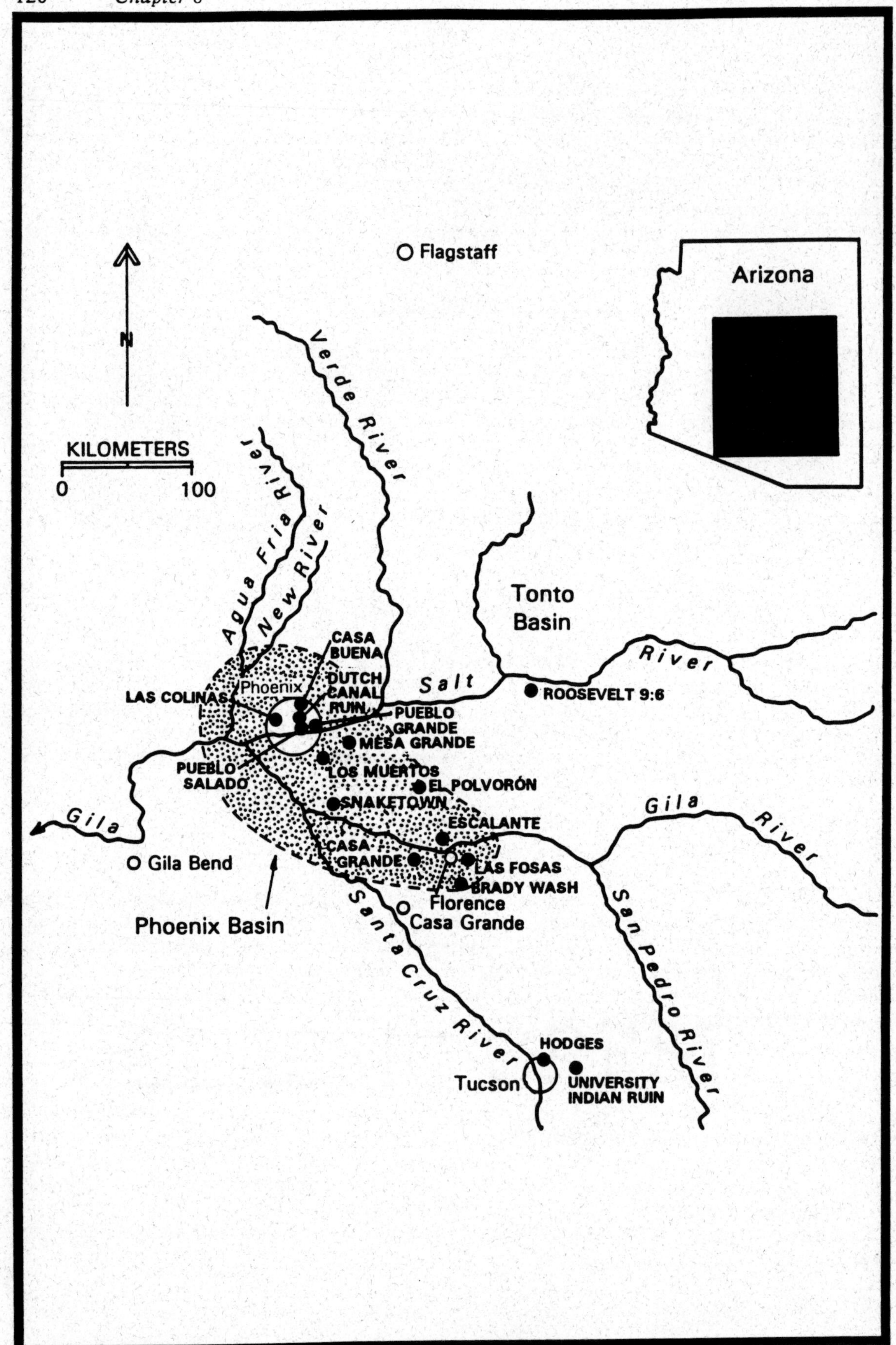

Figure 8.2. The geographic extent of the Phoenix Basin (from Doyel 1991b).

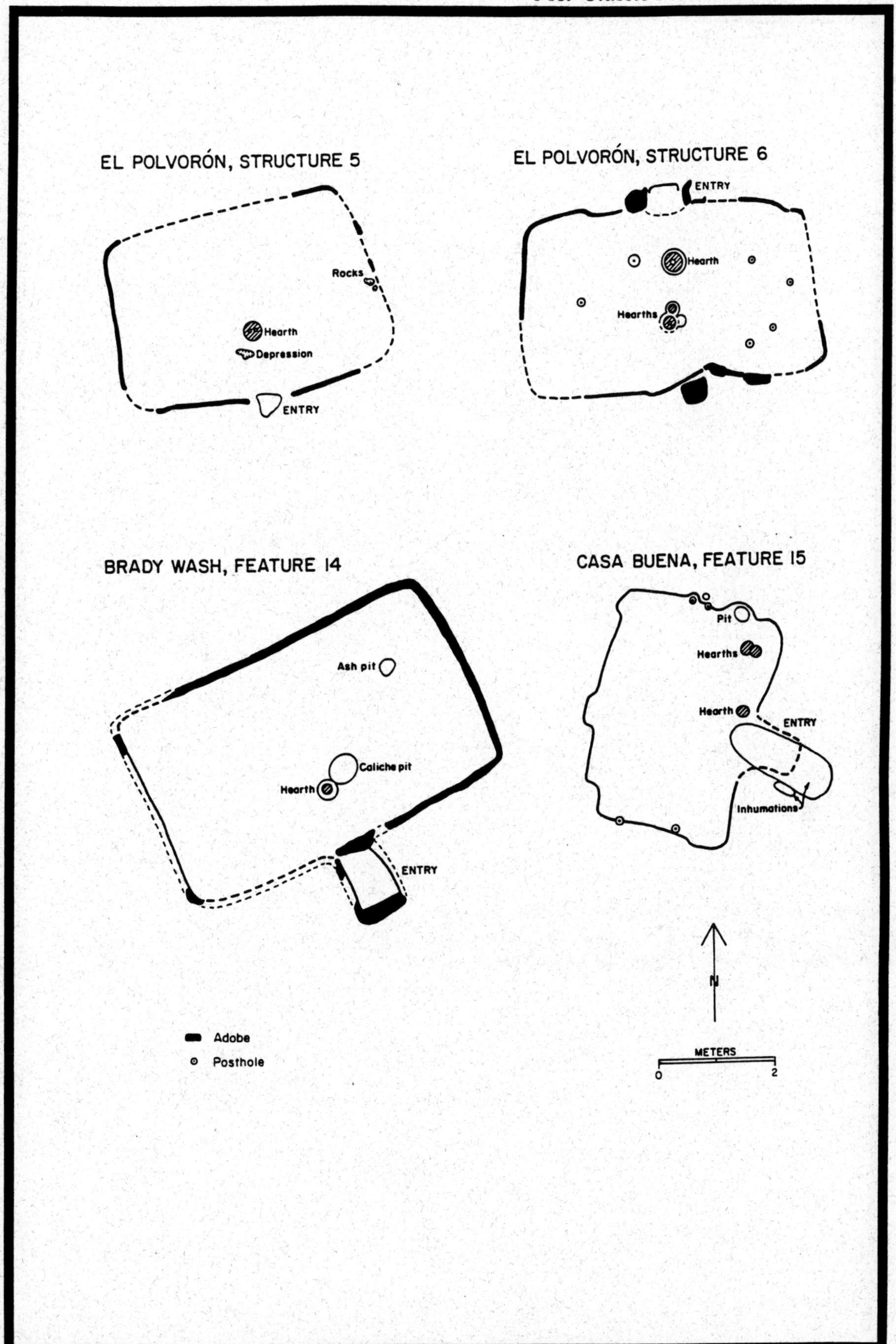

Figure 8.3. Plan views of selected Polvorón phase structures.

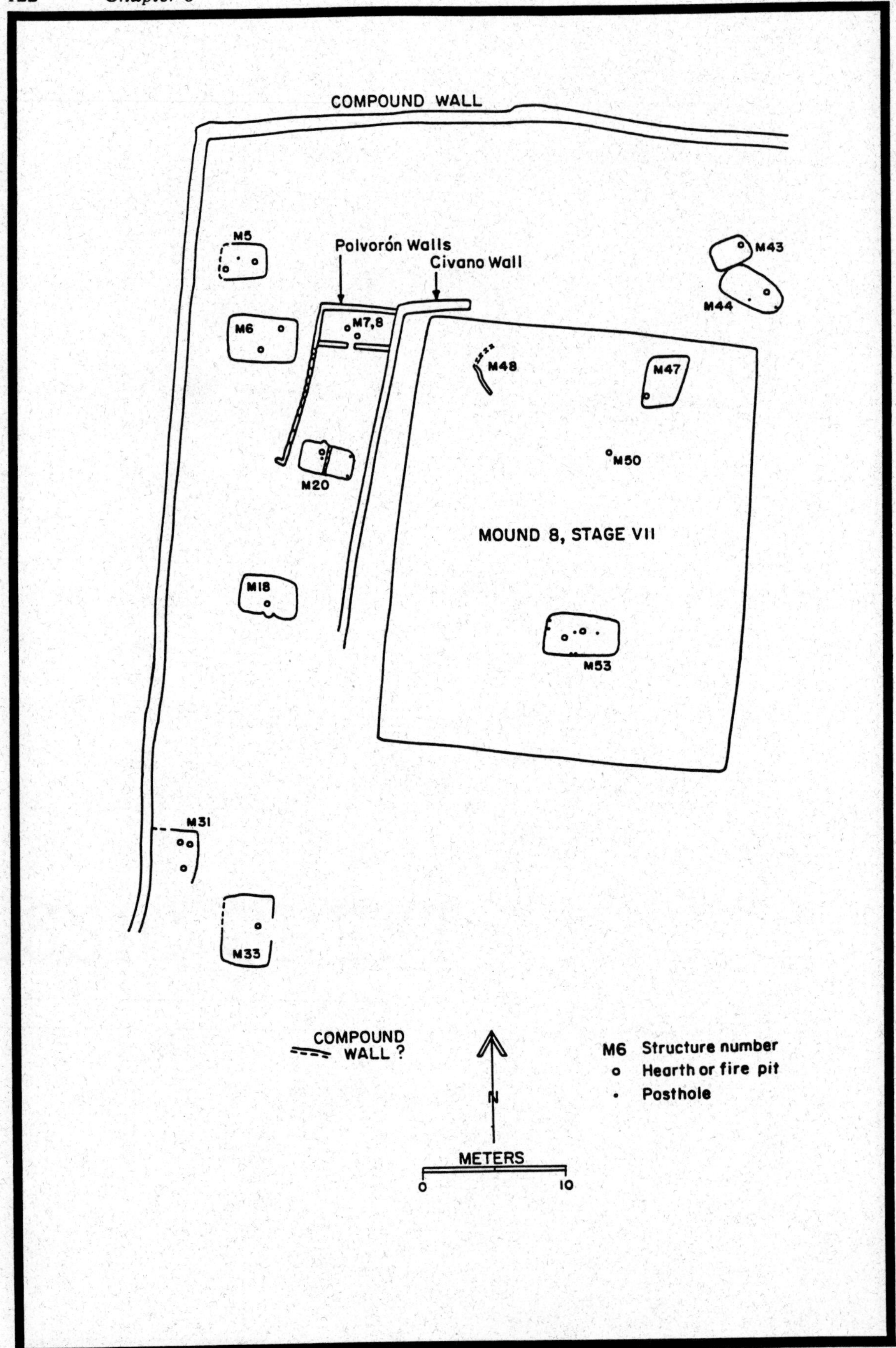

Figure 8.4. The post-Classic period Polvorón phase occupation at Mound 8, Las Colinas (after Gregory 1988).

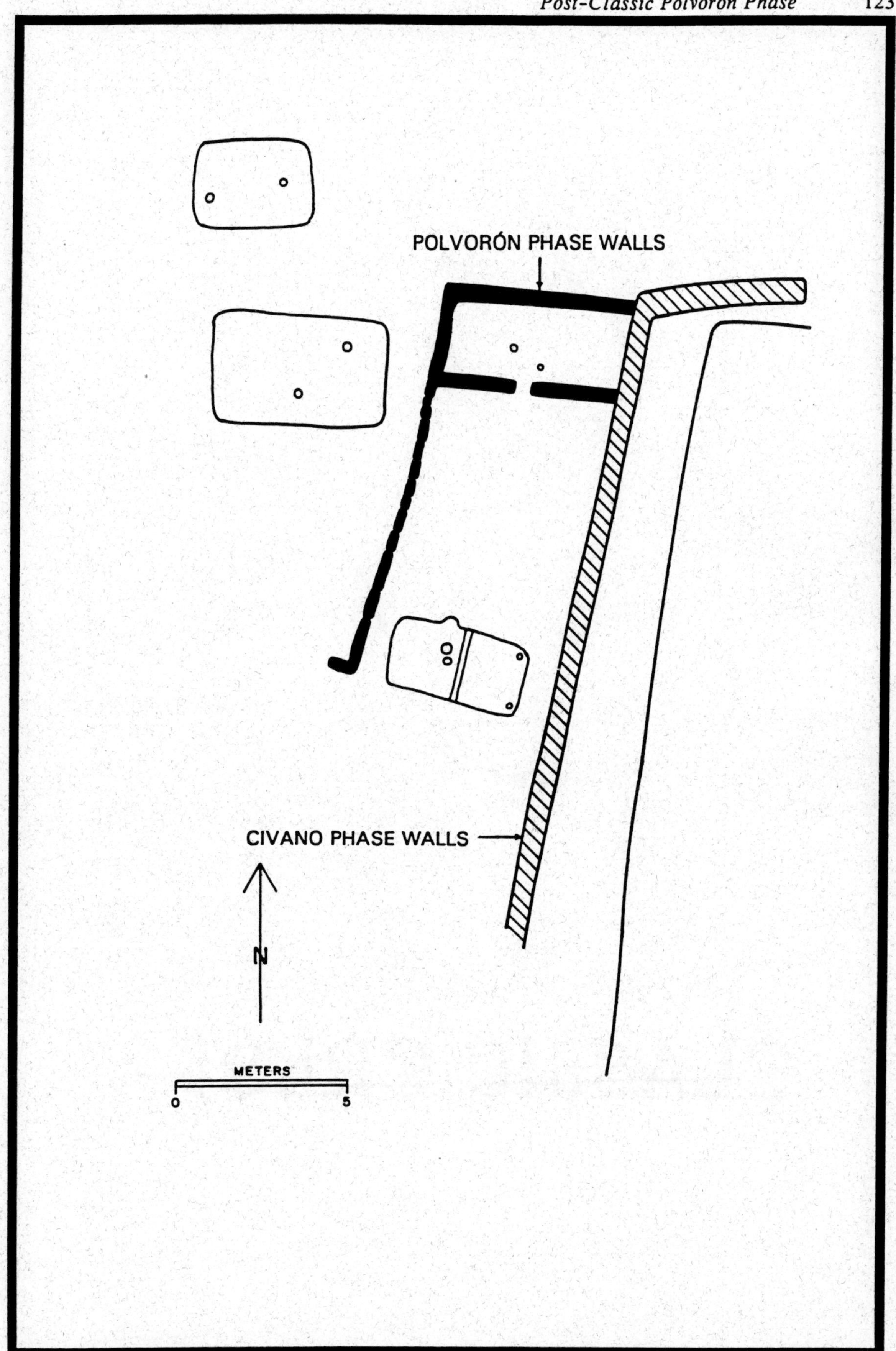

Figure 8.5. Civano phase walls used in Polvorón phase architecture at Mound 8, Las Colinas (after Gregory 1988).

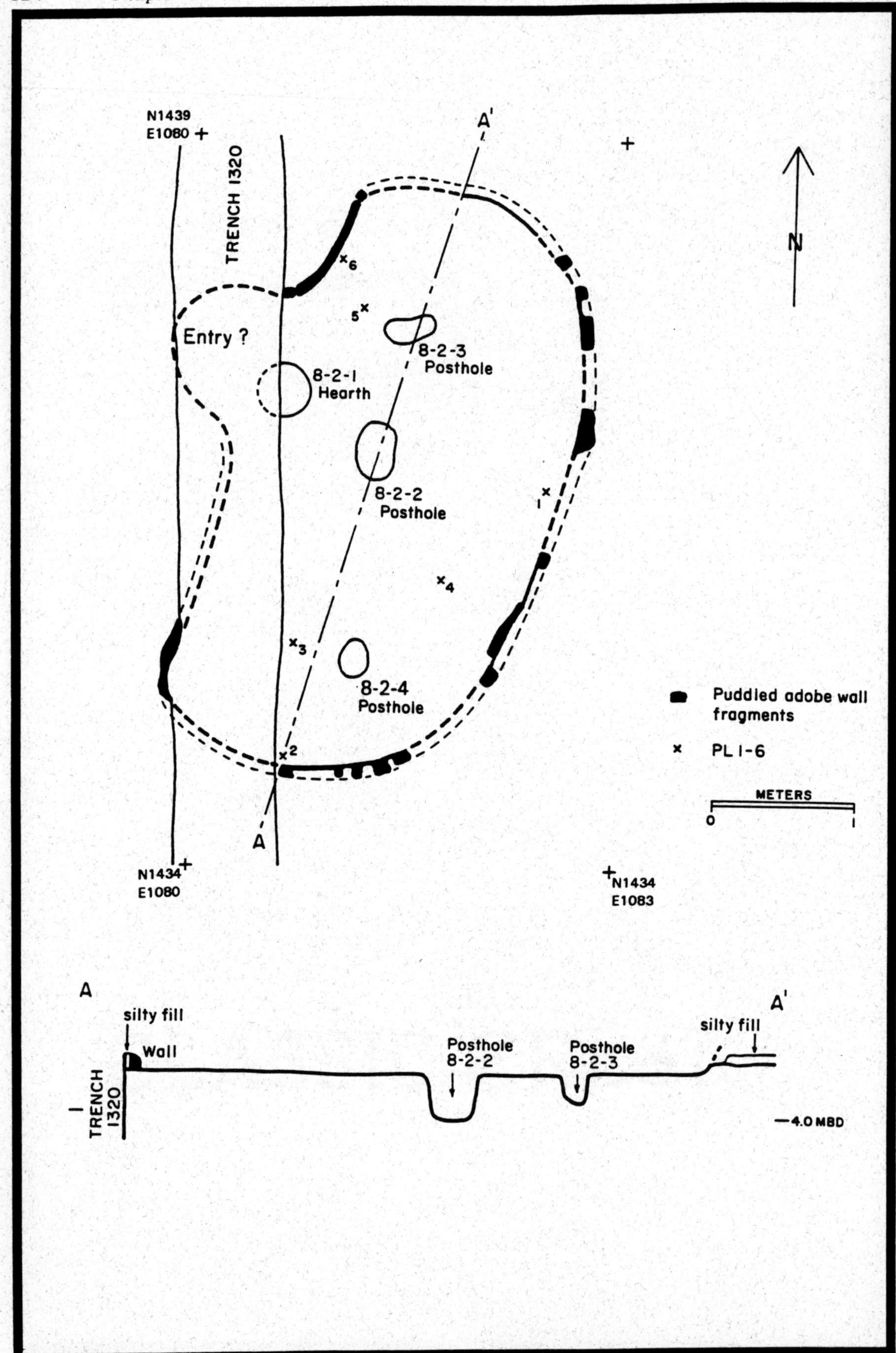

Figure 8.6. A Polvorón phase structure at Dutch Canal Ruin, Area 8 (Feature 8-2).

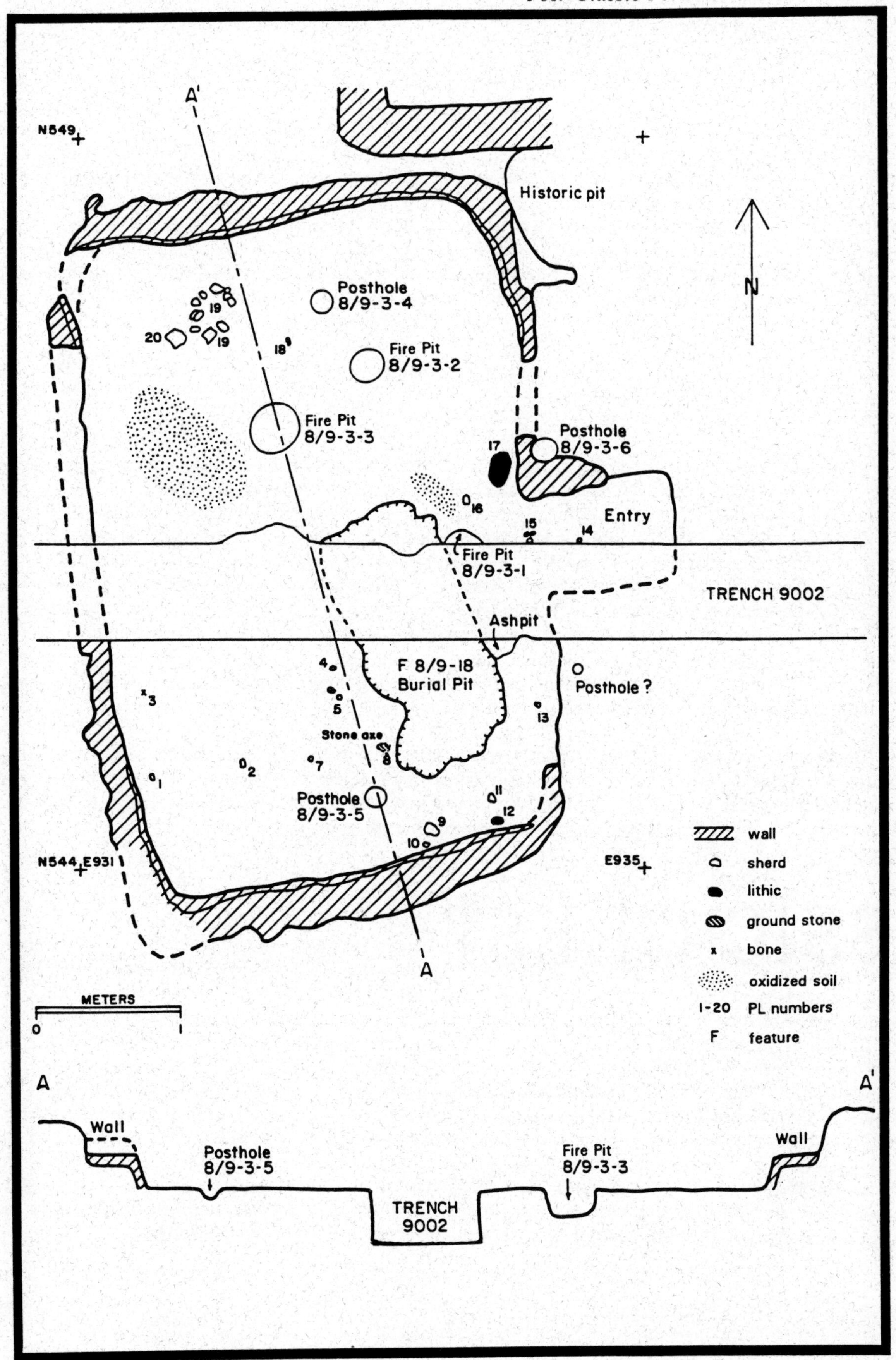

Figure 8.7. A Polvorón phase pit house (Feature 8/9-3) at Pueblo Salado, Area 8/9. The burial pit intruded into the fill and floor of the structure.

Sires (1984) concluded that Polvorón phase pit structures were arranged in small groups and retained the house cluster as the basic unit of organization (Figure 8.8). Gregory (1988:41) discussed the Polvorón phase structures at Las Colinas in terms of pairs. Examination of the layouts of Polvorón phase components at various sites suggests that the pairing of structures is the typical pattern, indicating that the size of the social units decreased during the post-Classic period. During the pre-Classic and Classic periods, the courtyard group, consisting of several structures, housed the primary social unit above the level of the individual household. During the Polvorón phase, however, the social unit appears to have been the individual household, consisting of a nuclear family or small extended family unit. The structures at Dutch Canal Ruin and Pueblo Salado seemed to fit this pattern and appeared to have been arranged in pairs.

Material Culture

Ceramics

Indigenous ceramics found in Polvorón phase components consist of very low frequencies of red-on-buff types (primarily Casa Grande Red-on-buff), high frequencies of redware and plainware, and what may be some locally manufactured red-on-brown wares (Volume 2:Chapter 12; Crown and Sires 1984). Roosevelt Red Ware types (Gila and Tonto polychromes) are abundant in Polvorón phase assemblages, usually representing 85–88% of total decorated ceramics. Some question remains, however, about whether Roosevelt Red Ware was produced by the Hohokam in the Salt-Gila Basin (Crown 1991) or was produced in other locations and traded into the basin. A study by Crown and Bishop (1987; see also Crown and Bishop 1991) indicates that the Roosevelt Red Ware polychromes were produced at multiple locations, and analysis of a small sample of sherds from Pueblo Grande by Peterson and Abbott (1991) suggests that Roosevelt Red Ware was produced outside the Phoenix Basin.

Decorated intrusive pottery from Polvorón phase contexts includes Hopi yellowware and orangeware, San Carlos Red-on-brown, and Tanque Verde Red-on-brown. As an example, field personnel recovered several sherds of Jeddito Black-on-yellow from Dutch Canal Ruin, Area 8, and Pueblo Salado, Area 8/9, along with several sherds of other intrusive types.

Flaked Stone

The main change evident in flaked stone assemblages between the Civano and Polvorón phases is an increase in the use of obsidian. Obsidian apparently was obtained mainly in the form of small nodules, commonly called "Apache tears," although Sires (1984:277) reported a variety lacking the water-rolled cortex of Apache tears. Obsidian nodules occur in a number of locations in Arizona, including the Superior area, the area around the Sauceda Mountains, the Vulture Mountain area, and other more distant locations (Shackley 1988; Slaughter et al. 1992). Obsidian accounted for approximately 30% of the flaked stone assemblage from El Polvorón but was less frequent at the Phoenix Sky Harbor Center sites. At Dutch Canal Ruin, 4.6% of the total flaked stone assemblage was obsidian. The assemblage from the Polvorón phase locus, Area 8, was 5.5% obsidian, and the pre-Classic component contained 0.2% obsidian. At Pueblo Salado, obsidian constituted 9.6% of the Polvorón phase flaked stone assemblage, 5.6% of the total site assemblage, 0.6% of the Area 14 assemblage, and 6.3% of the Area 8/9 assemblage. Small, triangular, side-notched projectile points common in Polvorón phase components were manufactured mainly from obsidian.

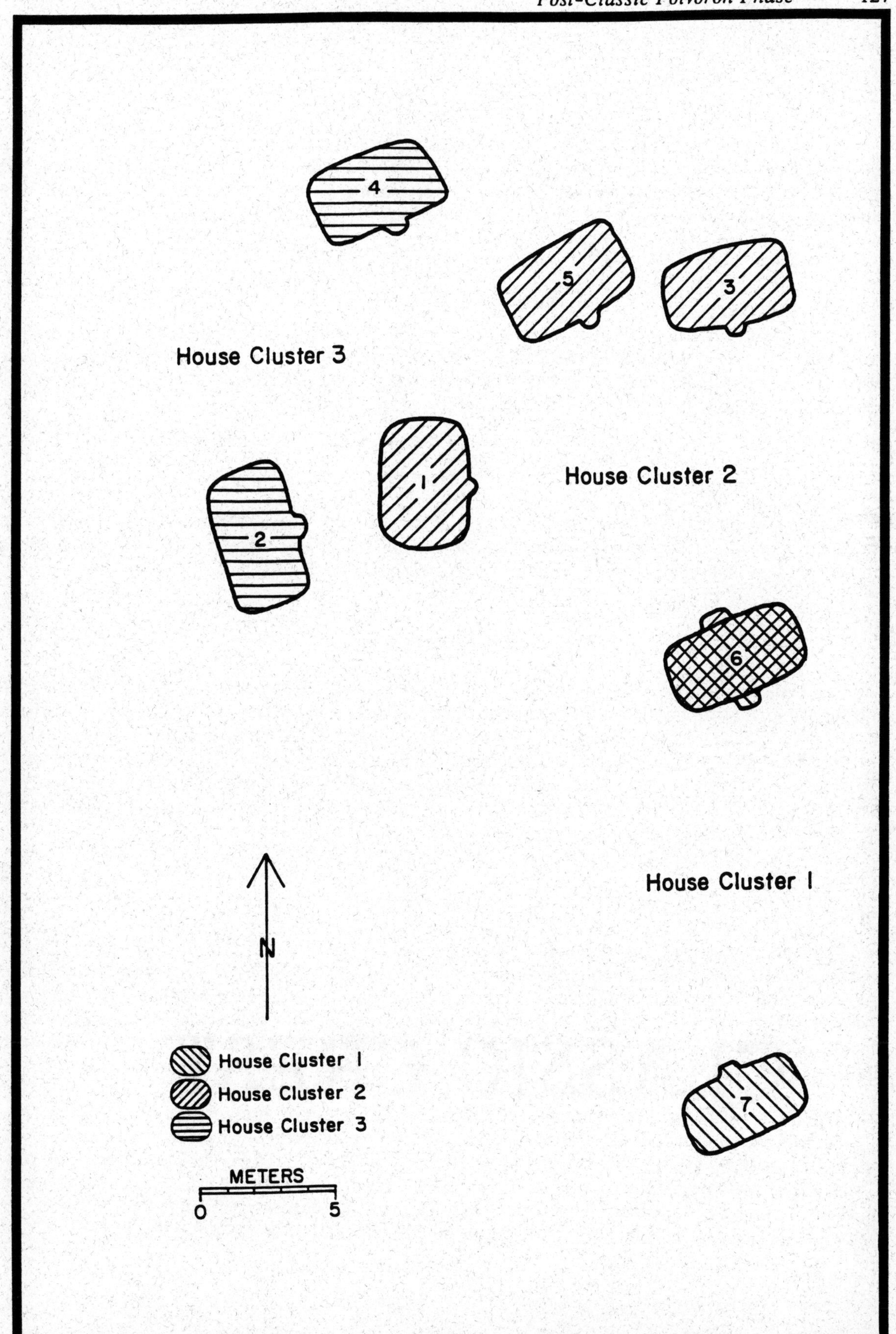

Figure 8.8. Household clusters at El Polvorón (from Sires 1984).

Ground Stone

According to Crown and Sires (1984), ground stone tools did not change noticeably from the Civano phase to the Polvorón phase. Polvorón phase assemblages include trough metates, slab metates, and shallow basin metates. Manos, three-quarter-grooved axes, polishing stones, stone rings, and stone bowls also occur in these assemblages.

Shell

Vokes (1984) noted a trend toward simplification of design in finished shell artifacts from Polvorón phase assemblages. He also found a decrease in the use of *Glycymeris* for production of bracelets, an artifact common in earlier Hohokam contexts. Zoomorphic and geometric cut shell is also largely lacking from Polvorón phase sites.

Clay Artifacts

Crown and Sires (1984) suggested that clay figurines had reappeared after an absence during the Classic period. Also, low-fired comales commonly occur and were recovered from Polvorón phase contexts at Pueblo Grande, Dutch Canal Ruin, and Pueblo Salado.

Burial Practices

The Hohokam of the post-Classic period apparently practiced both inhumation and cremation burials, favoring inhumations in the habitation areas and cremations in isolated plots (Crown and Sires 1984). The data from Dutch Canal Ruin and Pueblo Salado strongly supported this pattern. In Area 8 of Dutch Canal Ruin, two cremation burials lay approximately 10 m northwest of the structures, and three intrusive burials were present in probable Polvorón phase structures in Area 8/9 of Pueblo Salado. Those burials were almost certainly associated with the last-occupied pit structures at the site.

Subsistence

The pollen and carbonized remains of maize, cotton, beans, and squash provide evidence of cultivation of domesticated plants. Maize types found at El Polvorón include Onaveño, Harinoso de Ocho, and Maíz Blando (Sires 1984). Project investigators recovered another cultivated plant present in Polvorón phase contexts, little barley grass, from Dutch Canal Ruin, Area 8. Fibers from agave, believed to have been cultivated by the Phoenix Basin Hohokam (Volume 2:Chapter 15; Bohrer 1991), dominated the macrobotanical remains at Dutch Canal Ruin, Area 8.

A number of economically valuable wild plant species were also present at El Polvorón or Dutch Canal Ruin, Area 8, including Cheno-ams, prickly pear, buckwheat, wild gourd, wild tepary bean, spiderling, cattail, and possibly tobacco. Investigators found evidence for the use of cholla at Dutch Canal Ruin (Volume 2:Chapter 16) but not at El Polvorón (Sires 1984). Little change was apparent in the use of plant foods from the Civano phase through the Polvorón phase at Pueblo Salado, but the use of mesquite for fuel and construction material increased in the Polvorón phase.

ANALYSIS OF THE POLVORÓN PHASE CONCEPT

Interpretation One

The first and most basic issue to be examined is whether archaeologists can readily discern the Polvorón phase assemblages from those of the Civano phase. The interpretation of the supporting evidence can be stated as follows: *In the archaeological record substantial differences exist between remains assignable to the Civano phase and those assignable to the Polvorón phase.* Are the attributes suggested for the Polvorón phase different enough from those of the Civano phase to justify the formulation of a separate phase following the criteria of Willey and Phillips (1958)? They define a phase as

> an archaeological unit possessing traits sufficiently characteristic to distinguish it from all other units similarly conceived, whether of the same or other cultures or civilizations, spatially limited to the order of magnitude of a locality or region and chronologically limited to a relatively brief interval of time [Willey and Phillips 1958:22].

This interpretation implies that architectural remains, material culture, relative frequencies of artifacts, evidence of subsistence practices, burial practices, settlement patterns, and other aspects of the system differ enough from those of the preceding phase to warrant a different phase designation. In addition, a proposed phase should span a brief enough interval of time to fit the criteria for a phase and be compatible in length with the existing chronological sequence.

Architecturally, the late-dating pit structures found at the site of El Polvorón, and those found at some Classic period sites in the Phoenix Basin, differ greatly from the contiguous adobe-walled structures and compounds of the Civano phase. This difference is especially valid as an identifying criterion if one agrees with Doyel's (1991a) admonition that the Polvorón phase should not include assemblages with architecture traditionally assigned to the Civano phase simply because those remains postdate A.D. 1350. Such a late Civano phase component might best be thought of as a subphase of the Classic period rather than a separate phase (Willey and Phillips 1958:24). Civano phase architecture is typified by clusters of contiguous rooms, courtyards, and plazas surrounded by massive compound walls (Doyel 1991a; Haury 1945, 1976). Moreover, at some sites, the Hohokam had constructed platform mounds within the compound wall (Gregory 1987; Gregory and Nials 1985). Polvorón phase architecture, by contrast, consists of pit houses with jacal superstructures. The architectural units are noncontiguous, and the massive adobe-walled construction is not present. However, the shift in architectural style probably occurred in a transition period during which residents continued to use the adobe structures as well, in some cases renovating them (Volume 3:Chapter 2; Greenwald and Ciolek-Torrello 1988a).

Although it is entirely possible that adobe-walled Civano phase rooms were reoccupied following the Classic period Hohokam collapse, clear evidence for such a reoccupation would be difficult to find and attribute to the people of the Polvorón phase. For the purpose of defining the traits of the Polvorón phase, the only architectural types that can be assigned to the phase are late-dating true pit houses and adobe rooms that were clearly abandoned and then subsequently rebuilt and altered for reoccupation.

Site size also differs between the two phases; the number of pit structures is generally much smaller in Polvorón phase occupations than in Civano phase occupations. To date, archaeologists have found fewer sites assignable to the Polvorón phase and have not thoroughly defined the settlement patterns. Because of the ephemeral nature of the architectural remains of the Polvorón phase, excavators may overlook single-component occupations amid and beneath the modern urban features in the Phoenix Basin. Also, because they are more recent, Polvorón phase remains occur higher stratigraphically and are more likely to be disturbed by modern processes (Chapter 2; Schiffer 1987).

Cultural materials do not appear to differ significantly in type or morphology between the Civano phase and the Polvorón phase, but frequencies and, in some cases, raw material types vary. Peterson and Abbott (1991) showed that percentages of Roosevelt Red Ware pottery in assemblages of decorated ceramics attributed to the Polvorón phase range from around 60% to near 90%. Conversely, assemblages of decorated ceramics from the Civano phase contain less than 45% Roosevelt Red Ware, generally less than 30%. At Dutch Canal Ruin, Area 8, the percentage of Roosevelt Red Ware polychromes in the decorated ceramic assemblage was 86%. Intrusive wares in the Polvorón phase assemblages also include a few sherds of Hopi orangeware and yellowware, which are absent from Civano phase assemblages. More reliance on obsidian in relation to other lithic material types appears to be a hallmark of the Polvorón phase (Volume 2:Chapter 13; Volume 3:Chapter 6). Shell artifacts, too, exhibit temporal differences, with an overall decrease in amount, a simplification in artistic modification, and, as mentioned above, a decrease in the frequency of *Glycymeris* bracelets (Sires 1984). Investigators may reveal additional differences in cultural materials or aspects of the material assemblages through further research.

Subsistence practices, in contrast, appear to have remained the same. In Sires's words, "Most, if not all, of the traditional foods, both domestic and wild, are still in use" (1984:325). However, archaeologists also need to research this topic in more depth. According to Gasser and Kwiatkowski (1991), diachronic change in plant and animal use in the Phoenix Basin is minor, and even changes in subsistence between broad periods of time, such as between the pre-Classic and Classic periods, are not extensive (Mitchell 1989b). The problem is compounded with regard to the Polvorón phase, in that many of the components now thought to date to that phase were not recognized at the time of excavation and were lumped together with Civano phase components during analyses. To separate Polvorón phase components from those of the Civano phase, researchers will need to re-examine subsistence evidence on a feature-by-feature basis, if possible.

Evidence, then, seems to support this first interpretation: the material assemblages from Polvorón phase occupations are recognizably different from those of Civano phase occupations in some attributes. The most striking differences are in architecture and in occupation size, but analysts have also observed differences in material use and in artifact frequencies. In some categories, such as subsistence, changes are not clearly evident. However, this ambiguity might be due in part to the general problems with "fine tuning" the chronology and with the lumping together of assemblages from the later components at various sites. Concerning the temporal element, the Polvorón phase spans a period of time from A.D. 1350 to 1450, an interval sufficiently brief to conform to the criteria defined by Willey and Phillips (1958) and in keeping with the length of other phases in the Hohokam sequence.

Interpretation Two

A second interpretation can be stated as follows: *The Polvorón phase is a temporally distinct phase, not the representation of a separate cultural group coexisting alongside the late Classic period Civano phase Hohokam.* If this interpretation is correct, absolute dates for the Polvorón phase components will be consistently later (post-A.D. 1350) than those obtained for the Civano phase components; relative dates, such as those obtained through ceramic seriation and those indicated by cross-dating with intrusive wares, will be later for the Polvorón phase than for the Civano phase; and Polvorón phase components will be stratigraphically later than Civano phase components. The following discussion deals with the dates for possible Polvorón phase components at a number of sites in the Phoenix Basin. For a more detailed and systematic treatment of the chronometric data, see Chapter 9 of this volume.

At El Polvorón, analysts obtained archaeomagnetic dates for two of the eight structures excavated (Table 8.1). The temporal range for Structure 4 was from A.D. 1400 to 1450, and the range of dates for Structure 6 was from A.D. 1340 to 1390. Only one radiocarbon sample produced an acceptable date (obtained from Structure 4), of A.D. 1410–1650 (one sigma). One other radiocarbon sample produced a date of post–A.D. 1945.

Due to the poor preservation of the fire pits in the Dutch Canal Ruin structures, investigators recovered no archaeomagnetic samples from this site. Calibrated radiocarbon dates, however, did not refute the placement of the occupation at Area 8 in the Polvorón phase. Analysts obtained a radiocarbon date range of A.D. 1280–1405 from the fire pit in a pit structure (Feature 8-3) and date ranges of A.D. 1284–1431 and A.D. 1277–1405 from two extramural pits (Table 8.1). These dates yielded an average occupation range of A.D. 1280–1414. Although the early end of the range predates the proposed Polvorón phase, the late end of the range falls within the proposed Polvorón phase time span (A.D. 1350–1450).

At Locus E of the Brady Wash site, near the Picacho Mountains, four pit houses and an adobe structure had archaeomagnetic dates placing them in the temporal range of the Polvorón phase. Feature 14a was dated A.D. 1175–1425; Feature 39a was dated A.D. 1325–1400; Feature 57a was dated A.D. 1325–1425; Feature 58a, an adobe structure, dated A.D. 1375–1425; and Feature 60a was dated A.D. 1300–1425 (Table 8.1). A sixth feature, an adobe structure, had an archaeomagnetic upper range of A.D. 925–1025 but was assigned an A.D. 1400+ date based on other criteria (Greenwald and Ciolek-Torrello 1988b:353).

Two pit houses at Casa Buena had dates placing them within the temporal range for the Polvorón phase. Feature 15 was assigned a date of A.D. 1300–1400, and Feature 172 was dated to A.D. 1300–1425 (Howard 1988b) (Table 8.1). Several remnants of structures at the site also had dates within the A.D. 1350–1450 range, as did a number of the coursed-caliche structures.

At Las Colinas, nine archaeomagnetic dates from seven pit structures fit the morphology for Polvorón phase structures (Hammack and Sullivan 1981). The average early end of the date ranges for the Las Colinas structures is A.D. 1371, and the average late end is A.D. 1418 (Table 8.1).

Relative dating methods also indicate that Polvorón phase components date later in the Hohokam sequence than Civano phase components. As described above, the percentage of Roosevelt Red Ware relative to decorated pottery is higher in Polvorón phase assemblages than in Civano phase assemblages (Peterson and Abbott 1991). This continues an apparent trend of increasing frequency of Roosevelt Red Ware through time. For example, Roosevelt Red Ware was 43.8% of the decorated ceramic assemblage from the Civano phase component at Casa Grande, Compound F, and 63.5% of the decorated ceramics from the post-Classic period component (Andresen 1985; Peterson and Abbott 1991). At Pueblo Grande, the figures were 29.1% for the Civano phase and 66.9% for the Polvorón phase (Abbott, ed. 1994). At El Polvorón, which appeared to consist entirely of remains from the post-Classic period, 88.6% of the decorated sherds were Roosevelt Red Ware (Peterson and Abbott 1991).

This trend is concurrent with a general decrease in other Hohokam decorated ceramics beginning in the Sedentary period and continuing through the Classic period (Doyel 1980; Haury 1976). Amounts of plainware and redware increased throughout the Classic period, while decorated wares

Table 8.1. Chronometric Data for Features from Various Sites in the Phoenix Basin Assignable to the Polvorón Phase

Site	Feature	Date Range	Type
Brady Wash, Locus E	Feature 14a	A.D. 1175–1425	archaeomagnetic
Brady Wash, Locus E	Feature 39a	A.D. 1325–1400	archaeomagnetic
Brady Wash, Locus E	Feature 57a	A.D. 1325–1425	archaeomagnetic
Brady Wash, Locus E	Feature 58a	A.D. 1375–1425	archaeomagnetic
Brady Wash, Locus E	Feature 60a	A.D. 1300–1425	archaeomagnetic
Casa Buena	Feature 15	A.D. 1300–1400	archaeomagnetic
Casa Buena	Feature 172	A.D. 1300–1425	archaeomagnetic
Dutch Canal Ruin, Area 8	Feature 8-3-1	A.D. 1292–1405*	radiocarbon
Dutch Canal Ruin, Area 8	Feature 8-3-1	A.D. 1280–1405*	radiocarbon
Dutch Canal Ruin, Area 8	Feature 8-12	A.D. 1302–1431*	radiocarbon
Dutch Canal Ruin, Area 8	Feature 8-12	A.D. 1284–1423*	radiocarbon
Dutch Canal Ruin, Area 8	Feature 8-14	A.D. 1284–1405*	radiocarbon
Dutch Canal Ruin, Area 8	Feature 8-14	A.D. 1277–1405*	radiocarbon
El Polvorón	Structure 4	A.D. 1400–1450	archaeomagnetic
El Polvorón	Structure 4	A.D. 1410–1650*	radiocarbon
El Polvorón	Structure 6	A.D. 1340–1390	archaeomagnetic
Las Colinas	Feature 26	A.D. 1297–1373	archaeomagnetic
Las Colinas	Feature 34	A.D. 1392–1438	archaeomagnetic
Las Colinas	Feature 34	A.D. 1354–1426	archaeomagnetic
Las Colinas	Feature 35	A.D. 1383–1407	archaeomagnetic
Las Colinas	Feature 35	A.D. 1343–1397	archaeomagnetic
Las Colinas	Feature 40a	A.D. 1387–1433	archaeomagnetic
Las Colinas	Feature 41	A.D. 1383–1417	archaeomagnetic
Las Colinas	Feature 74	A.D. 1433–1467	archaeomagnetic
Las Colinas	Feature 114	A.D. 1363–1407	archaeomagnetic

*calibrated, one sigma

decreased (Crown 1981; Sires 1984). According to Sires (1984:302), plainware and redware increased in frequency from 91% to 95% during the Classic period.

Intrusive wares also indicate that Polvorón phase assemblages date late in the Hohokam sequence. Hopi orangeware from the Pueblo IV period occurred in many of the assemblages assigned to the Polvorón phase. For example, 48 sherds of Hopi orangeware and yellowware were present at El Polvorón (Sires 1984), 46 of them from one Bidahochi Polychrome (A.D. 1320–1400) bowl (Colton 1956). Also, a Bidahochi Polychrome vessel is reported to have been found at Las Colinas by a local family during unauthorized digging (Crown 1981). Investigators recovered 4 sherds of Jeddito Black-on-yellow (A.D. 1325–1600) (Colton 1956) from Area 8 of Dutch Canal Ruin (Volume 2:Chapter 12).

Examination of the stratigraphic context of Polvorón phase components indicates that they postdate the adobe-walled compound architecture assigned to the Civano phase. At Las Colinas, excavators found several pit houses on top of the mound, stratigraphically higher than Civano phase remains (Hammack and Sullivan 1981). At Compound F of Casa Grande, pit houses had been built into the debris of the collapsed compound wall (Hayden 1930). At Pueblo Grande, in a compound east of the platform mound, a Polvorón phase pit house was superimposed on a Civano phase adobe-walled compound room, and Downum and Bostwick (1994) reported the presence of Polvorón structures on top of the platform mound. Also, although the remains in Area 8 of Dutch Canal Ruin were not superimposed on earlier structures, their stratigraphic position was high within the project-area stratigraphy.

As the preceding discussion demonstrates, current evidence (particularly the stratigraphic evidence) generally supports the interpretation that the Polvorón phase is temporally distinct. In all multicomponent sites with pit structure architecture that appeared to conform to the Polvorón phase type, the Polvorón phase features intruded into, or were stratigraphically higher than, the features attributable to the Civano phase. In no known case was the opposite true.

Although pit structures corresponding to the Polvorón type tend to have temporal ranges later than Civano phase structures, some adobe-walled structures conforming to the Civano type also have absolute dates in the A.D. 1350–1450 range. At Escalante Ruin, for example, investigators found no pit structures resembling the typology of the Polvorón phase, but three adobe-walled rooms had archaeomagnetic dates later than A.D. 1400. Therefore, temporal placement alone does not appear to be an adequate means for distinguishing Polvorón phase remains from those of the Civano phase. Also, Dean (1991:73) examined the difficulty in distinguishing between adjacent phases using independent dates from Hohokam features, mainly because of the overlap in the ranges of dates. Similarly, in discussing periods and phases, Roberts stated, "It should be emphasized that these designations apply to the complex and not to a single element or series of years" (1935:33). Kluckhohn (1939:159), examining Anasazi ruins in Chaco Canyon, added "the various stages recognized by the Pecos classification (and very commonly referred to as 'periods') do not, necessarily, represent separate and clear-cut time periods, *even in the same geographical locality*" (emphasis in original).

Following the traditional concept of a phase, Doyel (1991a) appears to be correct in his contention that the adobe-walled compound sites of the Civano phase should not be included in the Polvorón phase, even when absolute dates of post-A.D. 1350 are obtained. As stated above, investigators can assign the late-dating Civano phase assemblages to a subphase of the Civano phase. Willey and Phillips (1958), who also advocated the use of the subphase unit rather than the phase when differences are present only in a few specific items or in variations in frequency, stated that "if it is impossible to present a sensible account of the culture of a unit except in terms of what went before or came after, it is probably better regarded as a subphase" (Willey and Phillips 1958:24).

The recommendation here, therefore, is that archaeologists use the Polvorón phase to denote the complex of traits described above under the heading "Characteristics of the Polvorón Phase" and that they date the phase primarily, but perhaps not exclusively, to the period from A.D. 1350 to 1450. The Civano phase, typified by coursed-adobe architecture, has been described elsewhere (Crown 1991; Doyel 1974, 1981, 1991a; Haury 1976). It may be possible to define a transition period, which may be considered a late subphase of the Civano phase, in which a reduced population continued to utilize the characteristic adobe-walled compound structures late in the Hohokam sequence.

Interpretation Three

Another interpretation is: *The Polvorón phase represents a choice to return to a former way of life involving smaller, more widely dispersed settlements and the construction and occupation of pit structures.* Implications include the post-Civano phase occurrence of cultural remains of a type found in pre-Civano phase or pre-Classic period contexts. This implies a knowledge by the prehistoric participants of their history, perhaps not an unreasonable assumption given the worldwide occurrence of oral histories, especially among nonliterate cultures. Alternatively, the Hohokam may have continued to use pit houses on a limited basis throughout the Classic period, thus retaining the knowledge of their use and construction.

Examination of the Polvorón phase data indicates support for this interpretation, at least at a general level. The Hohokam do appear to have returned to building and using pit houses rather than above-ground adobe structures; population levels were more like those of the early pre-Classic period; the data hint that site occupants may have had simpler levels of social organization; and settlement patterns may have been more dispersed than those during the Civano phase.

If, however, one looks at the specifics of the change in attributes from the Civano phase to the Polvorón phase, one can see that the process is one of culture change rather than of resurgence. The pit structures built during the Polvorón phase generally are different from those constructed during the pre-Classic and early Classic periods, which were houses-in-pits and not true pit houses (Howard 1988b). Therefore, the builders did not return exactly to the earlier architectural style. The change may have occurred because the simpler architecture was more appropriate to the size and level of complexity of the Polvorón phase system. The greatly reduced population level apparent for the Polvorón phase would not have provided a labor force large enough to maintain the more elaborate compound architecture and the extensive canal systems in operation during the Classic period. Of course, the smaller population during the Polvorón phase would also have required less food, thereby decreasing the need to maintain the irrigation systems at their previous level of complexity. Consideration should also be given to social and religious implications. As Bostwick (1992) suggested, loss of popular belief in the religious leaders' ability to control rainfall and flood events could have caused a shift in religious orientation. This change, in turn, could have altered social and organizational behaviors in ways that were reflected in architectural patterns.

Nor did the Hohokam revert to earlier types of material culture. For example, they did not resume production of red-on-buff ceramics but instead continued the use of polychromes, redware, and plainware. In other media, such as flaked and ground stone and shell, they also continued trends attributed to the Civano phase and did not return to the exact types of earlier periods.

In summary, the Hohokam appear to have adopted a lifestyle similar to that of the pre-Classic period but different in significant details. The change probably was dictated by a decrease in population and a resulting decrease in social complexity.

Interpretation Four

An interpretation similar to the preceding one can be stated as follows: *Remains dating to the Polvorón phase represent the continuation of the lifeways of a Hohokam lower class, which occupied pit structures and functioned within the larger Classic period system. When the upper and controlling classes abandoned the region, the lower-class peoples remained and continued their pit house-dwelling way of life.* The main implication of this interpretation is that investigators should find evidence of "lower-class" occupations contemporary with, and perhaps associated with, the coursed adobe-walled settlements of the Civano phase. The development and maintenance of that separate system should be traceable alongside the evidence of the Civano phase "upper-class" system. Material evidence would include the superpositioning of coursed-adobe structures over an occasional abandoned pit structure of the lower class, or the occupation of pit structures with dates placing them within the early Civano phase.

Further research into this topic is needed. Are there sites with contemporary pit house and adobe-walled components, for example? Is there evidence for pit house components contemporary with compound-architecture components but spatially separated?

Initial examination of the data suggests that whereas researchers have identified an occasional pit structure with a Civano phase date (A.D. 1300–1350) (e.g., Howard 1988b:70), they have found none intruded upon by compound architecture. Thus, perhaps the absolute dates for those structures do not accurately reflect their chronological relationship to the adobe-walled compound architecture.

Current data therefore do not support this interpretation. In fact, the apparent continuation of participation by a Polvorón phase population in a regional and interregional exchange system (see discussion below) suggests that the opposite might be true: the Polvorón phase might represent the continued occupation of the Phoenix Basin by persons who were the Hohokam elite, if such ranking existed in Hohokam society. Those individuals might have remained behind after the abandonment of the area by the general populace and maintained their ties with the elite of other regional systems, as evidenced by the exchange of Salado polychromes and the increased use and exchange of obsidian.

Interpretation Five

The possibility exists that Polvorón phase components represent limited activity sites, suggesting another interpretation: *The Polvorón phase components were habitation sites, not merely field houses occupied on a temporary basis by people with permanent habitations located elsewhere.* One implication of this interpretation is that evidence of long-term and possibly year-round occupation would be present. Such evidence would consist of substantial structures, abundant refuse, material culture consistent with habitation, burials (inhumation and cremation), and pollen and macrobotanical remains indicative of year-round processing of food resources. Another implication is that researchers would note an absence of contemporary sites of a more substantial and extensively used nature that could have functioned as permanent habitations.

Evidence from Dutch Canal Ruin (Area 8) and El Polvorón, both single-component sites, indicated that this interpretation probably is correct. Although architecture at the two sites was insubstantial, the structures would have provided adequate shelter, and other remains were indicative of long-term use. For example, trash deposits were extensive, and the artifacts recovered from the two sites indicated habitation rather than limited activity. Polychrome, redware, and plainware ceramics were abundant, and materials such as shell and obsidian also occurred. Investigators found two inhumations and one cremation at El Polvorón and two cremations at Dutch Canal Ruin, Area 8. Burials would not be likely in a limited activity, field house site. Even if an individual died in a field house location, he or she would probably be taken back to the permanent habitation for burial. Finally, macrobotanical evidence from Dutch Canal Ruin showed that the inhabitants had engaged in a wide range of subsistence-related activities indicating year-round occupation (Volume 2:Chapter 15). Although future work may change this interpretation, for the time being it appears that the Polvorón phase components are small habitation sites and not field houses.

Interpretation Six

A final interpretation deals with the ethnicity of people associated with the Polvorón phase: *The Polvorón phase represents a continuation of Hohokam culture and not the occupation of the area by a nonindigenous immigrant group*. Even in the absence of any known suggestions to the contrary, the possibility should perhaps be examined. Implications would be that continuation of trends evident in the early phases would be visible in manifestations such as architecture, material culture, and cultural practices (e.g., treatment of the dead). Another implication is that human physical characteristics would be similar to those of earlier Hohokam inhabitants.

Although architecture did change, investigators have identified a continuation of Hohokam material culture from the Classic period into the postulated Polvorón phase. Plainwares and redwares found in Civano phase contexts continued to be present, as did Roosevelt Red Ware. Characteristics found in flaked and ground stone artifacts during the Classic period also continued into the Polvorón phase. Likewise, burial practices appear to have changed little from earlier patterns and include both inhumation and cremation.

One of the most productive means of examining this interpretation is with the methods of physical anthropology. Analysis of the human remains recovered from Polvorón phase components can reveal physical similarities or differences compared to the remains found in Civano phase components. Specialists should initiate this type of study now that investigators are identifying Polvorón phase components at a number of sites in the Phoenix Basin.

The available data support the idea that the people of the Polvorón phase were the same culture group that occupied the Phoenix Basin during the Civano phase. Artifacts represent a continuum of previous trends along with an overall decrease in occupation size and complexity. As Sires states,

> there are few new attributes that distinguish the Polvorón phase from earlier Hohokam material assemblages. This phase is characterized more by the absence of elements common during the previous Civano phase than by the presence of distinctive and novel elements [1984:325].

HOHOKAM SOCIETY DURING THE POST-CLASSIC PERIOD

Evidence from the excavation of Polvorón phase components suggests that the Hohokam post-Classic period was a time of reduced population, simplified social structure, a more dispersed settlement pattern, a simplified agricultural production system involving less irrigation, simplified architectural construction and crafts production, and a probable change in ritual practices. Polvorón phase components are small, consisting of only a few contemporaneously occupied pit structures. The small component size indicates occupation by only a few family units or perhaps one or more extended families. The trend toward social differentiation visible during the Classic period disappeared with the Polvorón phase. Society was organized along more egalitarian lines, and the single family or small extended family was the main social unit. The labor force necessary for the construction and maintenance of elaborate canal systems was gone, but then so were the population and subsistence pressures that created a need for those systems. The collapse of the Classic period leveled Hohokam society, removing the need for a social hierarchy and the mechanisms that would support it.

One of the more intriguing aspects of the Polvorón phase for archaeologists is that despite the general societal decline, long-distance exchange continued and may have even increased relative to population size. Petrographic analysis of a sample of Gila and Tonto polychrome sherds from the Phoenix Sky Harbor Center sites indicated that the vessels had been produced from materials found outside of the Phoenix area (Schaller 1991). Peterson and Abbott (1991) obtained the same results with polychromes from Pueblo Grande. These two analyses suggested that Roosevelt Red Ware had been traded to sites in the Phoenix Basin from other locations, possibly in the Tonto Basin. Other studies have disagreed (Crown 1991; Crown and Bishop 1987), suggesting instead that the Hohokam produced Roosevelt Red Ware: "Although long believed intrusive to the Hohokam area, the Salado polychromes are now known to have been locally manufactured" (Crown 1991:152).

Further research into this topic throughout the Hohokam region is needed. Whether the Hohokam of the Phoenix Basin produced Roosevelt Red Ware polychromes locally or obtained them through trade with groups outside of the basin is important to our understanding of Hohokam society during the Classic and post-Classic periods. Recently, Crown stated that the primary result of studies of Roosevelt Red Ware is the realization that multiple production loci existed (Crown, personal communication 1991; Peterson and Abbott 1991). However, researchers have not yet determined precisely where those loci were.

Roosevelt Red Ware constitutes a large percentage of the decorated ceramic assemblage for Polvorón phase components. The remaining decorated ceramics consist of intrusive wares such as Hopi orangeware and yellowware, often present in contexts with Gila and Tonto polychromes (Rice, Lindauer, and Ravesloot 1990). Casa Grande Red-on-buff, which was almost certainly locally manufactured, occurs in very small amounts in Polvorón phase assemblages.

Intrusive materials were not limited to ceramics. As previously stated, obsidian use increased, and the Hohokam obtained obsidian from sources located outside of the Phoenix area. Amounts of marine shell decreased, but the use of some shell for jewelry continued. By contrast, materials of Mesoamerican origin (such as copper bells and macaws) that were present during the pre-Classic periods and to a lesser extent the Classic period seem to have been absent during the post-Classic period.

The above evidence indicates that during the Polvorón phase, interaction continued between the Hohokam in the Phoenix area and in neighboring areas, perhaps especially those to the east. Researchers have not determined the nature of the exchange system. However, a discussion of settlement clusters in the Southwest by Jewett (1989) provides a possible model. Alliances may have formed as insurance against subsistence stress caused by events such as localized drought and crop failure. In such an alliance, groups not adversely affected would have provided subsistence goods to those suffering shortages (Doyel 1991b). However, such an alliance need not have been highly organized or centrally controlled. Wilcox's conception of a Salado interaction sphere consisting of "a weakly integrated system of exchange among a large series of small-scale regional systems" (Wilcox and Sternberg 1983:244) is the most appropriate model. Within this interaction sphere, Gila and Tonto polychromes may have functioned not merely as trade items or as containers for exchanged substances, but also as symbols of participation in a regionwide alliance.

In a recent study, McGuire (1991) generally agreed with the Salado interaction sphere concept but stressed that core-periphery models should allow for the geographical shifting of system cores, and hence their peripheries, through time. In this view, the Phoenix Basin would have been on the periphery of the Saladoan system during the Classic period.

This model especially fits the pattern for the post-Classic period Polvorón phase. Following the collapse of the Phoenix Basin regional system (one of the small-scale systems in the larger Salado interaction sphere) at the end of the Civano phase, the remaining Hohokam would have continued to participate, with increasing dependency, in the Salado system. In this sense, the Hohokam during the Polvorón phase would have been truly peripheral to the core of the Salado system, in contrast to the major role they took during the Civano phase. When the Salado system collapsed in the mid to late A.D. 1400s, the Polvorón phase and the Hohokam culture would have fallen with it.

Some of the Hohokam abandoning the Phoenix Basin at the end of the Civano phase may have been absorbed by populations elsewhere in the Salado interaction sphere. This would have further strengthened the ties between those remaining in the now-peripheral Phoenix Basin and those in the Salado core, with kinship possibly being part of the formula. Of course, identification of Hohokam emigrants of the late Classic period in areas outside the Phoenix Basin would be very difficult, especially given the general similarities in material culture shared, for example, by the Phoenix and Tonto basins during the Classic period. Perhaps the extensive research now being conducted in the Roosevelt Lake area will identify site unit intrusions of Hohokam. However, assimilation may have been so complete as to obscure any cultural identity.

Both Wilcox (Wilcox and Sternberg 1983) and McGuire (1991) envisioned exchange within interaction spheres as occurring between elites. Whether or not this was the case for the Salado and the Hohokam of the Classic period is unclear. However, the Mesoamerican analog presented by McGuire (1991:371) appears to fit with the evidence for the Salado interaction sphere. In the Mesoamerican systems, the elite of different polities intermarried and formed alliances. These systems had two different levels of material culture. The individual groups had their own utilitarian ceramics and other material goods but shared a common style of polychrome pottery and possibly other items used by the elite (Blanton et al. 1981; McGuire 1991; Marcus 1973). The situation in the Hohokam and Salado regions fits this model, with Gila and Tonto polychromes being the shared polychrome style.

Drawing on a number of studies, Wilcox (1991) concluded that the Hohokam during the Civano phase had been organized in a ranked society in which an elite occupied residences on top of the platform mounds and in which sites were also hierarchically ranked. This complexity appeared to have ended with the advent of the Polvorón phase. Organization was probably at the site level, with sites formed by small house clusters consisting of one or two pairs of pit structures. Polvorón phase sites were probably inhabited by a single extended family (Sires 1984:311). Whether the inhabitants were formerly members of the lower class or the upper class is unclear. Because they continued to participate in a regional exchange system during the Polvorón phase, those who remained in the Phoenix Basin after the collapse of the Classic period may have been remnants of the Hohokam elite. The possibility also exists, however, that the exchange system was not elite based but was organized and controlled through other mechanisms. This would mean that Hohokam society during the Polvorón phase might have been composed of a mix of individuals or families formerly of different rank or of peoples of lower ranking.

Understanding the nature of this later society is closely tied to a better understanding of what took place during the Civano phase and of what caused its decline (Teague 1989). Part of the decline in complexity apparent at the end of the Civano phase includes the disappearance of the evidence for organized ritual. This is not to imply that ritual and ceremony ceased altogether but that large-scale community-integrative ceremony, as evidenced by architectural features such as ballcourts and platform mounds, no longer existed (Wilcox 1991).

Evidence from Dutch Canal Ruin, Area 8, and from El Polvorón suggested that cremation burial may once again have become the prevalent mode of treatment of the dead. However, excavators found inhumations at El Polvorón and the Polvorón phase components of Pueblo Salado as well. Other evidence of ritual and ceremony is largely missing from (or is undetectable in) Polvorón phase assemblages. The small size of Polvorón phase components, which probably consisted of individual extended families, suggests a lack of complexity in social organization.

The Polvorón phase represents the final recognizably Hohokam occupation in the Phoenix Basin. As such, this time period may hold clues to the causes for the decline of Hohokam society and the eventual abandonment of the Phoenix Basin sometime around A.D. 1450. Future research should examine the quantitative and qualitative changes in subsistence practices from the Classic to the post-Classic periods. Using data from late Civano phase and Polvorón phase components, researchers can test hypotheses concerning causal factors for the collapse of the Hohokam. Although an examination of the causes for the Hohokam collapse is beyond the scope of this chapter, two hypotheses can be presented as examples.

Teague (1989) suggested that Akimel O'odham (Pima) myths may provide an explanation of the collapse. The myths state that the leaders of the Hohokam, called the Sivanyi, who lived in the villages with platform mounds, became arrogant and were despised by many. The Akimel and Tohono O'odham, it is said, defeated the Sivanyi in a war, killing some and driving others away. According to the myths, some of the Hohokam fled to the Pueblos in the north (i.e., Hopi) (Teague 1989:157–158). This myth raises the possibility that warfare between the Hohokam and the "newly arrived" Akimel O'odham caused the collapse of the Hohokam system at the end of the Classic period. The question, however, of where the Akimel O'odham originated and when they arrived, if they were truly immigrants, is still unanswered. Teague (1989:166) proposed that the conflict might have been between dissident Hohokam and their leaders or between competing factions, rather than with outsiders. Therefore, archaeologists should not overlook the possibility that internal dissension and revolt caused the collapse.

Severe flooding also has been suggested as a cause for the collapse of Hohokam society. Nials, Gregory, and Graybill (1989) used climatic data records, streamflow records, and long-term upland precipitation data derived from tree-ring studies (Graybill 1989; Graybill and Nials 1989) to posit that flooding between A.D. 1356 and 1358 caused the Hohokam decline. The floods of the mid A.D. 1300s were preceded by a very dry period that would have caused a narrowing of the river channel. Adjustment of their canal system to offset those drying conditions could have made the Hohokam particularly vulnerable to flooding (Nials, Gregory, and Graybill 1989:69–70). Analysts thus have crafted a very convincing argument that flooding caused the collapse of the Hohokam system, or that it at least played a major role in the collapse, affecting irrigation systems (Doelle and Wallace 1990) and social structure. Flooding would, however, have been only the triggering mechanism. The Hohokam had survived droughts and floods before, as evidenced by data from earlier in the Hohokam sequence (Nials, Gregory, and Graybill 1989). Other factors must have helped bring the Hohokam to the point of collapse and prevented them from rebuilding after the floods of A.D. 1356–1358. For example, Bostwick (personal communication 1992) has suggested that disenchantment with their religious leaders, who failed to protect them from the destructive forces of nature, caused the Hohokam to revolt and the collapse to follow (see also Teague 1989).

In conclusion, the inhabitants of the Phoenix Basin during the Polvorón phase were a people who lived in pit houses, practiced agriculture, utilized wild plant and animal resources, and maintained membership in a regional or interregional exchange system. The insubstantial nature of Polvorón phase architecture suggests increased mobility, a decrease in the amount of time and effort available for construction and maintenance of facilities, or both. Although quantitative data are not yet available, a closer examination of the subsistence practices may reveal that the Hohokam of the Polvorón phase relied more heavily on hunted and gathered resources and exhibited a more mobile settlement system than did those of the Classic period.

CHAPTER 9

THE CHRONOLOGY OF THE POLVORÓN PHASE

M. Zyniecki

The post-Classic period Polvorón phase of the Hohokam was first named and described by Sires (1984; Crown and Sires 1984) based on his excavations at the site of El Polvorón. Although earlier investigators had recognized a post-Classic period phenomenon, which included many of the same traits that characterize the Polvorón phase, these earlier researchers did not formalize the phase as Sires did (A. E. Dittert, Jr., personal communication 1991). With time, Hohokam archaeologists have generally accepted the Polvorón phase, although some (e.g., Doyel 1991a) do not agree with all of the cultural manifestations that define the phase. Despite the objections, the previous chapter indicates that the Polvorón phase is an archaeologically recognizable entity.

The next logical question is, "When did the Polvorón phase occur?" As usual with questions of chronology for the Hohokam, this one is nettlesome. Dean (1991:87), in his analysis of the Hohokam chronology, recognized the Polvorón phase but rejected four of the five chronometric dates and thereby the possibility for chronometric dating. Moreover, he limited his analysis to those chronometric dates that were reported in the spring of 1988 (Dean 1991:71). Since that time, additional investigations have reported dates for the Polvorón phase (Volume 2:Chapters 9 and 18; Volume 3:Chapter 2; Howard 1988c; Mitchell 1989a; Teague 1988). Additionally, some of the dates that Dean ascribed to the Civano or late Civano phase have come to be recognized as part of the Polvorón phase. The intent of this investigation is to examine these dates in an attempt to develop an absolute chronology for the Polvorón phase.

METHODS

The methods involved in developing a Polvorón phase chronology were, for the most part, the same that Dean (1991:66–73) used. Weaknesses in Dean's approach have been noted elsewhere (Volume 2:Chapter 18), but it still seems a reasonable way to address the problem. As in Dean's analysis, radiocarbon dates were calibrated using the CALIB program, but a newer version, 3.03 (Stuiver and Reimer 1993), with expanded radiocarbon data sets. Archaeomagnetic dates represent the 95% confidence interval based on the Southwestern virtual geomagnetic pole (VGP) developed by Eighmy (Eighmy, Hathaway, and Counce 1987; Eighmy, Hathaway, and Kane 1985), when possible. Two points are relevant to this study. First, Dean (1991:Table 3.2) listed the mean age of radiocarbon dates, plus or minus one standard deviation (sigma), for every phase, along with the calibrated one sigma range; for the present study, the analyst did not follow this procedure. Because calibrated radiocarbon dates no longer have a central tendency (Stuiver and Becker 1993) such as might be expected with a normal distribution, the SWCA analyst averaged the radiocarbon dates using the CALIB program (Stuiver and Reimer 1993). This approach produces a calibrated range of one sigma dates but does not yield mean ages, plus or minus one sigma. Second, the archaeomagnetic dates used in this analysis differ from those used in the other discussions for the Phoenix Sky Harbor Center Project. Eighmy and Baker (1991:9) suggested the use of the visual date range rather than the 95% confidence interval range because the statistical method, in their opinion, produces date ranges that appear to be unlikely in some cases. At this time, archaeologists have reached no consensus as to which might be best (Eighmy and Baker 1991:9; Wolfman 1991). This study used the statistically derived date ranges so that the results would be comparable to Dean's (1991) analysis.

Another difference between Dean's (1991) analysis and this one is that, for the most part, Dean used the phase assignment of the excavator, while in this study some features have been reassigned to the Polvorón phase because some of the chronometric dates are from excavations that took place prior to the definition of the Polvorón phase (e.g., Hammack and Sullivan 1981). Other dates, such as the late Civano features from Brady Wash (Ciolek-Torrello 1988) were not assigned to the Polvorón phase because of the initial controversy surrounding the existence of the phase (D. H. Greenwald, personal communication 1992). Except in two instances, the reassignments of dates to the Polvorón phase were supported or suggested by other researchers.

This analysis used the same rejection criteria as those used by Dean (1991:70–73). However, the statistical methods used to develop the chronology differed in deleting from the statistical manipulations those archaeomagnetic date ranges that did not have definite endings (e.g., A.D. 1300–post 1400) but including them in graphical representations of the chronology.

DATA

As noted above, some of the data for this analysis resulted from the reassignment of features to the Polvorón phase. Included in this category are all the late Civano dates from the Brady Wash project conducted by the Museum of Northern Arizona (Ciolek-Torrello, Callahan, and Greenwald 1988). Investigators had originally assigned these features and dates as "late Civano" or "late Classic" to denote their late temporal affiliation and possible association with the Polvorón phase (D. H. Greenwald, personal communication 1992). As Ciolek-Torrello reported,

> No Polvorón phase assignments were made...although throughout the course of the project an attempt was made to distinguish late Civano phenomena which were believed to be comparable to the putative post-Classic Polvorón phase material as defined by Crown and Sires (1984) and Sires (1984)....The late Civano phase dates presented by Eighmy and McGuire (1988:23) are primarily Brady Wash samples. These appear to be indistinguishable from Polvorón dates, indicating that these two phases are temporally one and the same [1988:116].

Given the admission that the late Civano phase dates were contemporary with the Polvorón phase, as well as the fact that the material attributed to the late Civano phase phenomena was comparable to that of the Polvorón phase, the analyst elected to reassign these features to the Polvorón phase for this study.

Crown and Sires (1984) and Sires (1984) re-examined dates from the 1968 excavations at Las Colinas. In their initial descriptions of the Polvorón phase, they reported Civano phase pit houses in the vicinity of Mound 8 at Las Colinas as belonging to the Polvorón phase. Later investigators at Las Colinas upheld this premise (Teague 1988).

In the excavation report for the Brady Wash Site, Greenwald and Ciolek-Torrello (1988c) assigned Feature 83-A, a post-reinforced adobe structure at Locus C, to the Soho phase or possibly the early Civano phase based on dates from two hearths (Greenwald and Ciolek-Torrello 1988c:117). Reassignment of the structure to the Polvorón phase was based on the fact that the architectural descriptions fall within the acceptable characteristics for Polvorón architecture (Crown and Sires 1984; Sires 1984) and on the presence of an elevated relative proportion of Roosevelt Red Ware and the presence of Pimería Brown Ware within the structure, also characteristics of the Polvorón phase.

This analysis derived all data prior to 1988 from Dean 1991, Table 3.1 and all subsequent data from project reports. The data are presented as in the originals, except for radiocarbon dates. All radiocarbon dates were calibrated using CALIB version 3.03 (Stuiver and Reimer 1993), the newest version with more complete data sets, providing comparable calibrated dates using the most recent information available. For this reason, all radiocarbon dates in this chapter may differ from original project reports and from Dean 1991.

RESULTS

Fifty-one dates (38 archaeomagnetic and 13 radiocarbon) represented the Polvorón phase in this study (Table 9.1). For the purposes of the chronology 37 dates were acceptable. Of these, three radiocarbon dates from Grand Canal Ruins have been assigned to the late Civano/Polvorón phase (Cable and Mitchell 1989:Table 18.30) and can be considered transitional between the two phases. This analysis used the remaining 34 dates, 29 archaeomagnetic and 5 radiocarbon, for the Polvorón phase chronology. Figure 9.1 presents a graphic representation of the date ranges. The portion of the graph that represents the Polvorón phase exhibits the S shape that analysts expect to see in a near-normal distribution (Dean 1991:73).

The average of the three radiocarbon dates for the late Civano/Polvorón phase transition has a range of A.D. 1283–1449 and an average calibrated one sigma range of A.D. 1298–1397 (Table 9.2). The range of dates for the Polvorón phase is A.D. 1175–1660, with a midpoint range of A.D. 1275–1565 (Table 9.2). The one sigma range of the midpoint dates is A.D. 1323–1453; the average radiocarbon one sigma calibrated range is A.D. 1400–1424; and the one sigma range of the archaeomagnetic dates is A.D. 1321–1459. These dates exhibit the clustering that is desirable for an absolute chronology. The transition period, although based on only three dates, appears to date roughly to the period A.D. 1300–1400, and the Polvorón phase dates to the period circa A.D. 1330–1450. This date corresponds closely to the original estimates of A.D. 1350–1450 for the Polvorón phase (Crown and Sires 1984; Sires 1984). With the addition of dates from other excavations, future researchers can certainly refine this chronology further.

The three archaeomagnetic dates from Pueblo Salado are much later than the other dates as a whole, covering the period circa A.D. 1415–1660. The latter portions of these three date ranges are much later than would seem possible, since they extend into the historic period. These dates were retained in the statistical analysis because they met the other requirements for acceptable data. The dates from Pueblo Salado imply that the termination of the Polvorón phase may be later than the A.D. 1450 suggested by Sires (1984). The mean of the midpoints from the three date ranges is A.D. 1548 with a one sigma of 15 years, which results in a range of A.D. 1533–1563. These dates are too late for the end of the Hohokam culture, given traditional wisdom. However, the dates from Pueblo Salado suggest that an ending date near A.D. 1500 may be acceptable for the Polvorón phase.

Table 9.1. Archaeomagnetic and Radiocarbon Dates from the Polvorón Phase

No.	Site	Provenience	Nature of Sample	Association	Bias	C-14 Age	Dates	Acceptance Status	Reference
Late Civano/Polvorón									
1	Grand Canal Ruins, AZ T:12:14(ASU)	F 24 (cremation)	Fuel?	?	?	530±100	1310–1449	A	Cable and Mitchell 1989
2	Grand Canal Ruins, AZ T:12:14(ASU)	F 65 (cremation)	Fuel?	?	?	630±60	1295–1403	A	Cable and Mitchell 1989
3	Grand Canal Ruins, AZ T:12:16(ASU)	F 28 (inhumation)	?	?	?	680±60	1283–1391	A	Cable and Mitchell 1989
Polvorón									
4	Brady Wash Site, Locus C	F 83 A (adobe structure)	Hearth	Primary	0	—	925–1025 1250–1425*	A	Dean 1991
5	Brady Wash Site, Locus C	F 82 A (adobe structure)	Hearth	Primary	0	—	925–1025 1300–1425*	A	Dean 1991
6	Brady Wash Site, Locus E	F 11 A (adobe structure)	Hearth	Primary	0	—	700–720 925–1025	R[1]	Dean 1991
7	Brady Wash Site, Locus E	F 14 A (pit house)	Hearth	Primary	0	—	925–1025 1300–1425*	A	Dean 1991
8	Brady Wash Site, Locus E	F 60 A (pit house)	Hearth	Primary	0	—	925–1025 1300–1425*	A	Dean 1991
9	Brady Wash Site, Locus E	F 39 A (pit house)**	Hearth	Primary	0	—	925–1025 1325–1400*	A	Dean 1991
10	Brady Wash Site, Locus E	F 57 A (pit house)	Hearth	Primary	0	—	925–1025 1325–1425*	A	Dean 1991
11	Brady Wash Site, Locus E	F 58 A (adobe structure)	Hearth	Primary	0	—	925–1025 1375–1425*	A	Dean 1991
12	Brady Wash Site, Locus C	F 51 B (pit house)	Hearth	Primary	0	—	820–1070 1230–1450*	A	Dean 1991
13	Brady Wash Site, Locus C	F 45-B (pit house)	Hearth	Primary	0	—	880–1070 1180–1450	R	Dean 1991
14	Brady Wash Site, Locus C	F 44 C (pit house)	Hearth	Primary	0	—	910–1070 1300–1450*	A	Dean 1991
15	Las Colinas	F 34 (pit house)	Hearth B	Primary	0	—	940–1000 1400–post 1425*	A	Dean 1991
16	Las Colinas	F 34 (pit house)	Hearth A	Primary	0	—	950–1010 1325–post 1425*	A	Dean 1991
17	Las Colinas	F 40 (pit house)	?	?	?	—	940–1000 1400–post 1425*	A	Dean 1991
18	Las Colinas	F 39 (pit house)	?	?	?	—	950–1015 1325–1425*	A	Dean 1991

Table 9.1. Archaeomagnetic and Radiocarbon Dates from the Polvorón Phase, continued

No.	Site	Provenience	Nature of Sample	Association	Bias	C-14 Age	Dates	Acceptance Status	Reference
19	Las Colinas	F 26 (pit house)	?	?	?	—	1297–1393	A	Dean 1991
20	Las Colinas	F 114 (pit house)	?	?	?	—	975–1010 1325–post 1425*	A	Dean 1991
21	Las Colinas	F 41 (pit house)	Hearth	Primary	0	—	1383–1417	A	Dean 1991
22	Las Colinas	F 74 (pit house)	?	?	0	—	925–950 post 1425*	A	Dean 1991
23	Las Colinas	F 35 (pit house)	Hearth B	Primary	0	—	950–1010 1325–post 1425*	A	Dean 1991
24	Las Colinas	F 35 (pit house)	Hearth A	Primary	0	—	950–1015 1325–1425*	A	Dean 1991
25	El Polvorón	Structure 5 (pit house)	Hearth	Primary	0	—	700–900 1060–1450	R	Dean 1991
26	El Polvorón	Structure 4 (pit house)	Hearth	Primary	0	—	700–1060	R	Dean 1991
27	El Polvorón	Structure 4 (pit house)	Structure element	Primary	—	390±70	1441–1635	R	Dean 1991
28	El Polvorón	Structure 4 (pit house)	Structure element	Primary	—	post bomb[2]	—	R	Dean 1991
29	El Polvorón	Structure 6 (pit house)	Hearth	Primary	0	—	930–1070 1340–1450*	A	Dean 1991
30	Casa Buena	F 15 (pit house)	Hearth A	Primary	0	—	925–1025 1300–1400*	A	Howard and Cable 1988
31	Casa Buena	F 15 (pit house)	Hearth B	Primary	0	—	925–1025 1300–1400*	A	Howard and Cable 1988
32	Casa Buena	F 172 (pit house)	Hearth	Primary	0	—	975–1025 1300–1425*	A	Howard and Cable 1988
33	Grand Canal Ruins, AZ T:12:14(ASU)	F 12 (pit house)	Hearth	Primary	0	—	925–1025 1175–1375*	A	Cable and Mitchell 1989
34	Grand Canal Ruins, AZ T:12:14(ASU)	F 12 (pit house)	?	?	?	750±50	1250–1293	R	Cable and Mitchell 1989
35	Grand Canal Ruins, AZ T:12:14(ASU)	F 46 (pit house)	Hearth	Primary	0	—	975–1025	R	Cable and Mitchell 1989
36	Grand Canal Ruins, AZ T:12:14(ASU)	F 46 (pit house)	?	?	?	560±70	1310–1433	A	Cable and Mitchell 1989
37	Grand Canal Ruins, AZ T:12:14(ASU)	F 60 (pit house)	Hearth	Primary	0	—	925–1025 1175–1450	R	Cable and Mitchell 1989

Table 9.1. Archaeomagnetic and Radiocarbon Dates from the Polvorón Phase, continued

No.	Site	Provenience	Nature of Sample	Association	Bias	C-14 Age	Dates	Acceptance Status	Reference
38	Grand Canal Ruins, AZ T:12:14(ASU)	F 60 (pit house)	?	?	?	450±50	1430–1473	A	Cable and Mitchell 1989
39	Grand Canal Ruins, AZ T:12:14(ASU)	F 72 (pit house)	Hearth	Primary	0	—	925–1025 1150–1450	R	Cable and Mitchell 1989
40	Grand Canal Ruins, AZ T:12:14(ASU)	F 72 (pit house)	?	?	?	820±60	1174–1280	R	Cable and Mitchell 1989
41	Grand Canal Ruins, AZ T:12:14(ASU)	F 111 (pit house)	?	?	?	950±50	1022–1165	R	Cable and Mitchell 1989
42	Las Colinas	Structure 33	Hearth	Primary	0	—	1380–1450	A	Teague 1988
43	Las Colinas	Isolated fire pit	Hearth	Primary	0	—	1370–1450	A	Teague 1988
44	Dutch Canal Ruin	F 8-3-1 (fire pit)	Fuel	Primary	?	630±70	1292–1405	A	Volume 2, Chapter 9
45	Dutch Canal Ruin	F 8-12 (fire pit)	Fuel	Primary	?	580±80	1302–1431	A	Volume 2, Chapter 9
46	Dutch Canal Ruin	F 8-14 (fire pit)	Fuel	Primary	?	650±90	1284–1405	A	Volume 2, Chapter 9
47	Pueblo Salado	F 8/9-3-3	Hearth	Primary	0	—	580–725 920–1015 1480–1650*	A	Volume 3, Chapter 2
48	Pueblo Salado	F 8/9-3-2	Hearth	Primary	0	—	630–675 910–1015 1425–1660*	A	Volume 3, Chapter 2
49	Pueblo Salado	F 8/9-27-1	Hearth	Primary	0	—	915–1030 1300–1660	R	Volume 3, Chapter 2
50	Pueblo Salado	F 8/9-41-1	Hearth	Primary	0	—	915–1025 1330–1660	R	Volume 3, Chapter 2
51	Pueblo Salado	F 8/9-50-1	Hearth	Primary	0	—	630–675 920–1020 1415–1660*	A	Volume 3, Chapter 2

Note: After Dean (1991:Table 3.1)
A = Accepted
R = Rejected
*Accepted date range for multiple possibilities
**Typographical error in Dean (1991:Table 3.1) listed this as F 89 A
[1]Note that no preferred range is indicated for rejected dates.
[2]Per Dean 1991:Table 3.1. 1950 is generally accepted as the year beyond which samples can be expected to have been contaminated by fallout from nuclear weapons.

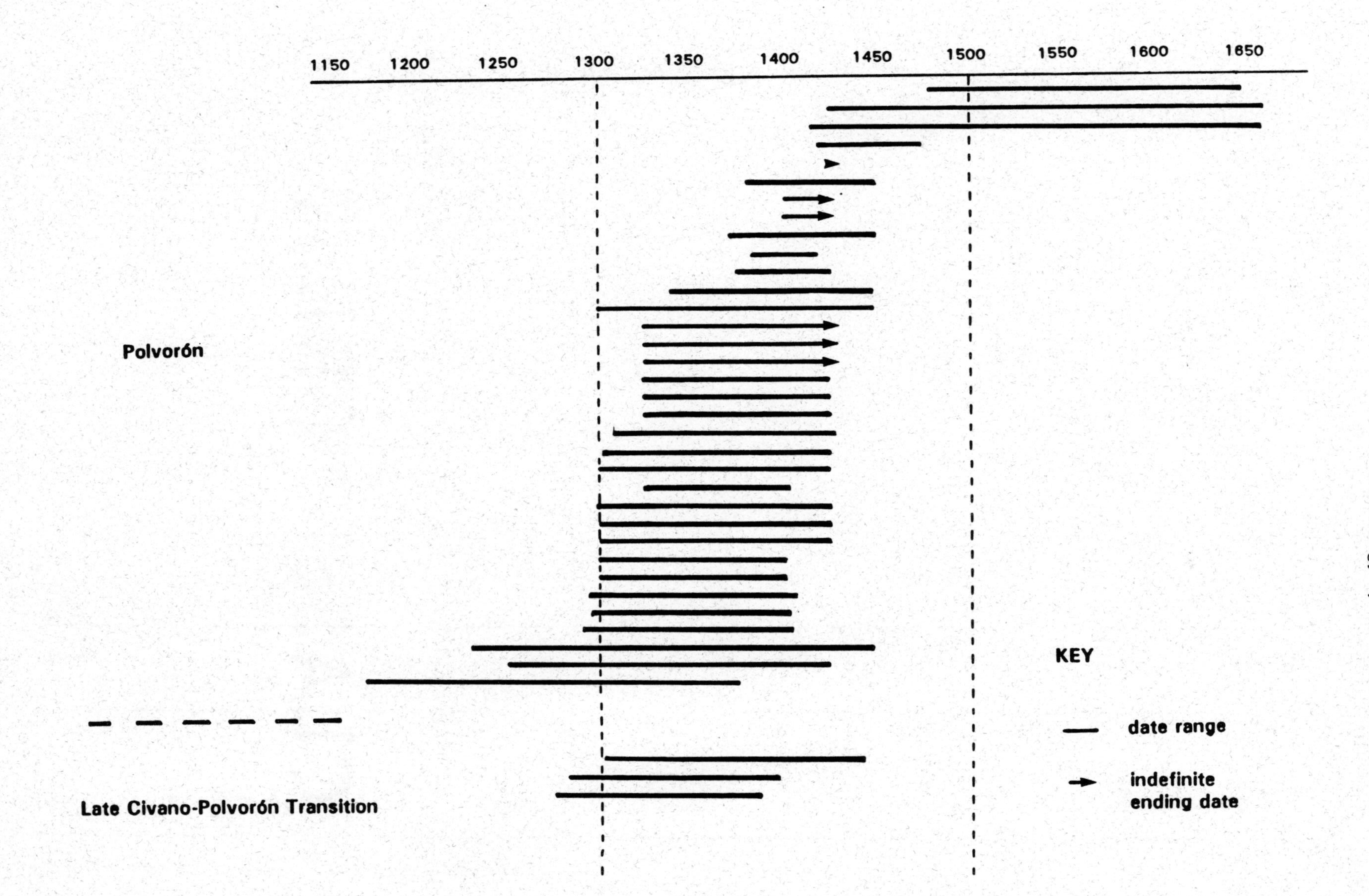

Figure 9.1. Archaeomagnetic and radiocarbon dates for the Polvorón phase.

Table 9.2. Summary Data for the Polvorón Phase

Phase	All Dates					Radiocarbon Dates		Archaeomagnetic Dates		
	N	Confidence Interval Range of Accepted Dates	Midpoint Range	Midpoint $\overline{X}$ ±1s	Midpoints 1s Range	N	Average Calibrated 1s Range	N	Midpoint $\overline{X}$ ±1s	Midpoints 1s Range
Late Civano/Polvorón	3	1283–1449	1337–1380	1355±22	1333–1377	3	1298–1397	—	—	—
Polvorón	28	1175–1660	1275–1565	1388±65	1323–1453	5	1400–1424	23	1390±69	1321–1459

CHAPTER 10

THE PUEBLO SALADO BURIALS: BIOARCHAEOLOGICAL CONSIDERATIONS

T. Michael Fink

In the past few years archaeologists have seen a significant increase in the number of Hohokam inhumations discovered. Not surprisingly, this increase is directly associated with the many contract archaeological projects undertaken in the Phoenix metropolitan area. Examination of mortuary remains has provided valuable insight into the level of Hohokam health and social complexity during the pre-Classic and Classic periods, and the recent excavations at Pueblo Salado have contributed to that growing data base.

The purpose of this chapter is to consider some of the biological and social aspects of the Pueblo Salado burial series based on a re-examination of dental remains and a more in-depth analysis of data presented in Volume 3, Chapter 11. Of particular importance to the study of Hohokam populations are (1) possible paleoepidemiological factors related to the distribution of enamel hypoplasias, (2) evidence for seasonality of mortality evident in the burial orientations, (3) burial history as indicated by the demographic profile of the burials, and (4) the level of social complexity suggested by an analysis of the mortuary remains.

PALEOEPIDEMIOLOGICAL CONSIDERATIONS

Of the 45 burial features excavated at Pueblo Salado, 34 were inhumations, 3 of them containing the remains of 2 individuals, 5 were cremations, and one was a concentration of human bone. Forty were from Area 8/9 (Volume 3:Chapter 2). Kathryn L. Wullstein conducted the initial osteological analysis of the remains of 43 individuals, which dated primarily to the late Civano phase (Volume 3:Chapter 11). Wullstein's analysis identified various physical conditions related to or resulting from the health of the Pueblo Salado populations. Of particular interest in understanding prehistoric health is the evidence for enamel hypoplasia in the Pueblo Salado dentition. Peak periods of hypoplasia development appear to reflect the synergistic interaction between infectious disease and malnutrition (Fink 1989b; Fink and Merbs 1991). The following discussion will focus on paleoepidemiological hypotheses suggested by the Pueblo Salado hypoplasia data base.

During the Civano phase, levels of infectious disease may have been critically high, a result of the lifestyle adopted during the Classic period. The inhabitants of the Salt River Valley tended to be aggregated in densely settled compound communities during this phase. These communities were often close to one another, with several sharing the same canal system as their main source of water. The combination of probable high population density and close proximity at both the inter- and intrasite level would have been sufficient to promote the spread of infectious diseases through person-to-person contact and mechanical transmission, for example by flies, food, or fomites (materials such as clothing capable of harboring and transmitting a disease-bearing organism). Unsanitary conditions probably already existing in Hohokam settlements would have increased during the Civano phase, and contamination of canals and reservoirs with human waste would have produced the added dimension of waterborne infections (Fink 1991). A preliminary analysis of the Pueblo Grande skeletons also tended to support the contention that the Hohokam of the Classic period may have been severely stressed (Amon and Karhu 1991).

Other causal factors may also have contributed to the paleoepidemiological variations that investigators noted in the Civano phase assemblage at Pueblo Salado in comparison with other Classic period burial assemblages. First, Pueblo Salado was not established until the Classic period and therefore would not have been subjected to the levels of contaminants from pre-Classic period occupations that accumulated at other locations in the valley. Second, because Pueblo Salado was on a single-village canal system, unlike most of the villages in Canal System 2, the water in the canals was less likely to have been polluted by upstream use. Furthermore, groundwater or water from other natural sources may have been easily accessible on the floodplain, reducing the use of canal water for domestic purposes.

One method of evaluating stress conditions that may be related to health problems is the occurrence of enamel hypoplasias. Enamel hypoplasias are deficiencies in dental enamel thickness caused by a disruption of amelogenesis and can be observed in both deciduous and permanent dentition. Even when exhibited in permanent teeth, the defects represent stress suffered in childhood, not adulthood.

Hypoplastic defects can be caused by trauma, hereditary conditions, infectious disease, and malnutrition. Researchers have recently proposed, however, that elevated levels of weaning stress exhibited in hypoplastic defects result from the synergistic interaction between infectious disease and nutritional inadequacies (El-Najjar, DeSanti, and Ozebek 1978; Goodman, Armelagos, and Rose 1984; Hutchinson and Larsen 1988; Rose, Condon, and Goodman 1985). This is the model currently prevailing among researchers working in the American Southwest (Martin et al. 1991; Stodder 1987) and has been used in the interpretation of enamel hypoplasias in Hohokam dentition (Fink 1989b; Fink and Merbs 1991).

Teeth were preserved in 26 individuals from the 37 inhumations excavated at Area 8/9 of Pueblo Salado; re-examination revealed a total of 34 enamel hypoplasias on teeth from 5 (19%) of these individuals (Figure 10.1). The frequent involvement of the canines, as seen in this mortuary population, is a common feature of hypoplasia development. However, the analyst had difficulty in composing an accurate picture of tooth involvement because of poor preservation in the sample.

Given the relatively short period of time that Pueblo Salado was inhabited, the working hypothesis for this analysis was that hypoplasia development in the Pueblo Salado population would be reduced when compared to other Classic period sites that had been occupied for a longer period of time and where sanitation problems could thus have become more acute. To tentatively test this hypothesis, dental data from the Classic period sites of Grand Canal Ruins and Casa Buena were compared with the data from the Area 8/9 burial population (Fink 1989b; Fink and Merbs 1991). The remains from Pueblo Salado exhibited hypoplastic bands in 19% of the burials with preserved dentition; the rates of involvement at the Grand Canal Ruins and Casa Buena were 26% (16 of 42) and 22% (9 of 41). Although this evidence conformed to the expectations of the hypothesis, the results were far from overwhelming. A Chi-square test (the null hypothesis stated that there would be no significant difference in enamel defects between sites) was not significant at the 0.05 level (X^2=0.36).

Although the argument that rates of infectious disease may have been less severe at Pueblo Salado than elsewhere in Canal System 2 remains plausible, the data from this study did not strongly support the notion. However, the hypothesis may have been overstated and may not have differentiated between alimentary or hematogenous infections as opposed to enteric or gastrointestinal conditions that are usually the cause of weanling diarrhea. Perhaps, then, weaning stress was an important cause of childhood morbidity, while other types of infections, dependent

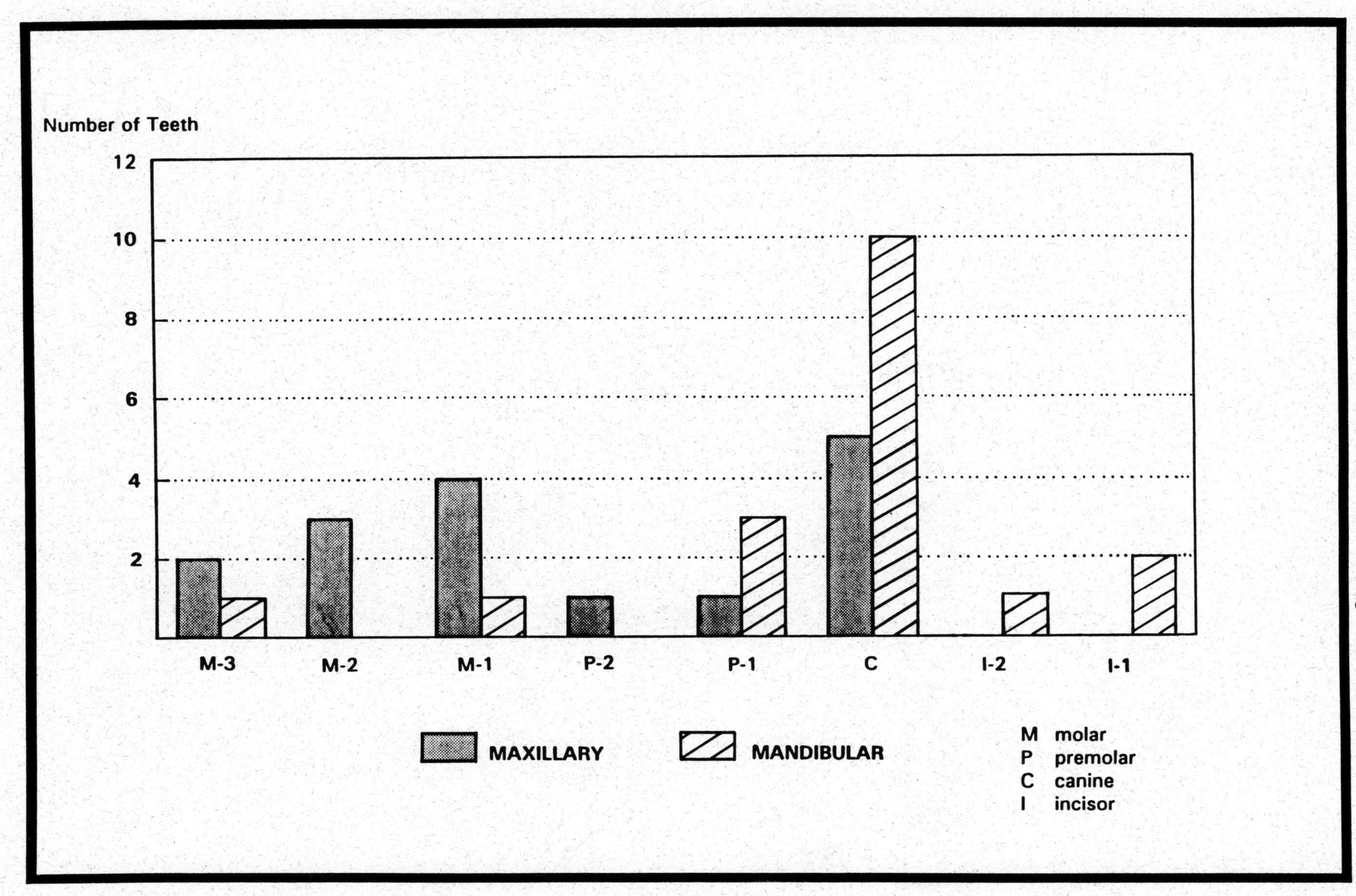

Figure 10.1. Enamel hypoplasia by tooth type in burials from Area 8/9 at Pueblo Salado.

on person-to-person contact, may have become more transitory, being spread from areas with greater contamination levels.

Alternatively, although rates of infectious disease may have occurred at a lower rate at Pueblo Salado, the overall effects may have been negated by an increase in nutritional inadequacies. This nutritional stress could have been caused by a change in the food menu due to decreased availability of plant foods or animal protein, an alternative not supported by the Pueblo Salado data, or it could represent alterations in the local subsistence base in response to changes in the sociopolitical or demographic structure. Kwiatkowski and Miller (Volume 3:Chapter 8) report less emphasis on cactus fruits and agave and probable increases in maize production at Pueblo Salado (perhaps facilitated by the geologic setting) in comparison with neighboring sites. The diminished use of cactus fruits would mean the loss of an important source of vitamin C, whereas an increase in maize consumption would add significantly to the carbohydrate load but minimally to vitamin and mineral stores.

One additional hypothesis concerns the possible influence of helminths (intestinal parasites) on the disease load of the Pueblo Saladoans. Because it was located in the Salt River floodplain, Pueblo Salado was situated in a pro-helminthic environment. The moist soils of the floodplain and riverbank would have been ideal for the survival and transmission of parasite eggs and larvae (Fink 1991; Reinhard 1985). Archaeologists have identified disease-causing helminths in Anasazi and Sinagua fecal deposits (Reinhard, Hevly, and Anderson 1987), and the Hohokam certainly could have likewise suffered from parasites.

The alternative hypotheses presented here can be tested by a variety of means and should be considered in future interpretations of hypoplasia development in Hohokam burial studies. Additional evidence of enamel hypoplasia in Classic and post-Classic period dentition would certainly improve the quantity and quality of the data base. Indeed, tentative evidence from Pueblo Grande (Amon and Karhu 1991) has suggested that weaning stress may have been a more serious problem during the Classic period than the remains from the Grand Canal Ruins and Casa Buena have indicated. Likewise, additional macrobotanical, pollen, and faunal evidence might prove useful to analysts attempting to discern possible shifts in the traditional Hohokam menu (Gasser and Kwiatkowski 1991) between the Classic and post-Classic periods or in the socioeconomic structure of the Hohokam. Even slight shifts might prove critical. Dietary data would be important in clarifying the role of synergy between infectious disease and malnutrition and the catalytic effect of either in producing weaning stress. Finally, the analysis of burial soils could provide evidence of helminthic parasites and help discern their role in producing weaning stress.

Although the frequency of enamel hypoplasia in the Classic period occupations did not appear to differ markedly between Canal System 2 and Pueblo Salado, the initial analytical hypothesis presented above predicted that the actual period of weaning, as evidenced by the peak period of defect development (Rose, Condon, and Goodman 1985), should be shorter or more condensed at Pueblo Salado than in Canal System 2. Again the basic premise was that infectious disease would have been less severe at Pueblo Salado and, therefore, that the analyst would see decreased interaction between the nutritive value of weaning foods and pathogenic microorganisms because of the decreased presence of these organisms in the local environment.

In determining the peak period of defect development, the author and Wullstein determined that 18 (53%) of the 34 Pueblo Salado hypoplasias were sufficiently demarcated to allow application of the odontometric scheme described by Rose, Condon, and Goodman (1985). This involved omitting one of the four affected individuals from the analysis. All but 3 (17%) of the 18 defects involved upper or lower canines; the others involved lower premolars. Table 10.1 presents the episodes of

Table 10.1. Stress Episodes Indicated by Enamel Hypoplasias, Area 8/9, Pueblo Salado

Feature	Age	Sex	Stress Episodes
8/9-5	25–30	Female	4.0–4.5 yrs 4.5–5.0 yrs
8/9-18	15–17	Male	4.5–5.0 yrs
8/9-60A	20–25	Male	3.5–4.0 yrs 4.5–5.0 yrs
8/9-60B	25–30	Male	3.5–4.0 yrs 4.0–4.5 yrs 4.5–5.0 yrs

insult suffered by the four individuals included in the odontometric analysis. Note that three of the individuals suffered recurrent insults. The ages at which these stress episodes occurred suggested that the weaning period at Pueblo Salado likely occurred between 3.5 and 5.0 years, a span of 1.5 years. The proposed weaning period observed at other Classic period sites also involved a span of 1.5 years but apparently between 2.5 and 4.0 years (Fink and Merbs 1991:306–307). Although the present data base does not support the hypothesis of a condensed period of weaning, it does suggest that the beginning of the weaning period at Pueblo Salado may have been postponed from late infancy (2.5 years) to early childhood (3.5 years). The fact that four of the eight insults (Table 10.1) occurred between 4.5 and 5.0 years can be interpreted as additional evidence for Pueblo Salado children having been weaned later than children at other Classic period sites.

The reason for the possible postponement of weaning at Pueblo Salado is problematic. It may have something to do with prolonged breast-feeding in response to low birth rates or high mortality. Researchers need a more extensive data base before they can properly assess the influence of weaning stress on childhood morbidity. Nevertheless, the hypoplasia evidence from Pueblo Salado has served to initiate the comparative study of health between autonomous settlements and those included in larger, integrated canal systems and is useful in generating hypotheses for future bioarchaeological research. Such studies expanded to populations of the Polvorón phase may be important in understanding the role infectious disease and malnutrition played in the dissolution of the Hohokam culture.

BURIAL ORIENTATION AND SEASONALITY OF DEATH

Hohokam inhumations are noted for exhibiting a general east-west orientation with the head located at the eastern end of the burial pit facing west (Effland 1988; Fink 1990; Merbs 1987; Mitchell, Fink, and Allen 1989). Merbs (1987; see also Merbs and Brunson 1987) recently suggested that this type of alignment represents an attempt by the Hohokam to orient the dead toward the sun at sunrise. He stated further that it may be possible to determine what time of the year a burial occurred from the side of the fall-spring equinox line on which the orientation lies. If the orientation lies north of 90°, the interment occurred in the summer between the spring and fall equinoxes. Orientations that fall along the 90° line represent spring or fall interments. This kind of analysis permits a theoretical plotting of the seasonality of death within a community.

Field investigators determined compass alignments for all 37 of the Area 8/9 inhumations (Volume 3:Chapter 2). These orientations have been grouped according to the 11° azimuth categories devised by Merbs (Merbs and Brunson 1987) and plotted in Figure 10.2a. Of the 37 orientations, 30 (81%) can be used to plot seasonality of death, as they fall between 12° NNE and 168° SSE and are, therefore, basically oriented in an easterly direction. Four (11%) others lie along the north-south axis, while 3 (8%) are oriented toward the west; since these orientations fall outside the prescribed eastern coordinates, they have been excluded from the following discussion.

Of the 30 east-west burials plotted in Figure 10.2a, 12 (40%) appear to be summer interments, one (3%) a winter burial, and 17 (57%) either spring or fall interments. If Merbs's sunrise orientation model is valid for the Hohokam, then the Pueblo Salado data differ markedly from those of other prehistoric Salt River Valley communities. In considering the possibility of a seasonal relationship, the analyst has plotted burial orientations for four Hohokam sites (Figures 10.2b-e): the pre-Classic period inhumations from La Ciudad (Merbs 1987) and the Classic period burials from Casa Buena, the Grand Canal Ruins (Merbs and Brunson 1987), and La Lomita (Fink 1990). Burials from Pueblo Salado differed from these interments in two ways. First, the pre-Classic and Classic period burials from the previous studies were overwhelmingly winter interments, whereas the Pueblo Salado burials showed a strong inclination toward summer. Second, unlike Casa Buena and La Lomita, Pueblo Salado showed a definite pattern of year-round burial (as did La Ciudad and the Grand Canal Ruins).

The orientations displayed by the pre-Classic and Classic period inhumations from previous studies differed from those of the Pueblo Salado burials in a pattern that was both striking and difficult to interpret. In his discussion of the Casa Buena data, Effland (1988) suggested that the paucity of summer interments may reflect a decrease in the number of individuals living at the site during the summer months. He postulated that some individuals "may have lived in smaller hamlets or farmsteads near fields" (Effland 1988:781). However, most "farmsteads" or field houses were fairly close to larger communities, and a corpse could therefore have been easily carried back for burial. It is more conceivable that some segments of the population were seasonally drawn away from their main habitation area so that they might participate in trading ventures, food gathering, rituals, or other activities routinely occurring in the summer. Indeed, this seasonal change would explain why three Classic period sites in Canal System 2—Casa Buena, Grand Canal Ruins, and La Lomita—had undergone simultaneous decreases in population and why Pueblo Salado differed from them, as its residents probably had remained within this microsystem (Chapter 3).

According to this hypothesis, the predominance of summer burials at Pueblo Salado would signal interment strategies different from those at the Classic period sites within Canal System 2, which may well have had their antecedents in the pre-Classic periods. Alternatively, the burial orientations may indicate seasonal use of discrete burial areas or loci within cemeteries. Hence, the burial series recovered from Pueblo Salado and the other communities may offer only a skewed view of seasonality of death. A third possible explanation is that the burial orientations do not indicate seasonality of any kind. However, Mixson and White's (1991) work at Hole-in-the-Rock in Phoenix's Papago Park strongly suggests that the Hohokam were concerned with the phases of the equinox (see also Bostwick 1992). Hence, Merbs's orientation model, though far from unequivocal, is not unfounded and is worthy of continued research.

Because archaeological evidence tentatively supports Merbs's (1987) orientation model, the analyst tabulated the Pueblo Salado data by age and sex to determine cohort variation in seasonality

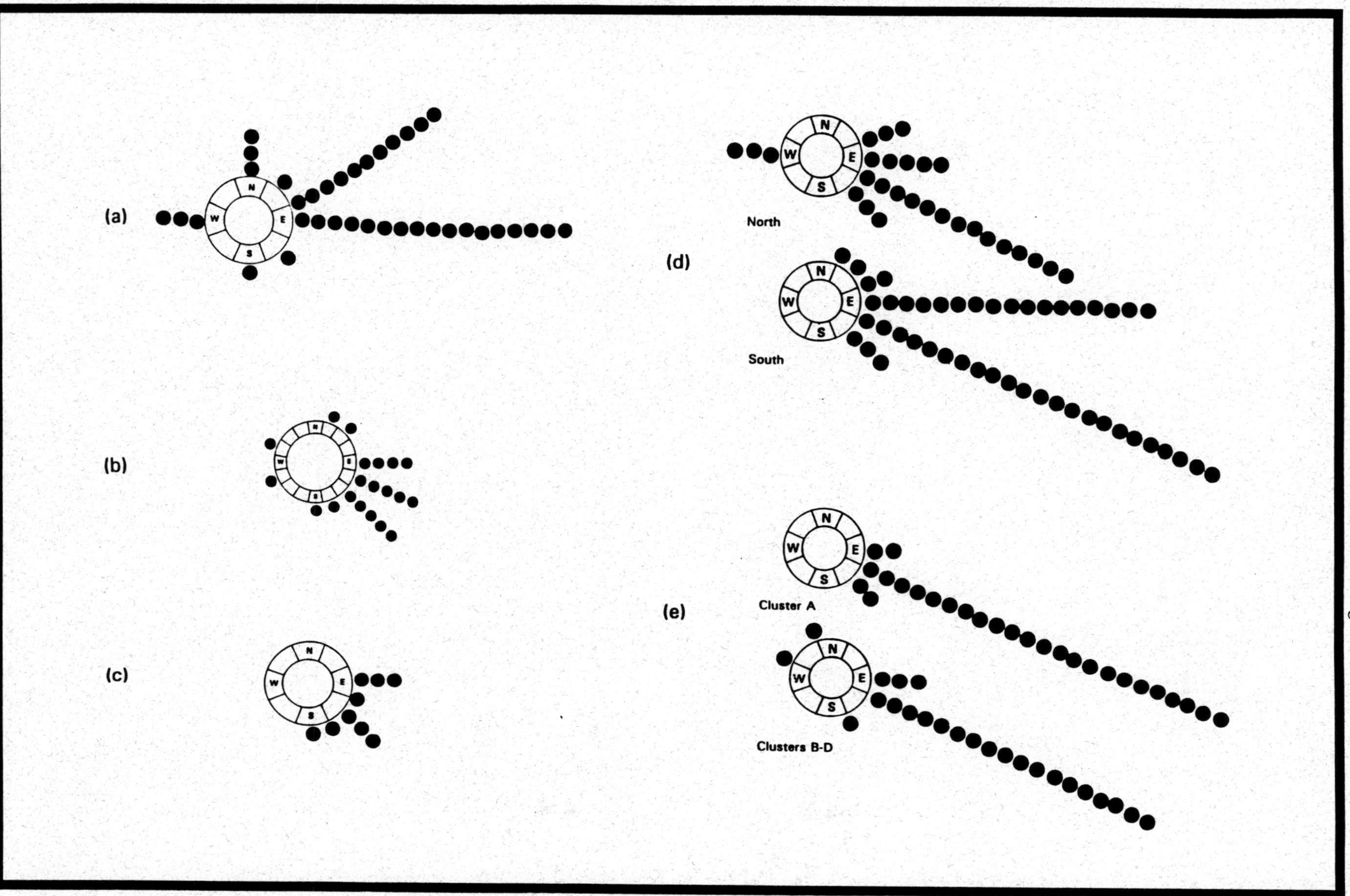

Figure 10.2. Burial orientations from five Hohokam sites: (a) Area 8/9 of Pueblo Salado; (b) La Ciudad (from Merbs 1987:209); (c) La Lomita (from Fink 1990:82); (d) Grand Canal Ruins (from Merbs and Brunson 1987); (e) Casa Buena (from Merbs and Brunson 1987).

of death (Tables 10.2, 10.3, 10.4). The following discussion deals solely with the patterns of summer and fall/spring mortality, as winter data are scarce. The analyst hypothesized that nonadults and females would consistently show higher mortality in each season examined, because nonadults generally exhibit greater susceptibility to infectious disease and females experience increased physiological demands associated with childbearing.

Apparent summer deaths in nonadults (Table 10.2) and females (Table 10.4) conformed to the hypothesis. Nonadults (infants, children, and adolescents) composed 75% of the summer deaths, and three of the four summer adolescent and adult deaths (Table 10.4) were females. High ambient summer temperatures may have interacted synergistically with infections to produce severe dehydration and diarrhea in the young and females already under chronic physiological stress during childbearing years. Furthermore, high ambient summer temperatures may have led to increased exposure to waterborne enteric infections through increased intake of river or canal water tainted with human waste (Fink 1991).

The same held true for the pattern of fall/spring deaths. Although adult deaths were well represented, nonadults accounted for 59% of the fall/spring deaths (Table 10.2). However, in contrast to the summer deaths, male and female deaths in the fall and spring were more nearly equivalent (Table 10.4). Again, despite the difficulty of discerning the cause or causes for fall/spring mortality, maladies actually contracted during summer or winter may have lingered on or induced complications that eventually brought about the death of an individual months later (i.e., in the fall and spring, respectively).

BURIAL HISTORY

Anderson (1986) recently argued that investigators can determine the history or longevity of Hohokam communities by examining the demographic profile of the burial areas. He wrote:

> Di Peso assumed that the total number of burials in a compound was a function of its age. In fact, it is more likely that the mean age of those interred is a function of the compound's duration. Given a normal settlement population of child-bearing households, a short-lived or new settlement should include a higher percentage of children than should an older or longer-lived settlement [Anderson 1986:199].

Anderson based his observation on sound paleodemographic principles, as infant and child mortality is frequently as high as 25–50% of a given burial series (Fink 1989b; Hinkes 1983:15–22). The young display greater risk of dying due to the complex interaction between the physiological demands of growth and the impacts of infectious disease, malnutrition, immunology, child-rearing practices, and maternal health (Fink 1985; Hinkes 1983). Therefore, the burial profile of a newly established settlement logically would be composed principally of nonadults. Elderly adults also would be present, as they too would have been at greater risk of dying. By contrast, all age groups should be well represented in the burial profile of a long-lived community.

The burial profile for the inhumations from Area 8/9 of Pueblo Salado at first appears to be of a fairly young burial series (Figure 10.3). That is, 38% of the individuals were under 10 years of age, while 19% died during their second decade of life. In other words, 57% of the Pueblo Salado burial series comprises individuals less than 20 years of age. The remaining 43% consists of fully adult individuals, primarily between the ages of 20 and 40 years. Only one individual could be determined to be 40 or more years of age, which probably represents not a virtual absence of

Table 10.2. Seasonal Distribution of Deaths by General Age Categories, Area 8/9, Pueblo Salado

Season Interred	Infant (0–2)		Child (2–12)		Adolescent (12–20)		Adult (20 +)		Total	
	No.	%	No.	%	No.	%	No.	%	No.	%
Winter	0	0.0	0	0.0	1	100.0	0	0.0	1	(3.3)
Summer	4	33.3	4	33.3	1	8.4	3	25.0	12	(40.0)
Fall/Spring	2	11.8	4	23.5	4	23.5	7	41.2	17	(56.7)
Total	**6**	**(20.0)**	**8**	**(26.7)**	**6**	**(20.0)**	**10**	**(33.3)**	**30**	**(100.0)**

Table 10.3. Seasonal Distribution of Adult Deaths, Area 8/9, Pueblo Salado

Season Interred	Young Adult 20–25 yrs		Middle Adult 25–40 yrs		Older Adult 40+ yrs		Age Unknown		Total	
	No.	%	No.	%	No.	%	No.	%	No.	%
Winter	0	0.0	0	0.0	0	0.0	0	0.0	0	(0.0)
Summer	0	0.0	1	33.3	0	0.0	2	66.7	3	(30.0)
Fall/Spring	1	14.3	3	42.9	1	14.3	2	28.6	7	(70.0)
Total	**1**	**(10.0)**	**4**	**(40.0)**	**1**	**(10.0)**	**4**	**(40.0)**	**10**	**(100.0)**

Table 10.4. Seasonal Distribution of Adolescent and Adult Deaths by Sex, Area 8/9, Pueblo Salado

Season Interred	Male		Female		Unknown		Total	
	No.	%	No.	%	No.	%	No.	%
Winter	1[1]	100.0	0	0.0	0	0.0	1	(6.3)
Summer	0	0.0	3	75.0	1[1]	25.0	4	(25.0)
Fall/Spring	4	36.4	6[2]	54.5	1[1]	9.1	11	(68.8)
Total	**5**	**(31.3)**	**9**	**(56.3)**	**2**	**(12.5)**	**16**	**(100.0)**

[1]adolescent [2]includes two adolescents

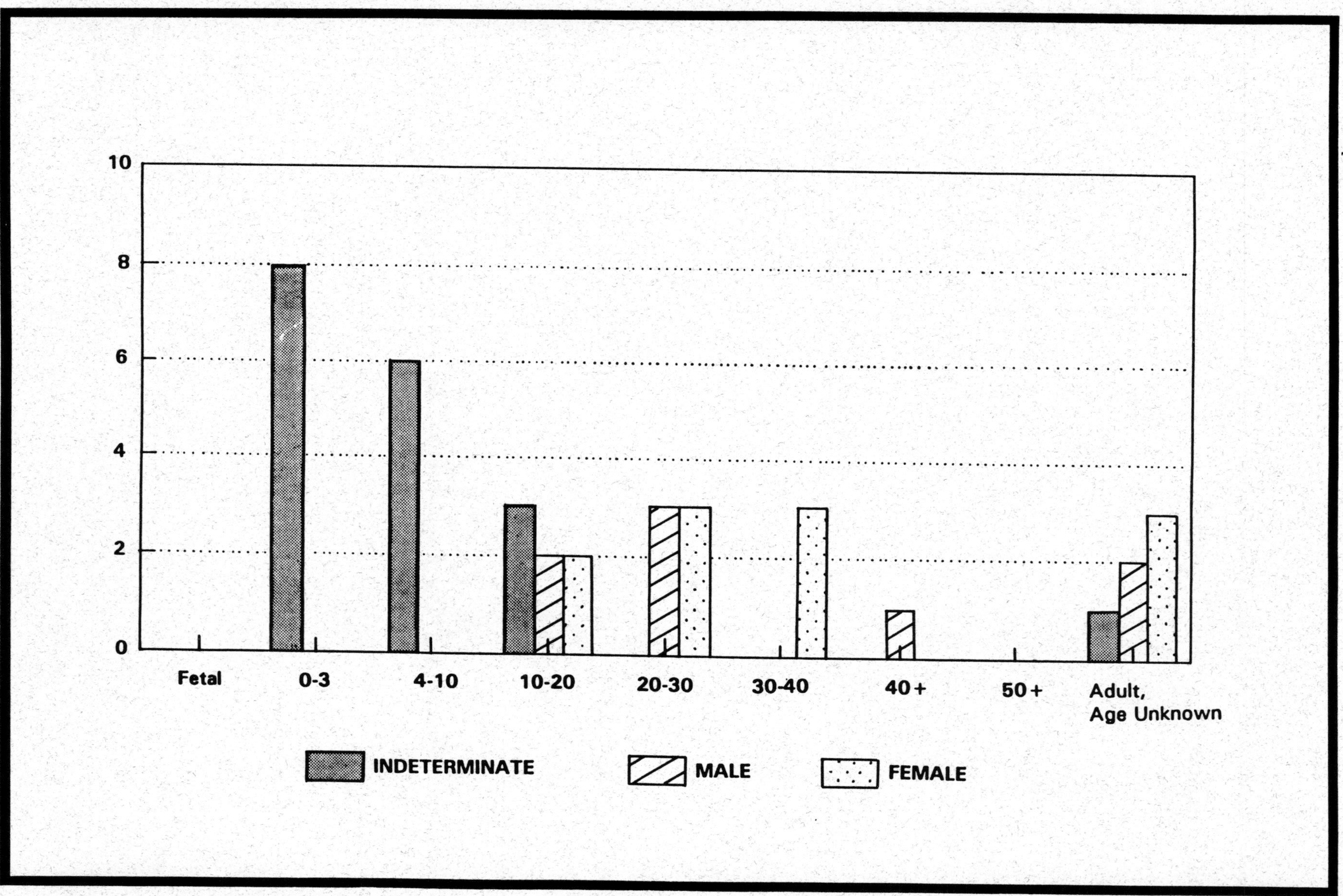

Figure 10.3. Demographic profile of inhumations from Area 8/9 of Pueblo Salado.

elderly adults at Pueblo Salado but instead the combined effects of poor preservation, sampling error, and difficulty in determining the age of older individuals (Willey and Mann 1986).

Three individuals included with the nonadults were 18-20 years of age (Table 10.2; Volume 3:Chapter 11) and could be considered "young adults." If these individuals were included with the adults instead of the nonadults, the nonadult and adult frequencies would change to 49% and 51%, respectively. Thus, the difference between adult and nonadult representation in the Pueblo Salado series is not overwhelming. The nonadult and adult frequencies indicate, then, that the Pueblo Salado burial series represents neither a newly established nor a long-lived settlement. Instead, it would appear to represent a community in existence for perhaps one or two generations.

MORTUARY ANALYSIS

In the last few years the number of mortuary analyses involving Hohokam cemeteries and funerary remains has increased significantly (e.g., Anderson 1986; Antieau 1981; Brunson 1989; Effland 1988; McGuire 1987; Mitchell, Fink, and Allen 1989; Schroeder 1990). The increase may be due to both a renewed interest in prehistoric mortuary studies in general and to the many recent contract archaeological projects, which have excavated a sizable number of Hohokam burials. The study of these collections has usually focused on the use of mortuary data to discern a settlement's level of social complexity and organization.

In his original conception of this study, the investigator envisioned analyzing the Pueblo Salado burials, both inhumations and cremations, in the same methodological fashion as the remains recovered from La Ciudad (McGuire 1987), Casa Buena (Effland 1988), or the Grand Canal Ruins (Mitchell, Fink, and Allen 1989). However, tabulation of the distribution of the Pueblo Salado grave goods (Table 10.5) as identified during the analyses reported in Volume 3 made it clear that the series exhibits both a low absolute number of items and a wide array of artifact types (Volume 3:Chapters 2 and 3). Furthermore, 22 (46%) of the 48 burials of individuals from all areas at Pueblo Salado contained no grave goods beyond ceramic sherds, flaked stone, or shell debris. Consequently, a complex statistical analysis could not be properly applied to the Pueblo Salado data base. Instead, this study describes the distribution of the artifact types and draws inferences wherever possible, illustrating salient points with tabulations of pertinent data. In spite of the limitations of the data, the evidence suggests that social complexity at Pueblo Salado may have decreased in comparison to other Classic period sites. The three inhumations and two cremations recovered from Dutch Canal Ruin (Volume 2:Chapters 2 and 9 and Appendix E) are discussed as well.

Pueblo Salado Burials

Table 10.5 presents a distribution of the artifact types recovered from the Pueblo Salado burials. The analyst considered all artifacts recovered within the definable boundaries of a burial pit to be associated materials, even though some could have come from the matrix surrounding and covering the burial. Tables 10.6, 10.7, and 10.8 present the distribution of the flaked and ground stone artifacts.

The distribution of flaked stone in Table 10.6 seems unusual in two ways. First, 50% of the inhumations containing flaked stone were adult interments. Second, the highest flaked stone counts in inhumations occurred in the burials of two females (Features 8/9-51 and 8/9-104) and a child (Feature 8/9-124). The other inhumations, including those of males, contained only small quantities of the material. In fact, of all the adult inhumations with flaked stone, 75% were female interments

Table 10.5. Distribution of Grave Goods and Other Associated Artifacts in the Pueblo Salado Burials by Artifact Category

Artifact Category	Number of Burials Containing Artifact Type	Percentage of Burials Containing Artifact Type
Grave Goods		
Salt Red Bowl	10	21
Salt Red Jar	1	2
Plainware Bird Effigy Jar*	2	4
Salt Red Pitcher	1	2
Phoenix Red Bowl	4	6
Phoenix Red Jar	1	2
Gila Plain Bowl	3	6
Gila Plain Scoop	2	4
Squaw Peak Plain Bowl	2	4
Lower Colorado Buff Ware Bowl	1	2
Ceramic Pipe	1	2
Projectile Point	3	6
Obsidian Whole Flake	1	2
Handstone	1	2
Grinding Slab	1	2
Ground Stone Jar Lid/Rasp	1	2
Grinder NFS	1	2
Stone Beads	3	6
Kelletia Shell	1	2
Glycymeris Shell Bracelet	3	2
Shell Beads	2	2
Turtle/Tortoise Carapace	1	2
Burials with No Grave Goods**	22	46
Other Associated Materials		
Plainware Sherds	20	43
Redware Sherds	22	47
Polychrome Sherds	6	13
Flaked Stone	25	51
Ground Stone	6	11
Burials with No Grave Goods or Other Associated Materials	11	23

Notes: Total number of individual burials = 48; data compiled from Volume 3, Chapters 2, 3, 4
*Identified as redware in the field.
**Items directly associated with the remains; does not include sherds, flaked stone, fragments of ground stone, or noneconomic shell or shell debris found in the feature matrix.

Table 10.6. Distribution of Flaked Stone from Inhumations by Sex and Age, Pueblo Salado

Burial No.	Sex/Age	Material Type																													Total
		Basalt					Obsidian					Igneous					Quartz				Quartzite				Chert					Metamorphic	
		W	*U*	*F*	*P*	*AD*	*W*	*U*	*P*	*C*	*PP*	*W*	*U*	*F*	*P*	*AD*	*U*	*F*	*P*	*AD*	*W*	*F*	*P*	*AD*	*W*	*F*	*P*	*B*	*PP*	*F*	
8/9-5	F/25-30			2								2																			4
8/9-18	M/15-17			3	1	1				1		1		1																	8
8/9-20	?/14-18						1					3		1																	5
8/9-51	F/18-20	2		19	3	1	2	2	1			2	1	7			1	1		1		2				1					46
8/9-60A	M/20-25																												1		1
8/9-74	?/14-15	1		1							1			2							2					1					8
8/9-75	F/adult			2														1													3
8/9-77	M?/adult				1								1	1					1												4
8/9-78	F/30-35			4																		1									5
8/9-79	F/>25			5											1						1					1					8
8/9-80B	?/infant	3		1								2		1										1							8
8/9-83	F/adult			1	1																		1				1				4
8/9-84	?/child			3										1								1								1	6
8/9-91	F/>35										1																				1
8/9-93	?/infant		1		2																										3
8/9-94	?/infant						1																								1
8/9-95	?/infant			1																							1	1			3
8/9-96	?/child			1																											1
8/9-97	?/infant	3																													3
8/9-103	M/25-35			1										2																	3
8/9-104	F/<30	2		1	2			1				1		2		1		1							1	2				1	15
8/9-124	?/child			3										4		3					1				1	1					13
14-5	?/child			1								2																			3
14-14	F?/>20											1																			1
Total		11	1	49	10	2	4	3	1	1	2	14	2	22	1	4	1	3	1	1	4	4	1	1	2	6	2	1	1	2	156

W = whole
U = used
F = fragment
P = probable
AD = angular debris
C = core
B = biface
PP = projectile point

Table 10.7. Distribution of Flaked Stone from Cremations by Sex and Age, Pueblo Salado

Burial No. Cremation	Sex/Age Sex/Age	Basalt *W*	Basalt *U*	Basalt *F*	Basalt *P*	Basalt *AD*	Obsidian *W*	Obsidian *U*	Obsidian *F*	Obsidian *P*	Obsidian *AD*	Igneous *W*	Igneous *F*	Igneous *P*	Igneous *C*	Igneous *AD*	Metasediment *F*	Metasediment *P*	Quartzite *F*	Quartzite *P*	Chert *F*	Metamorphic *F*	Total
8/9-117	?/adult	13		22	2	3	8	4	13	2	5	5	12		1	2			2	1	1	1	97
8/9-121	?/adult		1	2			2		1			1									1		8
8/9-122	?/adult	6		23	8	2	3					1	5	1	2	1	3	2	1				58
Total		19	1	47	10	5	13	4	14	2	5	7	17	1	3	3	3	2	3	1	2	1	163

Material Type: Basalt, Obsidian, Igneous, Metasediment, Quartzite, Chert, Metamorphic

W = whole
U = used
F = fragment
P = probable
AD = angular debris
C = core

Table 10.8. Distribution of Ground Stone from All Burials by Sex and Age, Pueblo Salado

Burial No.	Sex/Age	Burial Type	Artifact Type	Material Type	Total
8/9-18	M/15–17	inhumation	1 grinder NFS	scoria	1
8/9-51	F/18–20	inhumation	1 handstone 1 grinding slab 1 grinder NFS	quartzite quartzite greenstone	3
8/9-76	M/16–18	inhumation	8 beads [bracelet]	sedimentary	8
8/9-117	?/adult	cremation	ground stone fragments 1 bead 2 beads	quartzite and sandstone sedimentary metamorphic	3+
8/9-124	?/child	inhumation	1 manuport	metasediment	1
14-14	F?/>20	inhumation	1 bead	sedimentary	1

and 25% were male. Although women certainly must have used flaked stone implements, the distribution of this lithic type at Pueblo Salado does not follow the expected gender norm (i.e., archaeologists more often have assumed that lithic artifacts were associated with men). The inability to determine the sex of the cremated remains (Table 10.7) sheds no additional light on this phenomenon. Hence, both the distributions of flaked stone relative to age and sex and the low amount of flaked stone present in most of the graves suggest that this artifact class may not represent burial offerings in every case. Although flaked stone associated with female burials may represent components of tool kits, in some cases the flaked stone may be intrusive. Because of the long history of habitation at Pueblo Salado and the ease with which the Pueblo Saladoans probably produced and discarded stone tools, ample amounts of lithic trash could have become incorporated in burial fill, especially if cemetery areas were in or near trash middens.

Although the burial assemblage was small, the analyst noted a distinct difference in the flaked stone distribution between inhumations and primary cremations (Tables 10.6 and 10.7): overall, the primary cremations exhibited markedly higher numbers of flaked stone items per burial. The Pueblo Saladoans deposited flaked stone items in the cremation pits following completion of the cremation process, as demonstrated by the absence of alteration to the flaked stone from exposure to burning (e.g., heat spalls, warping, sooting, and crazing). Utilitarian deposition (i.e., trash disposal) or ritual deposition following the cremating of the remains (relating to ceremonial practice, perhaps indicating status) may have contributed to the increased number of items in primary cremations. According to Gregory (1991:167), cremation areas at platform mound villages tend to occur outside the compound walls, while inhumations tend to occur within the compound. Investigators generally noted this pattern at Pueblo Salado (Volume 3:Figure 2.1). Zyniecki (1993) suggested that such a dichotomy reflects a difference in status between the individuals whose remains were buried and those who were cremated.

The pattern of ground stone tools (Table 10.8) seemed logical only in the context of Feature 8/9-51, in which utilitarian grinding implements used by this female could have been included as burial offerings in her grave. This may or not have been the case in the other examples. If these artifacts held any meaning, they were probably utilitarian items that had been used or owned by the deceased, as with the ground stone beads in Features 8/9-76, 8/9-117, and 14-14.

Field personnel found projectile points with three burials: a young adult male (Feature 8/9-60A); an adolescent (14–15 years) of indeterminable sex (Feature 8/9-74); and a middle-aged adult female (Feature 8/9-91). In the first and possibly in the second case, the presence of a projectile point may indicate the role of hunter and fledgling hunter, respectively. Other investigators have suggested that the chert side-notched point in the thoracic region of the adult male (Feature 8/9-60A) may be evidence of violence (Volume 3:Chapter 2). However, the point was not embedded in bone and originally may have been placed on the individual's chest and settled into the thoracic region as the body decomposed. The significance of the artifact for the female was difficult to determine. Perhaps it was an offering from the woman's husband, son, or some other male relative or was a keepsake retained or found. Its presence in the burial matrix might also suggest that the artifact was intrusive. Nevertheless, the utilitarian nature of these objects is difficult to contradict.

Forty percent of the burials at Pueblo Salado contained ceramic vessels as funerary offerings, whereas virtually all of the inhumations and cremations from Casa Buena (Effland 1988) and the Grand Canal Ruins (Mitchell, Fink, and Allen 1989) contained some form of ceramic vessel. The difference between these assemblages and that of Pueblo Salado may be related to individual status or may reflect the temporal setting of the site. Although status may have a bearing on the burial assemblage, few of the current data allow assessment of burial practices during the late Civano and

Polvorón phases. However, a decrease in burial goods may correspond to other changes in lifestyle that occurred at the time of the Hohokam collapse.

Abbott and Schaller (1991) recently reported that ceramic temper types differ between Hohokam communities north and south of the Salt River. Vessels from southern production areas are tempered with micaceous schist, South Mountain granodiorite, and Estrella gneiss, whereas phyllite, Camelback granite, and Squaw Peak schist are present in ceramics produced north of the river. From this kind of evidence researchers can generate hypotheses regarding the level of interaction between communities or possible corporate groups north and south of the river located along Turney's (1929) Canal Systems 1 and 2.

The temper analysis of the whole vessels from Pueblo Salado burials (Chapter 11) indicated that 24% came from southern production areas and 40% from northern production areas (Table 10.9); the analysis did not provide evidence for the origin of the other 36%. While the burial vessels from Area 14, representing the Soho phase, were evenly divided between northern and southern production areas, all but one of the later Area 8/9 burials contained vessels from southern production areas. These data suggest that in the Classic period the inhabitants of Pueblo Salado may have had a strong association with communities or corporate groups south of the Salt River despite being located on the northern floodplain. The nature of or cause for this affiliation is problematic (Chapter 3). Future research into this problem should include skeletal evidence of biological or genetic affinity. The combination of evidence from material culture and biological remains would provide a much stronger argument for the nature of sociopolitical "alliances" within or between canal systems.

Polychromes (Roosevelt Red Ware varieties) were noticeably absent from the Pueblo Salado burial vessel assemblage. Walsh-Anduze (Chapter 11) reports that redwares were the preferred burial offering, although plainware was also commonly present. Several of the intact vessels showed some evidence of prior use (Chapter 11), and the interment assemblage appeared to consist of utilitarian or personal items. The absence of polychromes from the mortuary assemblage was most likely a temporal phenomenon. Motsinger (Volume 3, Chapter 5) noted that polychromes were rare at Pueblo Salado until the Civano phase and that their frequency markedly increased toward the end of the Civano phase and especially during the Polvorón phase. This trend suggests that many of the burials probably dated to the early Civano phase, not that the Pueblo Saladoans excluded polychromes from the burial ritual. Moreover, Crown's (1984a) study associated with the Salt-Gila Aqueduct Project that included examination of mortuary vessels from Los Muertos found no significant relationship between use wear and ceramic ware, including polychromes. Thus the presence of utilitarian items in burials at Pueblo Salado cannot be attributed to contextual differences in ceramic ware (Walsh-Anduze, personal communication 1992).

Items of personal adornment were present in five of the Pueblo Salado graves. Field personnel recovered shell jewelry from three graves: two *Glycymeris* shell bracelets on the left humerus of an infant (Feature 8/9-94), marine shell beads in Feature 8/9-121, and a fragment of a *Glycymeris* shell bracelet in Feature 14-14. Also, the graves of three adults (Features 8/9-76, 8/9-117, and 14-14) contained ground stone beads (Table 10.8). The overall pattern of personal adornment at Pueblo Salado seemed rather simple, lacking in elite symbolism. Of course, the number of observations was small, and social rank or status could have been expressed in other ways that have not survived in the archaeological record (e.g., clothing).

Finally, almost all (33, or 89%) of the Area 8/9 inhumations were in a supine position (Volume 3:Chapter 2). Two (5%) were placed in a prone position, one (3%) was in a seated position, and one (3%) was lying on the right side. Excavators could not determine the position of the remaining three

Table 10.9. Production Area Affiliation of Whole Vessels from Pueblo Salado Burials

Burial	**Sex**	**Age**	**Whole Vessel Type**	**North**	**South**	**Unknown**
8/9-20	?	14-18	redware bowl			X
8/9-48	?	18-20	plainware scoop		X	
8/9-51	F	18-20	redware bowl			X
8/9-74	?	14-15	redware bowl			X
8/9-79	F	>25	redware bowl		X	
			plainware bowl	X		
8/9-82	M	40	redware bowl			X
			plainware bowl			X
			plainware bowl			X
			plainware bowl			X
8/9-93	?	infant	redware bowl		X	
8/9-94	?	infant	plainware bowl		X	
			red/plain bowl		X	
8/9-117	?	adult	plainware effigy			X
14-5	?	child	plainware bowl	X		
			plainware bowl	X		
			red/plain bowl	X		
14-13	?	child	plainware bowl	X		
			plainware bowl	X		
			plainware effigy			X
14-14	F?	>20	redware jar		X	
			redware bowl		X	
			redware bowl		X	
			redware bowl		X	
			redware bowl		X	
Total				6	10	9

**Note:* The whole vessel analysis (Chapter 11) included only ceramics determined in the laboratory to represent one-third or more of a vessel. In some cases the field identification of ware or form, reflected in feature descriptions, was corrected.

burials. This pattern does not deviate from that observed at other Classic period sites (Effland 1988; Mitchell, Fink, and Allen 1989). Researchers have not determined possible meanings associated with body positioning in Hohokam graves, although recent research has suggested that body position may not have social significance (Brunson 1989; Savage 1991).

From the five cremation features excavated at Pueblo Salado, analysts could not make a formal comparison between this mode of burial and inhumation. Sprague (1968) argued that cremation represents primarily a reduction of the corpse as opposed to an additional mode of interment. The single secondary deposit of calcined bone may have been derived from any of the primary cremations. Research recently completed (Fink 1996) indicates that so-called primary cremations rarely contain enough bone to represent a complete individual; portions of the deceased appear to have been removed from the pit for burial elsewhere. Consequently, although they may contain human bone, features deemed "primary cremations" may actually be crematory pits where the dead were cremated rather than the primary locus of interment (Fink 1989a).

Although evidence for ascribed status appeared to be absent from the individual artifact categories, investigators expected that possible evidence for elite graves would be more distinct in the combined assemblage for each burial. In fact, six burials (Features 8/9-51, 8/9-82, 8/9-117, 14-5, 14-13, and 14-14) tended to stand out in the combined assemblage. Features 8/9-51 and 8/9-117 both contained large amounts of flaked stone; the former also yielded several grinding implements, and the latter contained ground stone beads. Features 8/9-82, 14-5, 14-13, and 14-14 were the only burials that contained more than two ceramic vessels, and Features 8/9-117 and 14-13 contained the only ceramic effigies. However, with the possible exception of the effigy vessels, most of these artifacts were utilitarian items used or worn by these individuals when they were alive and could hardly be considered exotic. In this respect, the artifacts from these four burials did not seem to differ from the offerings found in the other Pueblo Salado burials. The "commonness" of the Pueblo Salado burial assemblage was its most striking feature.

The combined data offered little evidence for ascribed status in the burials at Pueblo Salado. Although Features 8/9-51, 8/9-82, 8/9-117, 14-5, 14-13, and 14-14 may have differed from the rest of the burials in the quantity of lithic and ceramic artifacts they contained, they did not appear to have differed in quality. Indeed, turquoise jewelry, carved shell pendants and effigies, polychrome vessels, and other elaborate grave goods were completely lacking from the Pueblo Salado burials.

Dutch Canal Ruin Burials

Field personnel excavated five burial features at Dutch Canal Ruin. During the testing phase they identified and excavated two inhumations dating to the Classic period (Features 1021 and 1029), and during full-scale excavations they recovered two cremations dating to the late Civano or Polvorón phase (Features 8-13 and 8-15) and one inhumation (Feature 1-3) that could not be temporally assigned.

Excavators recovered five redware vessels from Feature 1021; none were intact, and only three met the criterion of being at least one-third complete for inclusion in the whole vessel analysis (Chapter 11). A single redware scoop from Feature 1029, and no apparent grave goods from Feature 1-3. The skeletal remains in all three interments represented adults. Although analysts could not determine sex for Feature 1021, the remains were those of an adult, perhaps 35-45 years of age, the remains from Feature 1029 were those of a female, 30–40 years of age (Volume 2:Appendix E), and the remains from Feature 1-3 were of a male, 25–35 years of age (Volume 2:Chapter 2).

The cremated remains appeared to be adults, but analysts could not determine the sex or the age more precisely (Volume 2:Chapter 9). The remains for Feature 8-13 had been placed in a partly restorable plainware jar. The feature also contained 1 shell disc bead, 1 worked fragment of shell, 2 lithic cores (1 basalt, 1 quartz), and 25 fragments of flaked stone (6 quartzite, 1 chert, 15 basalt, and 3 igneous). Other than the plainware cremation vessel covered by an overturned bowl, Feature 8-15 contained only sherds.

The small size of the Dutch Canal Ruin burial series precluded making any substantial comparisons with other mortuary data sets. However, the two inhumations uncovered at this site during testing were typical of Classic period interments, based on their burial accompaniments, and did not differ from the burials found at Casa Buena (Effland 1988) and the Grand Canal Ruins (Mitchell, Fink, and Allen 1989). The third inhumation may also have been from the Classic period.

SUMMARY AND CONCLUSIONS

The skeletal remains and grave offerings recovered from the Pueblo Salado burials have permitted an initial look at the paleoepidemiology and level of social complexity of Pueblo Salado, a site probably representing an autonomous canal system. The analyst initially hypothesized that enamel hypoplasias would occur less frequently in autonomous systems where single villages occurred and that contamination of water supplies by pathogenic organisms may have been less severe than in larger, more densely populated irrigation communities. Although enamel hypoplasias occurred less frequently in the burial assemblage from Pueblo Salado than in burial assemblages from sites in the larger canal systems during the Classic period, the difference was not statistically significant. However, the peak period of hypoplasia development in the Pueblo Salado dentition suggested that children had been weaned later at this site than at other Classic period sites.

The orientation of the inhumations at Pueblo Salado also differed from the pattern at other Classic period sites in Canal System 2. Application of Merbs's (1987) model to the Pueblo Salado burial orientations suggested a high incidence of summer deaths, whereas other sites showed orientations that suggested greater rates of winter deaths. The reason for this difference, however, is difficult to discern; residents of villages in Canal System 2 may have seasonally spent more time away from the permanent village than did Pueblo Saladoans, resulting in fewer summer burials at those sites. Nevertheless, a demographic analysis of the burial orientations indicated that nonadults and young females had been more likely than other segments of the population to die in the summer. Nonadults showed a relatively equal incidence of mortality during the summer and during the fall and spring.

The evidence presented regarding burial orientation and seasonality of death can be considered only tentative. The skeletal series from Area 8/9 at Pueblo Salado was too small to test specific hypotheses to any significant degree. Nevertheless, the seasonal distribution of these burials was provocative and suggested that future analyses involving larger numbers of burials may produce substantial results. Critical to this kind of analysis, however, is greater knowledge of the demographic composition of communities. Therefore, it is essential to examine the variation present in a larger sample of sites. Perhaps one method of further examining seasonality would be to closely scrutinize associated botanical remains.

The demographic profile of the Pueblo Salado burials suggested that the site had been inhabited for at least a generation. The archaeological data, moreover, suggested that the site had been occupied considerably longer, especially since evidence of occupation during both the Civano and the Polvorón phase was present (Volume 3:Chapter 2). Analyses of several burial assemblages from other sites in the Phoenix Basin have shown that the majority of vessels exhibit use-related wear patterns (Crown 1984a; Walsh-Anduze 1994), indicating that the vessels were in circulation prior to their interment as burial offerings. No ceramic evidence indicated that the level of social complexity at Pueblo Salado was less developed than elsewhere; instead, the general utilitarian nature of the Pueblo Salado assemblage demonstrated, along with the preference for redware, a general homogeneity in the burial ritual that identified various smaller groups as Hohokam (Walsh-Anduze, personal communication 1992). Noticeably absent, however, were polychrome vessels and exotic goods such as turquoise and the carved shell pendants and effigies generally found in Classic period graves (Brunson 1989; Effland 1988; Mitchell, Fink, and Allen 1989).

Social status at Pueblo Salado was probably based more on achievement than on social rank. Although archaeologists have not yet evaluated burial assemblages from other sites that exhibit a high degree of autonomy, they may find in future studies that small irrigation systems never developed status stratification that could be recognized beyond that of the "common" level, at least as reflected in the kinds of artifacts that have survived. Additional research concerning social organization and complexity in large and small canal systems is needed for a better understanding of these differences. The contrast between the burial assemblage from Pueblo Salado and assemblages from other sites may be related to its having been occupied later in the Hohokam sequence and may reflect social, economic, and political reorientation (D. H. Greenwald, personal communication 1992).

CHAPTER 11

CERAMIC TEMPER STUDIES AND WHOLE VESSEL ANALYSIS

Mary-Ellen Walsh-Anduze

This chapter is divided into two sections describing the temper and whole vessel analyses of the plainware and redware ceramics from Dutch Canal Ruin and Pueblo Salado. The first section, the temper analysis, discusses inferences about production and exchange based on spatial and temporal differences in temper type. The classification of ceramics by temper type rather than typology (Haury 1945, 1976; Schroeder 1940) is not meant to diminish the value of the traditional type classifications but instead to provide a means to examine one aspect of ceramic variability (production sources) in greater detail. It is also a more valid approach for this study, in which the analyst did not consider surface treatment, which is an important typological attribute. The second section, the whole vessel analysis, then uses the temper study in conjunction with an examination of surface finish attributes to consider more fully the issues of Hohokam ceramic production and exchange during the Classic and post-Classic periods. Specific issues addressed are based on the research by SWCA at Dutch Canal Ruin and Pueblo Salado (Volumes 2 and 3).

The first issue involves a consideration of land use in the pre-Classic period component of Dutch Canal Ruin, which included seasonally occupied field houses and farmsteads and has been defined as an agricultural zone (Volume 1:Chapter 1). Investigators have proposed two related models: both suggest that residents of permanent village sites located nearby (e.g., Los Solares, La Ciudad, Pueblo Grande, Pueblo Patricio, La Villa) may have been involved in agricultural activities at Dutch Canal Ruin (Volume 2:Chapter 19). If this premise is true, then the material culture at the site should reflect some type of interaction, for example, in the presence of ceramics manufactured at those sites.

The second issue concerns the identification of changes in Hohokam ceramic procurement strategies between the Classic and post-Classic periods. Comparisons of temporal differences in temper type can provide information on changing strategies and the relationship of Pueblo Salado to Canal Systems 1 and 2.

THE TEMPER ANALYSIS

Most studies of Hohokam ceramics have focused on chronology building; more recent work has included the spatial distribution of ceramic types. However, during the past decade archaeologists have increasingly recognized a high degree of variability, especially among plainware and redware types (Abbott 1984; Abbott and Gregory 1988; Cable and Gould 1988; Doyel 1980; Doyel and Elson 1985; Lane 1989), which challenges the traditional typologies (Gladwin et al. 1937; Haury 1945, 1976; Schroeder 1940). Recently the results of an extensive and detailed investigation of Hohokam ceramics and temper source areas within the Salt River Valley (Abbott 1992, 1994a, ed. 1994; Abbott and Schaller 1991; Abbott, Schaller, and Birnie 1991; Bostwick and Downum 1994; Schaller 1993) have provided a better understanding of ceramic variability. The study demonstrated that Hohokam Plain Ware and Red Ware production throughout the lower Salt River Valley comprised distinct ceramic traditions rather than homogeneous developments of unified traditions as previously assumed. Further, temporal changes in production were neither uniform nor widespread (Abbott 1992; Abbott and Walsh-Anduze 1991; Walsh-Anduze 1994).

Building the Temper Typology

Nine sand composition zones that have been defined in the lower Salt River Valley (Figure 11.1) are separated by mutually distinguishable rock types (Abbott and Schaller 1991; Abbott, Schaller, and Birnie 1991). Results of the Pueblo Grande analysis, determined by chemical assays of the clay fraction in the pottery by electron microprobe analysis and a modified version of inductively coupled plasma spectroscopy (ICPS), indicated that Hohokam potters had used locally available materials (Abbott, ed. 1994; Abbott, Schaller, and Birnie 1991).

The petrographic analysis of 30 sherds from Dutch Canal Ruin and Pueblo Salado indicated that five of the nine sand composition zones were represented in ceramic temper at these sites, exhibiting the rock types South Mountain granodiorite, Estrella gneiss, Camelback granite, Squaw Peak schist, and micaceous schist (Appendix C). Paradoxically, both Dutch Canal Ruin and Pueblo Salado were situated on the geologic floodplain on the north side of the Salt River and not within any of the defined zones (Figure 11.1). However, the majority of the petrographic samples (both plainware and redware) were tempered with South Mountain granodiorite (Appendix C), which is likely to have important implications for identifying exchange relationships in the Salt River Valley. The rock types identified by petrographic analysis (Abbott, ed. 1994; Schaller 1993) are described below.

Camelback Granite

This rock type is distinguished from other granites by the presence of rose quartz crystals, which usually occur in minor amounts. Angular quartz and feldspar grains generally are the major constituents when Camelback granite is identified as temper in ceramic analysis. The quartz crystals tend to be translucent and fairly lustrous; feldspar is often white, but sometimes pink. The source of Camelback granite is the Phoenix Mountains.

Estrella Gneiss

Spheroidal hornblende grains are diagnostic of Estrella gneiss. Often, the hornblende particles occur as part of a larger matrix consisting of feldspar and quartz, which also is characteristic of South Mountain granodiorite (defined below). A higher proportion of hornblende distinguishes Estrella gneiss from South Mountain granodiorite. Estrella gneiss outcrops on the western slope of the South Mountain range.

Micaceous Schist

Fragments are platy and lustrous, and they appear to be stacked. Black spots of tourmaline are diagnostic of this rock type. Large platelets of muscovite occur in abundance; additional inclusions are quartz and feldspar. The only known source areas within the (immediate) region are along the Gila River at Pima Butte, Gila Butte, and the Santan Mountains.

South Mountain Granodiorite

Although South Mountain granodiorite has a granitic structure like that of Camelback granite, its grain size is smaller. Additionally, quartz and feldspar particles occur in zones and are identified

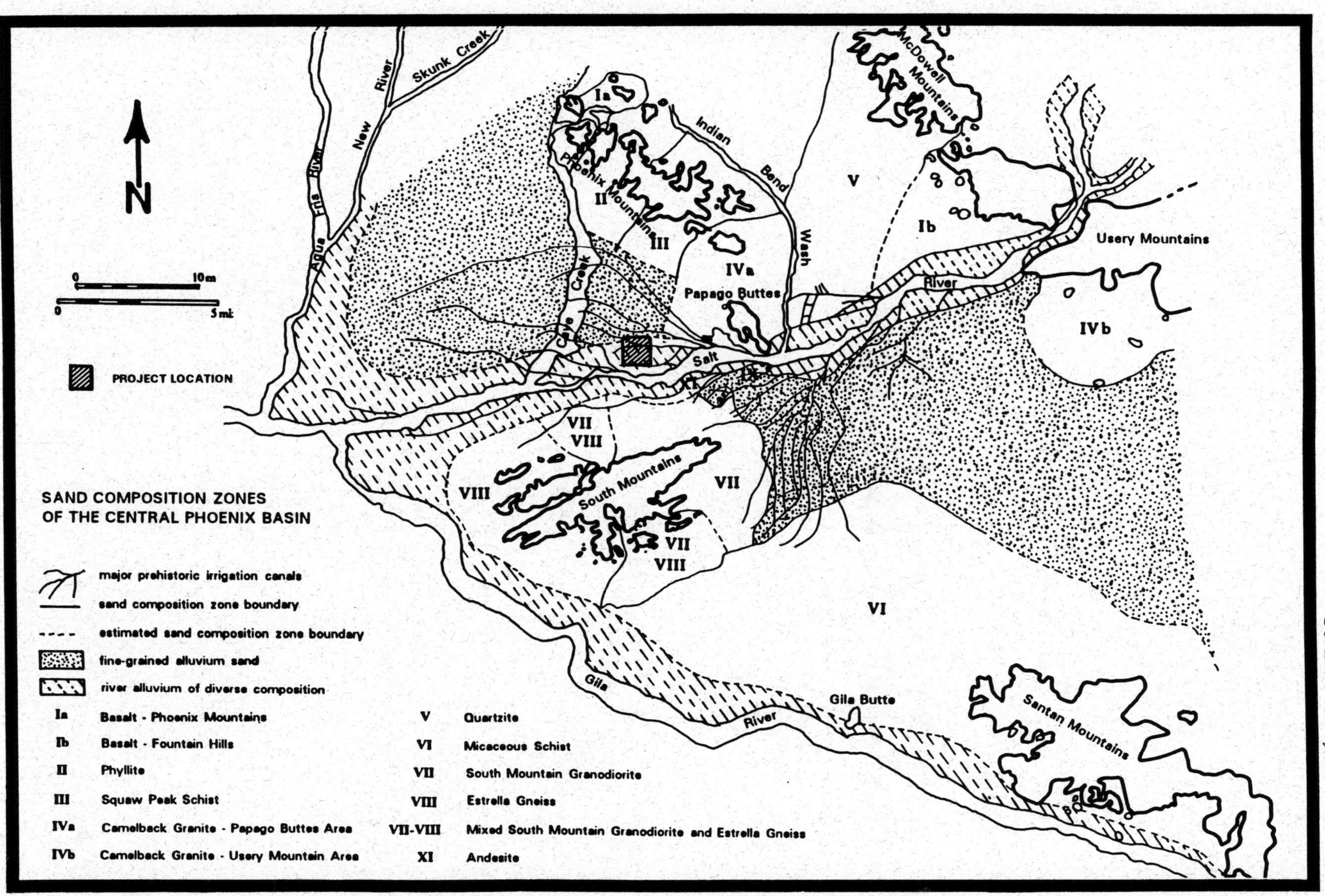

Figure 11.1. Sand composition zones in the lower Salt River Valley.

by contrasting colors (gray quartz and chalk-white feldspar). Hornblende is a common inclusion but should be less than 20% of the matrix; a larger percentage would indicate Estrella gneiss. South Mountain granodiorite outcrops on the eastern side of the South Mountain range.

Squaw Peak Schist

Depending on foliation (well developed vs. poorly developed), fragments have either a platy or a grainy appearance. Squaw Peak schist has a silvery white color with green and gray-green tints. It is further distinguished from phyllite and micaceous schist by its waxy luster and visible banding of larger-sized particles. The source area is the southern half of the Phoenix Mountains.

Phyllite

This rock type was not identified in the petrographic analysis, but it did occur as temper in small amounts of pottery from Dutch Canal Ruin and Pueblo Salado. Phyllite is distinguished from micaceous schist and Squaw Peak schist by its platy appearance and relatively large size (5 mm in length). Additionally, phyllite has a dull luster; its color is variable, but lead gray is most common. The suggested source area for phyllite-tempered ceramics in the Phoenix Basin is the northern half of the Phoenix Mountains.

Methods

The 6181 sherds microscopically examined constituted approximately 12% of the Dutch Canal Ruin and Pueblo Salado ceramic assemblages. The ceramics, selected from floor fill and floor contexts of architectural features, pits, hornos, and other trash deposits, seemed to be representative of the total assemblage from both sites, covering the span of Hohokam occupation previously discussed (Volume 2:Chapter 12; Volume 3:Chapter 5). Since the samples used for addressing particular questions were relatively small, the results of this study are preliminary and could be used as a basis for future research.

Motsinger's study of the assemblage from Dutch Canal Ruin (Volume 2:Chapter 12) identified Areas 1-7 as pre-Classic period components, whereas Area 8 had a post-Classic period or Polvorón phase occupation. Additionally, the ceramic data supported the contention that Pueblo Salado had been occupied no earlier than the Soho phase of the Classic period (Area 14), while Area 8/9 had evidence of Civano and Polvorón phase occupations (Volume 3:Chapter 5). These data provided the temporal framework in the current study.

In the temper analysis, each of the identified rock types was used to define a discrete temper category. Temper type identification with a 30X stereoscope closely followed the Pueblo Grande study (Abbott, ed. 1994; Abbott, Schaller, and Birnie 1991), with additional temper type categories based on the co-occurrence of two materials. The first material listed was the dominant temper, with the exception of Camelback granite, identified only when the analyst observed rose quartz. The final temper type categories (Table 11.1) were created by cross-tabulating primary and secondary temper materials. Recoding of some categories kept the data as comparable as possible with the Pueblo Grande assemblage. However, some notable differences still appeared. Other variables recorded for each of the sherds included ware, sherd temper (the inclusion of crushed material), and count.

Table 11.1. Identified Temper Types

0.	Fine-paste (natural inclusions)
1.	Micaceous schist
2.	South Mountain granodiorite
3.	Estrella gneiss
4.	Phyllite
5.	Squaw Peak schist
6.	Camelback granite
7.	Arkosic sand*
8.	Probable South Mountain granodiorite
9.	South Mountain granodiorite/Estrella gneiss
10.	Phyllite/Squaw Peak schist
11.	Mixed sand**/micaceous schist
12.	Mixed sand/Squaw Peak schist
13.	Camelback granite/Squaw Peak schist
14.	South Mountain granodiorite/mixed sand
15.	Squaw Peak schist/mixed sand
16.	Mixed sand
17.	Sherd
18.	Indeterminate

*Arkosic sand is defined as undifferentiated granitic rock, primarily quartz and feldspar; its source area is unknown.

**Mixed sand refers to generally small-sized and indeterminate temper particles. The distinction between arkosic sand and mixed sand in this analysis was based on differences in grain size and the inclusion of black particles, believed to be hornblende or basalt, in the mixed sand. Additionally, arkosic sand grains tend to be angular, which suggested procurement close to the source area. Conversely, mixed sand grains tend to be round or subangular, which suggested that the material had been gathered some distance away from its original source, possibly within a riverbed.

Hohokam Interaction during the Pre-Classic Periods

To address the question of Dutch Canal Ruin's external relationships during the pre-Classic period, the analyst examined 49 sherds. Eighteen sherds (36.7%) were plainware, 30 sherds (61.2%) were buffware, and 1 sherd (2.0%) was a brown-paste variant. The plainwares were distributed among four temper types (Table 11.2). The majority of these would traditionally be defined as Gila Plain, Salt Variety (Haury 1976) because of the high proportions of sand temper; most (67%) of this sample was arkosic sand. Although Abbott (personal communication 1992) believes this category represents a locally produced plainware at Pueblo Grande (during the Classic period), presently the distinction between arkosic sand found north and south of the Salt River is not clear-cut.

The existing data base provided too little information for investigators to adequately test the hypothesis that Dutch Canal Ruin had been seasonally occupied by residents of one or more permanent village sites within Canal System 2 (Volume 2:Chapter 19). The success of this study relied on the identification of ceramics locally produced at the permanent village sites, but researchers have not yet applied a similar sourcing analysis to the ceramics from the sites in question (with the exception of Pueblo Grande). Additionally, the size of the ceramic sample from Dutch Canal Ruin was much too small and variable to test the hypothesis adequately. Nevertheless,

Table 11.2. Plainware Temper Types at Dutch Canal Ruin

Temper Type	Frequency	Percent
South Mountain Granodiorite	1	5.6
Phyllite	2	11.1
Arkosic Sand	12	66.7
Mixed Sand with Indeterminate Schist	1	5.6
South Mountain Granodiorite with Mixed Sand	1	5.6
Indeterminate	1	5.6
Total	**18**	

the following highly speculative ideas concern the potential relationship between Dutch Canal Ruin and Las Colinas, La Ciudad, and Pueblo Grande.

Las Colinas

Las Colinas, near the western end of Canal System 2, was approximately 6 km from Dutch Canal Ruin. The site was first occupied in the late Colonial period as a small farmstead (Teague 1988:121). Population increased and the site grew proportionally in size and complexity during the Sedentary and Classic periods. Whereas investigators considered the Colonial period occupation at Las Colinas to be "limited in scope and spatial extent" (Gregory 1988:319) and no ceramic data are available for the early components, they thoroughly studied the later components (permanent habitation areas). The results of the plainware study (Abbott 1988:177–178) suggested that phyllite-tempered pottery (Wingfield Plain), which occurred in high frequencies compared to sand-tempered plainware (Gila Plain, Salt Variety) during the Sedentary period, had been locally produced. X-ray fluorescence analysis supported this proposition, although perhaps not conclusively (Crown, Schwalbe, and London 1988:29-71). More recently, Abbott's (1992) analysis of a ceramic sample from Las Colinas, using electron microprobe assays, strengthened the argument for the local production of phyllite-tempered pottery.

At Dutch Canal Ruin, Motsinger (Volume 2:Chapter 12) proposed that high frequencies of sand-tempered ceramics were indicative of local production. Sherds tempered with phyllite, by contrast, accounted for less than 1% of the total assemblage and had probably been brought to the site through trade or exchange. The sample data for the current temper analysis, although limited, were comparable; only two sherds were tempered with phyllite.

If, indeed, the high frequency of phyllite-tempered pottery is associated with the Hohokam Sedentary period as suggested by Abbott (1988, 1992), then the lack of phyllite-tempered pottery in Areas 1–7 at Dutch Canal Ruin could indicate that the Hohokam occupied these areas prior to the Sedentary period. Another interpretation may be that phyllite-tempered vessels were not involved in the trade and exchange network at Dutch Canal Ruin; however, the high frequencies of Santa Cruz Red-on-buff ceramics compared to later buffwares (Volume 2:Chapter 12) also tend to support the idea that Dutch Canal Ruin was occupied prior to the permanent habitation at Las Colinas. Although the field house components of the two sites were contemporary, project investigators assumed that they had not been used by a single population, given their functional similarity and their geographic positions within Canal System 2. However, this proposition cannot be tested without comparable data from Las Colinas.

La Ciudad

La Ciudad was approximately 1.6 km from Dutch Canal Ruin. Kisselburg (1985) found that a high frequency of undifferentiated sand-tempered pottery dominated the plainware assemblage of the Colonial period, and she suggested that these ceramics had been locally produced, although researchers have identified no other supporting data. However, this lack of data leaves the door open for future studies using comparable methods to determine if the sand-tempered pottery at Dutch Canal Ruin and La Ciudad were made at the same location and to test the proposition that the Hohokam of La Ciudad had used the field houses at Dutch Canal Ruin on a seasonal basis.

Pueblo Grande

Pueblo Grande, located near the heads of Canal System 2, was approximately 7.0 km from Dutch Canal Ruin. Recent excavations by Soil Systems, Inc. (SSI), resulted in an extensive and detailed ceramic study focusing on ceramic production and exchange (Abbott, ed. 1994) that served as the basis for the current study design. The ceramic assemblage, however, was primarily derived from the Classic period component and thus could not be compared with the current data base for Dutch Canal Ruin. Although early excavators had recorded a large Colonial period occupation at the site (Bostwick 1994), the ceramic data focused on diagnostic buffware. Two recent testing projects conducted by SSI at the Pueblo Grande Cultural Park (Birnie and Walsh-Anduze 1991) and on Washington Street (Birnie 1991) did produce comparable data, although the assemblages from both projects were small.

Nevertheless, the dominant tempering materials in plainware (both Pueblo Grande assemblages combined) were Camelback granite and unidentified arkosic sand with major amounts of Squaw Peak schist (Walsh-Anduze 1991a, 1991b, 1991c). Abbott's study of the Classic period component determined that vessels tempered with these materials had been locally produced, at least during the Classic period. If (and this possibility has yet to be tested) the Hohokam of Pueblo Grande also manufactured these materials locally during the Colonial period, then they were probably not related to the seasonal inhabitants of Dutch Canal Ruin. Neither of these temper types was present in the Dutch Canal Ruin sample, and ceramics tempered with Squaw Peak schist were not especially common in the total assemblage (Volume 2:Chapter 12).

Summary

The use of a detailed temper analysis to examine possible relationships between the seasonally occupied field houses at Dutch Canal Ruin and permanent village sites located within Canal System 2 was hampered by a very small ceramic assemblage and the lack of comparable data. Under such circumstances, the use of a traditional typological approach would have sufficed and yielded similar results. However, one goal of this research was to broaden the newly formed data base begun with the Pueblo Grande ceramic analysis (Abbott, ed. 1994). As such, the analysis was useful for determining that the permanent habitation areas at Las Colinas were not contemporary with the field house occupation at Dutch Canal Ruin. Also, the inhabitants of Dutch Canal Ruin and Pueblo Grande during the Colonial period appeared to have had little or no relationship with each other. The assemblage from La Ciudad needs to be evaluated further using a comparable approach.

Ceramic Production during the Classic and Post-Classic Periods

As stated earlier, another objective of this study was to examine changes in procurement strategies during later Hohokam prehistory and to determine, if possible, the place of Pueblo Salado within the Phoenix Basin. Analyses of Classic and post-Classic period plainware and redware from Pueblo Salado and Dutch Canal Ruin provided some relevant data. However, such an approach considers only a limited aspect of ceramic production technology, the procurement of raw resources, and therefore provides only the baseline data for additional studies, which should consider other technological and social parameters that guided prehistoric ceramic production.

Dutch Canal Ruin, as previously stated, was primarily occupied during the late Colonial Period. However, project investigators also defined a Polvorón phase occupation (small hamlet) in Area 8 based on architectural morphology and the associated material culture (Volume 2:Chapter 9).

During data recovery at Pueblo Salado, investigators determined that the earliest occupation of the site was at Area 14, which dated to the late Soho and early Civano phases (Volume 3:Chapter 2). During this time, the site was a continuously occupied but dispersed habitation area or small hamlet. By the Civano phase, the site had grown in size and complexity as evidenced by compound architecture in Area 8/9. Excavators also recorded a small Polvorón phase component representing a residual population of the earlier Civano phase occupation in Area 8/9.

The ceramics from Pueblo Salado (Table 11.3) were typical of those from most later Hohokam sites, with plainware dominating the assemblage (81%) and redware accounting for approximately 13%. As expected, decorated ceramics occurred only infrequently. The proportion of plainware to redware was very similar in the assemblages from the Soho and Civano phases (6:1 and 7:1, respectively) but decreased in the Polvorón phase assemblage (5:1). A comparison of proportions of redware to buffware (including brown-paste variants) through time showed a slight decrease between the Soho (22:1) and Civano (19:1) phases at Pueblo Salado but a dramatic increase by the Polvorón phase (193:1). At Dutch Canal Ruin, the proportion of plainware to redware in the Polvorón phase assemblage was only slightly higher than at Pueblo Salado. The ratios of redware to buffware were high at both sites, although the ratio at Dutch Canal Ruin (20:1) was considerably lower than that at Pueblo Salado. This difference may be related to one or more of the following: sample size, temporal variability, functional variability, or differences in population size between the two sites.

Plainware

Table 11.4 shows the distribution of all plainware temper types, with arkosic sand-tempered ceramics dominating both the Pueblo Salado and Dutch Canal Ruin assemblages (47.4% and 50.7%, respectively). Analysis did not reveal the source area or areas of this material; moreover, the high percentage recorded may reflect a problem with distinguishing between South Mountain granodiorite and arkosic sand. Microscopically, South Mountain granodiorite was identified by the contrasting or zonal white and gray colors of the feldspar and quartz constituents. Although Schaller's (Appendix C) petrographic analysis suggested that the majority of plainware was tempered

Table 11.3. Distribution of Ceramic Wares at Pueblo Salado and Area 8, Dutch Canal Ruin

Ceramic Ware	Pueblo Salado				Dutch Canal Ruin
	Soho (Area 14)	Civano (Area 8/9)	Polvorón (Area 8/9)	Total	Polvorón (Area 8)
Plainware	928 *22.9 **84.3	2206 54.5 81.4	914 22.6 76.1	4048 80.8	823 **80.3
Redware	154 23.2 14.0	317 47.7 11.7	193 29.1 16.1	664 13.3	133 13.0
Red/Plain	12 7.2 1.1	91 54.8 3.4	63 38.0 5.2	166 3.3	28 2.7
Buffware	7 41.2 0.6	9 52.9 0.3	1 5.9 0.1	17 0.3	5 0.5
Brown-paste Variants		8 100.0 0.3		8 0.2	2 0.2
Roosevelt Red Ware		78 72.2 2.9	30 27.8 2.5	108 2.2	34 3.3
Total[1]	1101 *22.0	2709 54.1	1201 24.0	5011 100.0	1025 100.0

*Row %
**Column %
[1]Three sherds of unidentified ware not included in table.

with South Mountain granodiorite, a microscopic examination of the remnant sherds from those samples as part of the current study did not yield the same confident results. Because the problem was unresolved in the current study, the following discussion is based on relative proportions of temper types other than arkosic sand (Table 11.5).

Leaving out the sherds with arkosic sand temper, sherds occurring with the highest frequency at Pueblo Salado were those containing mixed sand temper (Table 11.5). Although the source area or areas of this material is not known, the characteristics of both paste (silty) and temper (round or subangular) suggested a riverine origin. The frequency of plainware tempered with mixed sand increased through time at Pueblo Salado, and by the Polvorón phase it accounted for approximately 46% of the plainware.

Among plainwares from known source areas (Table 11.5), ceramics tempered with South Mountain granodiorite constituted 20.3% of the Soho phase assemblage from Pueblo Salado; the proportion increased to 24.3% in the Civano phase assemblage and 34.1% in the Polvorón phase. In comparison, pottery tempered with Squaw Peak schist represented 23.5% of the Soho phase assemblage, decreasing to 10.6% in the Civano phase and 5.3% in the Polvorón phase.

Table 11.4. All Plainware Temper Types through Time

Temper Type	Pueblo Salado				Dutch Canal Ruin
	Soho (Area 14)	Civano (Area 8/9)	Polvorón (Area 8/9)	Total	Polvorón (Area 8)
Micaceous Schist	2 *22.2 **0.2	6 66.6 0.3	1 11.1 0.1	9 **0.2	2 **0.2
South Mountain Granodiorite	94 17.5 10.0	309 57.5 14.0	134 25.0 14.7	537 13.3	264 32.1
Estrella Gneiss	8 15.1 0.9	39 73.6 1.8	6 11.3 0.7	53 1.3	1 0.1
South Mountain Granodiorite/ Estrella Gneiss	38 15.8 4.1	170 70.8 7.7	32 13.3 3.5	240 5.9	17 2.1
Phyllite	38 69.1 4.1	13 23.6 0.6	4 7.3 0.4	55 1.4	1 0.1
Squaw Peak Schist	109 41.1 11.7	135 50.9 6.1	21 7.9 2.3	265 6.5	46 5.6
Phyllite/Squaw Peak Schist	6 50.0 0.6	4 33.3 0.2	2 16.7 0.2	12 0.3	—
Camelback Granite	3 16.7 0.3	5 27.8 0.2	10 55.5 1.1	18 0.4	12 1.5
Camelback Granite/ Squaw Peak Schist	—	2 100.0 0.1	—	2 0.0	3 0.4
Arkosic Sand	464 24.2 50.0	932 48.6 42.2	521 27.2 57.0	1917 47.4	417 50.7
Mixed Sand	105 14.0 11.3	470 62.8 21.3	174 23.2 19.0	749 18.5	30 3.6
Mixed Sand/Squaw Peak Schist	46 29.7 5.0	101 65.2 4.6	8 5.2 0.9	155 3.8	30 3.6
Other/Unknown	15 41.7 1.6	20 55.6 0.9	1 2.7 0.1	36 0.9	—
Total	928 *22.9	2206 54.5	914 22.6	4048 100.0	823 100.0

*Row %
**Column %

Table 11.5. Plainware Temper Types through Time, Excluding Arkosic Sand

Temper Type	Pueblo Salado				Dutch Canal Ruin
	Soho (Area 14)	Civano (Area 8/9)	Polvorón (Area 8/9)	Total	Polvorón (Area 8)
Micaceous Schist	2	6	1	9	2
	*22.2	66.6	11.1	**0.4	**0.5
	**0.4	0.5	0.3		
South Mountain Granodiorite	94	309	134	537	264
	17.5	57.5	25.0	25.2	65.0
	20.3	24.3	34.1		
Estrella Gneiss	8	39	6	53	1
	15.1	73.6	11.3	2.5	0.2
	1.7	3.1	01.5		
South Mountain Granodiorite/ Estrella Gneiss	38	170	32	240	17
	15.8	70.8	13.3	11.3	4.2
	8.2	13.3	8.1		
Phyllite	38	13	4	55	1
	69.1	23.6	7.3	2.6	0.2
	8.2	1.0	1.0		
Squaw Peak Schist	109	135	21	265	46
	41.1	50.9	7.9	12.4	11.3
	23.5	10.6	5.3		
Phyllite/Squaw Peak Schist	6	4	2	12	—
	50.0	33.3	16.7	0.6	
	1.3	0.3	0.5		
Camelback Granite	3	5	10	18	12
	16.7	27.8	55.5	0.8	3.0
	0.6	0.4	2.5		
Camelback Granite/ Squaw Peak Schist	—	2	—	2	3
		100.0		0.1	0.7
		0.2			
Mixed Sand	105	470	174	749	30
	14.0	62.8	23.2	35.1	7.4
	22.6	36.9	44.3		
Mixed Sand/Squaw Peak Schist	46	101	8	155	30
	29.7	65.2	5.2	7.3	7.4
	9.9	7.9	2.0		
Other/Unknown	15	20	1	36	—
	41.7	55.6	2.7	1.7	
	3.2	1.6	0.3		
Total	464	1274	393	2131	406
	*21.8	59.8	18.4	100.0	100.0

*Row %
**Column %

Overall, plainwares tempered with micaceous schist, Estrella gneiss, phyllite, and Camelback granite occurred in the Pueblo Salado assemblage during each phase but relatively infrequently; cumulatively these temper types, alone or in combination, accounted for approximately 18.3% of the total from known sources. These types were also infrequent at Dutch Canal Ruin during the Polvorón phase. In addition, the proportion of South Mountain granodiorite to Squaw Peak schist in plainware was high at Dutch Canal Ruin, as it was at Pueblo Salado; the analyst had no available data with which to examine this proportion through time. Finally, one significant difference between the Polvorón phase assemblages from Pueblo Salado and Dutch Canal Ruin was in the proportion of South Mountain granodiorite to mixed-sand temper. The ratio was approximately 1:1 at Pueblo Salado compared to nearly 9:1 at Dutch Canal Ruin. Presently, there is no explanation for this difference.

Redware

Ceramics containing arkosic sand temper accounted for nearly 19% of the total redware assemblage from Pueblo Salado (Table 11.6); again, because of the identification problems described above, they are not included in this discussion. The next most common temper material was South Mountain granodiorite, identified in 51% of the redware from Pueblo Salado (Table 11.7). The proportion of sherds tempered with South Mountain granodiorite compared to all other temper types increased through time, although not significantly.

Sherds tempered with mixed sand (30.0%) also constituted a major portion of the Pueblo Salado redware assemblage (Table 11.7). As with South Mountain Granodiorite, the frequency of mixed sand temper was slightly higher during the Polvorón phase than in earlier time periods, but not significantly so. As in the plainware assemblage, micaceous schist and Estrella gneiss temper were infrequent in the Pueblo Salado redwares. Phyllite and Squaw Peak schist temper were virtually absent from the redware assemblage, and only two sherds contained Camelback granite temper.

At Dutch Canal Ruin, redware tempered with South Mountain granodiorite accounted for 88% of the Polvorón phase assemblage (Table 11.7), a significantly greater frequency than at Pueblo Salado. The difference between the assemblages can be attributed to the proportion of redware containing mixed sand temper, which was only 7.5% at Dutch Canal Ruin.

Discussion

Analysts examining ceramics from Pueblo Grande (Abbott, ed. 1994) determined that, during the Classic period, plainware ceramics had been manufactured primarily with materials found north of the Salt River (phyllite, Squaw Peak schist, and Camelback granite), whereas the materials for redware ceramics (particularly South Mountain granodiorite) had come from source areas to the south. Additionally, plainware tempered with Camelback granite and arkosic sand with major amounts of Squaw Peak schist had been locally produced. Because it was located north of the Salt River, a similar pattern might be expected at Pueblo Salado even though the site was not located within any of the defined sand composition zones (Appendix C).

Because the current study did not include sherds tempered with arkosic sand, the analyst can only speculate that Pueblo Salado had interacted more heavily with areas south of the Salt River throughout the Classic period, based on the relatively high frequencies of South Mountain granodiorite temper among both plainware and redware. Moreover, the predominance of South Mountain granodiorite may indicate that relationships between Pueblo Salado and Pueblo Grande

Table 11.6. All Redware Temper Types through Time

Temper Type	Pueblo Salado				Dutch Canal Ruin
	Soho (Area 14)	Civano (Area 8/9	Polvorón (Area 8/9)	Total	Polvorón (Area 8)
Micaceous Schist	—	2 *100.0 **0.6	—	2 0.3	—
South Mountain Granodiorite	47 17.1 30.5	133 48.4 42.0	95 34.5 49.2	275 41.4	94 **70.7
Estrella Gneiss	8 38.1 5.2	10 47.6 3.2	3 14.3 1.6	21 3.2	—
South Mountain Granodiorite/ Estrella Gneiss	20 57.1 13.0	5 14.3 1.6	10 28.6 5.2	35 5.3	3 2.3
Camelback Granite	2 100.0 1.3	—	—	2 0.3	1 0.8
Arkosic Sand	33 26.6 21.4	65 52.4 20.5	26 21.0 13.5	124 18.7	26 19.5
Mixed Sand	30 18.3 19.5	75 47.2 23.7	54 34.0 28.0	159 23.9	5 3.8
Mixed Sand/ Squaw Peak Schist	—	1 33.3 0.3	2 66.7 1.0	3 0.5	3 2.3
Other/Unknown	14 32.6 9.1	26 60.5 8.2	3 7.0 1.6	43 6.5	1 0.8
Total	154 *23.2	317 47.7	193 29.1	664 100.0	133 100.0

*Row %
**Column %

and other sites (e.g., Las Colinas, La Ciudad) within Canal System 2 were minimal. Consequently, the relatively low frequencies at Pueblo Salado of plainware and especially of redware with temper from other southern production sources, such as micaceous schist and Estrella gneiss, were unexpected. Analysts have not yet tested whether this pattern was a result of site specialization or differential exchange patterns of ceramics with particular temper types. However, because ceramics tempered with micaceous schist, especially redware, are temporally sensitive, the absence of such ceramics was probably related to the period when Pueblo Salado was occupied.

Finally, the examination of Polvorón phase assemblages from Dutch Canal Ruin and Pueblo Salado offered additional questions. At both sites, the frequency of sherds tempered with South Mountain granodiorite was high, which may indicate that residents of the two sites had shared a close social relationship. However, significant differences between the two sites in the proportion

Table 11.7. Redware Temper Types through Time, Excluding Arkosic Sand

Temper Type	Pueblo Salado				Dutch Canal Ruin
	Soho (Area 14)	Civano (Area 8/9)	Polvorón (Area 8/9)	Total	Polvorón (Area 8)
Micaceous Schist	–	2 *100.0 **0.8	–	2 0.4	–
South Mountain Granodiorite	47 17.1 38.8	133 48.4 52.8	95 34.5 56.9	275 50.9	94 **87.9
Estrella Gneiss	8 38.1 6.6	10 47.6 4.0	3 14.3 1.8	21 3.9	–
South Mountain Granodiorite/ Estrella Gneiss	20 57.1 16.5	5 14.3 2.0	10 28.6 6.0	35 6.5	3 2.8
Camelback Granite	2 100.0 1.7	–	–	2 0.4	1 0.9
Mixed Sand	30 18.3 24.8	75 47.2 29.8	54 34.0 32.3	159 29.4	5 4.7
Mixed Sand/ Squaw Peak Schist	–	1 33.3 0.4	2 66.7 1.2	3 0.6	3 2.8
Other/Unknown	14 32.6 11.6	26 60.5 10.3	3 7.0 1.8	43 8.0	1 0.9
Total	121 *22.4	252 46.7	167 30.9	540 100.0	107 100.0

*Row %
**Column %

of South Mountain granodiorite to pottery with mixed-sand temper, among both plainware and redware, are not easily explained. The variation in the assemblages may reflect differences in function (e.g., cooking, storage), in period of occupation, or in longevity of occupation (seasonal vs. year-round). The variability also could reflect the dispersion or aggregation of populations from one site to the other; some of the ceramic differences, therefore, might be related to site function.

Conclusions

Comparative studies of Hohokam ceramics using a detailed temper analysis approach, rather than a typological one, are useful for identifying spatial and temporal differences in production. Although the results of the current study were tenuous because of the small sample and the problem of distinguishing between arkosic sand and South Mountain granodiorite, a typological approach would not have identified these differences (or potential differences) in production; both tempers would have been subsumed under a single type such as Gila Plain, Salt Variety or Salt Red. In the current study, the analyst used the detailed examination of temper to address specific questions about intraregional relationships.

The first step was to attempt to define relationships between Dutch Canal Ruin and permanent habitation sites north of the Salt River. While the conclusions reached were not definitive, the evidence suggested that use of the (Colonial period) seasonal field houses at Dutch Canal Ruin had not been related to permanent occupation at Las Colinas or Pueblo Grande. The relationship between Dutch Canal Ruin and La Ciudad remains to be tested.

Second, the analyst used ceramic data from Pueblo Salado to examine Classic and post-Classic period production strategies. The results showed that the inhabitants of Pueblo Salado had favored ceramics tempered with South Mountain granodiorite throughout the Classic period (based on known temper source areas). However, during the Civano phase, mixed-sand temper may have replaced Squaw Peak schist in plainware, as Squaw Peak schist occurred infrequently by the Polvorón phase. The relationship between South Mountain granodiorite and plainware containing mixed-sand temper also changed through time, but not significantly. These patterns may have major implications for understanding Classic period production and exchange within the Phoenix Basin, but they could not be further investigated with the current data set.

With regard to the possibility of local production at Pueblo Salado, the temper study indicated that a large portion of the plainware assemblage had been manufactured with arkosic sand and possible river sands, both of which could have been obtained locally rather than from one of the nine defined temper zones (Abbott, Schaller, and Birnie 1991). The presence of polishing stones (Chapter 12) supported the premise of local ceramic manufacture, although excavators did not discover features such as kilns and mixing basins at the site.

The relationship between Pueblo Salado and Dutch Canal Ruin during the Polvorón phase remains uncertain because of the high proportion of South Mountain granodiorite at both sites among both plainware and redware ceramics. Additionally, the relatively low proportion of pottery exhibiting mixed-sand temper at Dutch Canal Ruin, compared to Pueblo Salado, may reflect a temporal or functional difference between the two sites; perhaps, too, the high proportions of South Mountain granodiorite and mixed sand together define a late Polvorón phase assemblage. This possibility should be examined further with an attribute analysis; observation of this material suggests that mixed-sand temper with matte or wiped surfaces in both plainware and redware may be characteristic of later ceramics.

Third, future research should evaluate the question of the role of Pueblo Salado within the Phoenix Basin canal systems after analysts have resolved the arkosic sand problem. However, the preliminary indications are that Pueblo Salado may have functioned differently than other sites, such as Pueblo Grande, particularly in the production and distribution of plainware. At Pueblo Salado, unlike Pueblo Grande, comparatively few ceramics were tempered with Squaw Peak schist, Camelback granite, and phyllite; these materials had been derived from production areas north of the Salt River. From this difference, researchers may infer that Pueblo Salado had stronger social connections with southern production areas than did Pueblo Grande. Although redware ceramics from southern production areas dominated the assemblages at both sites, plainware and redware apparently were differentially exchanged (Abbott 1984); moreover, the distribution of redware at sites in northern and southern production areas is apparently related to Hohokam mortuary customs (Brunson 1989; Crown 1984a; Walsh-Anduze 1994). In conclusion, the results of the temper studies, although encouraging, should be considered preliminary until analysts have examined larger samples and formulated more comparable data sets.

WHOLE VESSEL ANALYSIS

This section presents the analysis of whole vessels from Dutch Canal Ruin and Pueblo Salado, incorporating the results of the temper analysis. The objective of this research was to expand the data base for exploring questions about ceramic production and exchange within the Phoenix Basin (Abbott 1992, 1994b; Walsh-Anduze 1994). This analysis considered vessels only, not other ceramic forms, and only vessels judged to be one-third or more complete. In some cases field identifications of ware or form were corrected.

Dutch Canal Ruin

During data recovery three partial, reconstructible vessels were recovered from the Classic period component at this site; all three were plainware and provided only a limited amount of information. Two of the vessels were jars tempered with unidentified arkosic sand. The form of the other vessel was coded as indeterminate, although the lack of interior surface finish suggested that it, too, was a jar. It was tempered with South Mountain granodiorite.

Pueblo Salado

Analyzed ceramics from Pueblo Salado included 59 items, both whole vessels and reconstructible partial vessels. The majority (n=44) were from Area 8/9; 15 vessels were from Area 14. Fourteen of the vessels from Area 8/9 and 11 of those from Area 14 were from mortuary contexts; the rest of the assemblage was from various contexts, including structures, courtyards, pits, and caches. Because of the relatively small number of vessels within each context, the assemblage was analyzed as a whole.

Ware and Form

Most (76%) of the vessels in this collection were plainware and redware bowls and jars; 2 scoops, 2 pitchers, 3 (duck) effigy vessels, and 2 (partial) vessels of unknown form completed the inventory (Table 11.8). Although excavators found no red-on-buff vessels, they did recover two polychrome vessels. The focus of this study was on the plain and red bowls and jars because of the larger sample of these wares.

A comparison of the redware and plainware vessels showed that jars occurred significantly more often as plainware than as redware (.05 confidence level); bowls occurred more often as redware, but not significantly. Because of the small assemblage, the analyst did not place vessels into specific form groups, although previous analyses have demonstrated significant relationships between specific vessel forms and ceramic ware (Brunson 1989; Crown 1984a; Walsh-Anduze 1994). However, the majority of bowls at Pueblo Salado had unrestricted hemispheric or vertical shapes; only one vessel fit the definition of a wide-mouthed (Walsh-Anduze 1994) or Group 2 (Crown 1984a) jar, which has a zone of restriction at or near the orifice. Comparative use-wear analyses have demonstrated that these vessels functioned more like bowls than (necked) jars (Crown 1984a; Walsh-Anduze 1991a, 1994), so their classification as bowls is not misleading. The Pueblo Salado jars were predominantly necked vessels, although a few vessels had an incurved (neckless) profile similar to that of a seed jar.

Table 11.8. Relationship between Ware and Vessel Form, Pueblo Salado Ceramic Assemblage

Vessel Form	Plainware	Redware	Red/Plain	Salado Polychrome	Total
Bowl	11 *32.3 **37.9	19 55.9 82.6	2 5.9 40.0	2 5.9 100.0	**34** ****57.6**
Jar	13 81.3 44.8	2 12.5 8.7	1 6.2 20.0		**16** **27.1**
Scoop	1 50.0 3.4	1 50.0 4.3			**2** **3.4**
Pitcher		1 50.0 4.3	1 50.0 20.0		**2** **3.4**
Effigy Vessel	2 66.7 6.9		1 33.3 20.0		**3** **5.1**
Unknown	2 100.0 6.9				**2** **3.4**
Total	**29** ***49.1**	**23** **39.0**	**5** **8.5**	2 3.4	**59** **100.0**

*Row %
**Column %

Temper

The relationship between ware and temper, for bowls and jars combined, is shown in Table 11.9; examination of all temper types for plainware and redware vessels revealed no significant association. However, when the temper types (excluding arkosic sand, the source of which could not be determined) were collapsed into two groups—representing production areas north and south of the Salt River—a significant association (.05 confidence level) between production area and ceramic ware existed: redware vessels had been manufactured with temper materials found south of the Salt River (South Mountain granodiorite, Estrella gneiss, and micaceous schist).

These findings were similar, in part, to the results of the Pueblo Grande analysis, where a significant number of redware vessels, compared to plainware, were from southern production areas. However, the majority of plainware at Pueblo Grande either had been produced locally or had come from northern production areas. Although the interpretation is tentative (because of the arkosic sand temper problem), the data from Pueblo Salado indicated that plainware, like redware at this site, had come from southern production areas. However, because the form classification schemes were not identical, the apparent difference between the two sites cannot be tested.

Table 11.9. Relationship between Ware and Temper, Pueblo Salado Ceramic Assemblage

Temper Type	**Plainware**	**Redware**	**Total**
Micaceous Schist		1	**1**
		*100.0	****2.2**
		**4.8	
South Mountain Granodiorite	7	7	**14**
	50.0	50.0	**31.1**
	29.2	33.3	
South Mountain Granodiorite/ Estrella Gneiss		4	**4**
		100.0	**8.9**
		19.0	
Estrella Gneiss		1	**1**
		100.0	**2.2**
		4.8	
Phyllite	2		**2**
	100.0		**4.4**
	8.3		
Squaw Peak Schist	2		**2**
	100.0		**4.4**
	8.3		
Unidentified Arkosic Sand	11	7	**18**
	61.1	38.9	**40.0**
	45.8	33.3	
Indeterminate	2	1	**3**
	66.7	33.3	**6.7**
	8.3	4.8	
Total	**24**	**21**	**45**
	***53.3**	**46.7**	**100.0**

*Row %
**Column %

Surface Finish

Interior and exterior surface finish attributes of plainware and redware vessels from Pueblo Salado are shown in Table 11.10. The lack of interior finish on jars, regardless of ware, was expected, because jars have a significant zone of restriction at or near the orifice. In contrast, the interior surfaces of nearly 78% of the plainware bowls were either polished or floated; exterior surfaces were less often finished. Interior surfaces of redware bowls tended to be polished (and smudged) but only seldom slipped. Exterior surfaces of redware bowls were slipped but not polished; the majority had a matte finish, which may be temporally diagnostic (i.e., late). The analyst also examined the relationships between ware and smudging, patterned fire clouds, and patterned polishing.

Table 11.10. Surface Finish of Pueblo Salado Ceramic Assemblage

Surface Finish	Plainware						Redware					
	Interior			Exterior			Interior			Exterior		
	Bowl	*Jar*	*Total*	*Bowl*	*Jar*	*Total*	*Bowl*	*Jar*	*Total*	*Bowl*	*Jar*	*Total*
None	2 *11.8 **22.2	15 88.2 100.0	17 70.8	3 33.3 30.0	6 66.7 42.9	9 37.5	1 33.3 5.3	2 66.7 100.0	3 14.3	1 50.0 5.6	1 50.0 50.0	2 **10.0
Nonlustrous with visible striations	2 100.0 22.2		2 8.3	1 50.0 10.0	1 50.0 7.1	2 8.3	13 100.0 68.4		13 61.9			
Floated (unslipped surface)	5 100.0 55.6		5 20.8	6 46.2 60.0	7 53.8 50.0	13 54.2						
Slipped; lustrous with visible striations							2 100.0 10.5		2 9.5	6 100.0 33.3		6 30.0
Matte (slipped surface)							3 100.0 15.8		3 14.3	11 91.7 61.1	1 8.3 50.0	12 60.0
Total	9 *37.5	15 62.5	24 100.0	10 41.7	14 58.3	24 100.0	19 90.5	2 9.5	21 100.0	18 90.0	2 10.0	20 100.0

*Row %
**Column %

Smudging

Smudging, the deliberate blackening or carbonization of a vessel's surface (Shepard 1980), correlated closely with redware vessels (.05 confidence level). Smudging was present on 14 (66.7%) of the 21 redware vessels but only 1 (4.2%) of the 24 plainware vessels. This association supported the results of previous studies in the region (Crown 1984a; Walsh-Anduze 1994). Additionally, because the majority of redware was produced south of the Salt River, smudging is likely to be more strongly associated with southern production areas.

Fire Clouds

Fire clouds are defined as localized deposits of carbon that occur on a vessel's surface when fuel or another object makes contact with the vessel during firing (Shepard 1980:92). Schroeder (1940) noted that patterned fire clouds occurred in association with Hohokam pottery identified as Salt Red, and Haury (1945) confirmed the relationship in his analysis of the Los Muertos assemblage. However, although the association between redware and patterned fire clouds is real, another assessment of the Los Muertos assemblage (see Crown 1984a) found that patterned fire clouds occurred in only 10% of the redware vessels. Similarly, a study of vessels from Pueblo Grande also showed a significant association between redware and so-called patterned fire clouds, although, again, fire clouds occurred on a small number of vessels. Consequently, additional analyses of the Pueblo Grande assemblage were conducted to test the proposition that redware with patterned fire clouds was valued more highly than other redware. An examination of use wear and context found no significant results, which led to the conclusion that patterned fire clouds occurred unintentionally during the firing process, probably when relatively few vessels were fired together (Walsh-Anduze 1994). If patterned fire clouds were, indeed, an attribute of Salt Red pottery, then a more significant relationship should appear.

At Pueblo Salado, the relationship between patterned fire clouds and redware was not significant (.05 confidence level). The majority of plainware (23, or 96%) and redware vessels (17, or 81%) exhibited fire clouds, but patterned fire clouds were present on only 5 vessels (2 plainware, 3 redware), 11% of the jars and bowls. Although probably influenced by sample size, the data tended to support the idea that patterned fire clouds were unintentional results of the firing process.

Patterned Polishing

Vessels exhibiting patterned polishing were first recorded in the Los Muertos collection by Haury (1945), who found their presence strongly associated with Classic period redware, especially Gila Red ceramics. Subsequent analyses (Crown 1984a; Lane 1989) have confirmed this relationship and identified other correlations between specific patterns and vessel forms. The Pueblo Grande study also showed that patterned polishing correlated with plainware tempered with micaceous schist during the Soho phase of the Classic period (Walsh-Anduze 1994).

At Pueblo Salado, however, patterned polishing rarely occurred, regardless of ceramic ware. Only 2 (8.3%) plainware vessels and 1 (4.8%) redware vessel exhibited patterned polishing. These figures may indicate that the assemblage represented manufacture or use after the Soho phase or possibly that Pueblo Salado was not part of the network (i.e., did not participate in Canal System 2) that received these vessels. Sample size is another consideration.

How the Attribute Study Relates to the Ceramic Typology

The initial study of ceramics from Dutch Canal Ruin (Volume 2:Chapter 12) and Pueblo Salado (Volume 3:Chapter 5) involved the classification of ceramics by ware and type names based on the examination of several attributes, including temper and surface finish treatments, among others. Among plainware and redware ceramics, temper was the most important criterion distinguishing between types. In the current study, examination of temper alone (Table 11.11) demonstrated that more than one material may be correlated with a traditional (or recently defined) ceramic type. A ceramic type approach, therefore, would mask important spatial differences in production as recently defined.

Of course, analysts should probably develop a ceramic typology based on the covariance of two or more attributes, rather than on a single variable. The results of this study (and others) have demonstrated that surface finish attributes such as smudging are strongly associated with specific tempers derived from source areas south of the Salt River. Although not demonstrated in this analysis, a similar correlation may exist between patterned polishing and southern production areas. Consequently, if archaeologists utilize the traditional typology, they cannot examine the production and distribution of these ceramics in as much detail, since the traditional typology could mask important differences in Hohokam trade and exchange networks and social relationships. However, the development of a behaviorally meaningful typology, incorporating this data, was beyond the scope of this research. An attempt to create or reform the Hohokam typology should probably wait until additional studies are conducted and critically evaluated. In the meantime, the traditional type classification system remains a valid approach for addressing other, perhaps more broadly defined questions related to Hohokam occupation in the Phoenix Basin.

Results and Conclusions

The attribute relationships identified above for the Pueblo Salado whole vessel assemblage appear to have implications for trade and exchange within the Phoenix Basin. Undoubtedly, they are also influenced by social behaviors, especially those related to Hohokam mortuary customs (Brunson 1989). Although excavators recovered the assemblage from various contexts, redware ceramics were more prevalent among burials at Pueblo Salado, as apparent in the ratio of redware to plainware compared between mortuary (9:1) and trash (1:1) contexts (derived from sherd data). Similar results from the Los Muertos (Brunson 1989) and Pueblo Grande (Walsh-Anduze 1994) assemblages partially supported Abbott's (1984) hypothesis that the Hohokam had valued redware ceramics more highly than plainware.

Additionally, the high frequency of redware bowls in mortuary contexts at Pueblo Salado suggested that they may have been selected over other vessel forms, although this cannot be demonstrated due to the absence of a comparable data set from trash deposits. Interestingly, however, the Hohokam mortuary ritual at Pueblo Grande apparently favored wide-mouthed redware jars. If the Pueblo Saladoans favored unrestricted bowls, then the differences in vessel form may help analysts distinguish between social groups and further reconstruct exchange relationships within the Phoenix Basin.

Although the temporal factor was not controlled in this study, other than to note that the vessels were associated with the Classic period of Hohokam occupation, some patterns suggested that the majority of the assemblage was associated with the Civano and possibly the Polvorón phases. First, the only decorated vessels were Gila and Tonto polychromes, which had been manufactured primarily during the Civano phase. Second, only one redware vessel was tempered with micaceous

Table 11.11. Relating Temper Type with the Traditional Typology, Hohokam Plainware and Redware

Traditional Type	**Temper Type**								
	South Mountain Granodiorite	Camelback Granite	Arkosic Sand	Estrella Gneiss	Mixed Sand	Micaceous Schist	Phyllite	Squaw Peak Schist	Other/ Indeterminat e
Gila Plain, Gila						X			
Gila Plain, Salt	X	X	X	X	X				?
Wingfield Plain							X		
Squaw Peak Plain								X	
Other Hohokam Plain									X
Gila Red						X			
Salt Red	X	X	X	X	X				?
Wingfield Red							X		
Squaw Peak Red								X	
Phoenix Red									X

schist. This was the equivalent of Gila Red, as defined by Schroeder (1940), which had primarily been produced during the Soho phase. The absence of Gila Red vessels also correlated with the lack of vessels with patterned polishing.

Finally, the distribution of redware vessels from southern production sources to villages located north of the Salt River, including both Pueblo Salado and Pueblo Grande, does not contradict the idea that Pueblo Salado had a unique position within the Phoenix Basin. It does, however, support the theory that in addition to the high social value of redware compared to plainware (Abbott 1984), in the burial ritual the Hohokam in the Phoenix Basin favored redware pottery from southern production areas over its northern counterparts (Walsh-Anduze 1994). Moreover, the relationship of Pueblo Salado to Canal System 2 and its apparent relationship with sites south of the Salt River along Canal System 1 emphasize the regional cohesiveness of Hohokam mortuary customs, although groups along different canal systems may have favored one vessel type over another for burial offerings.

CHAPTER 12

INTRAREGIONAL COMPARISONS OF FLAKED STONE AND GROUND STONE ASSEMBLAGES FROM TEMPORARILY AND PERMANENTLY OCCUPIED SITES

Dawn M. Greenwald

Hohokam flaked and ground stone studies usually focus on the description of selected characteristics that the project analyst deems adequate to categorize the technological and functional attributes of the assemblage. Descriptions vary due to the background of different analysts as well as changes through time in the state of the art. For example, early Hohokam studies did not record the amount and type of flaked stone debitage recovered (e.g., DiPeso 1951; Haury 1976; Weaver 1977), emphasizing instead the type and variety of tools in the assemblage. The attention that archaeologists later paid to debitage grew out of important advances in the interpretation of flake attributes that result from flake production (Magne 1981; Phagan 1980; Pokotylo 1978; Sheets 1978). Traces left on a flake from the force, angle, and location of a blow by a percussor to a core and the size and preparation of the platform prior to the application of force can supply direct information about the kind of percussor that was used, the nature of the force that was applied, the general abilities of the knapper, and the stage of core and flake reduction that was taking place. All of these pieces of information taken together from a collection of debitage are diagnostic attributes that contribute to an understanding of prehistoric technological behavior.

Although investigators now record debitage counts in analytical reports, they tend to gather data on flake types and attributes sporadically and differentially. The most likely reason for this inconsistency is the seemingly predictable nature of Hohokam flaked stone reduction and the lack of variation through time or among activity areas or sites with structures (Huckell 1988:175, 199; Rozen 1984:426–427). According to most lithic studies, Hohokam flaked stone technology can be summarized as an expedient technology of core reduction using locally available materials. However, some analysts have detected variability in Hohokam assemblages, usually as a matter of degree rather than as discrete differences in kind or number. They most often note variation between site categories, particularly between habitation and other site types such as field houses and resource procurement or processing areas (Bernard-Shaw 1983; Greenwald 1988a; Hoffman and Doyel 1985; Swidler 1989). Other recorded differences include material type preference or use based on local variation in resources (Allen and Cable 1984; Bernard-Shaw 1983; Greenwald 1988a; Hoffman and Doyel 1985). Analysts for the Phoenix Sky Harbor Center Project noted such differences in degree in the assemblages from Dutch Canal Ruin and Pueblo Salado (Volume 2:Chapter 13; Volume 3:Chapter 6).

In the Dutch Canal Ruin and Pueblo Salado flaked and ground stone assemblages, the analysts examined intrasite distributions by area, by feature type, and by temporal groupings. However, to understand how these sites compared to sites of similar type and location within the regional system, a more thorough evaluation was needed. Part of the goal of this study was to compile data from other sites and projects to produce a data base for technological, site type, and intraregional comparisons. This approach has proved successful in studies where data are comparable, such as the Murphy's Addition site (Allen and Cable 1984), the Tucson Aqueduct Project, Phase A (Greenwald 1988b), the Baccharis site (Greenwald 1988c), and New River (Hoffman and Doyel 1985), and in a study comparing data from several projects (Greenwald 1990). Patterns in the data reflected the relative frequency and nature of activities at various types of sites and differences in the emphasis on food production compared to other types of economic activity between core and peripheral sites.

The Phoenix Sky Harbor Center assemblages were then compared with general regional patterns for a better understanding of their place within the Hohokam cultural system.

METHODS

Collecting and combining data that have been assembled by various analysts is a difficult task. Analysts use not only different nomenclature but also typologies, manner of recording data, and judgments of what is deemed worthwhile to include as basic descriptive information from a project or site area. Other factors—such as the sampling techniques that investigators used during field excavations and for analysis, as well as differing research objectives—contribute to variations between the work of different researchers such as those encountered during data collection for this study. To use these data for intraregional comparisons in the current project, the lithic analyst first had to sift through the mounds of artifact descriptions, definitions, and tables that were available and then integrate the relevant information.

The first stage in the process of data collection was to target particular projects for inclusion in this study and compile an adequate sample of field house, farmstead, hamlet, and village sites, including a cross section of sites from the core and peripheral areas of the Hohokam region. Since most projects were assigned to the Colonial or Sedentary periods, adequate refinement of site types and intraregional groupings by temporal designation was not possible. Thus, after searching the archaeological literature, the project analyst designated each project by site function and location. Most of the data was available in published project and site reports. When information that seemed too valuable to omit was not in these sources, or when raw data were not included in the published reports, the analyst contacted the repository that housed the project information. In one such case, after compiling information about each flaked and ground stone category by tallying statistics from reams of computer paper printouts, the SWCA analyst discovered that the published counts for individual artifact classes did not coincide with the printout totals. The data from that study were thus not considered reliable for comparative purposes.

The next stage of the process involved the actual recording of flaked and ground stone attributes. The analyst for this study evaluated definitions of each artifact type to determine how best to fit the numerous classifications into one system. Most categories, particularly the ground stone types, could be placed into a category that was comparable to that of other analysts, often by lumping categories together. Classes such as blades, pressure flakes, and bifacial thinning flakes had to be condensed into a general debitage category, since most analysts do not use these separate morphological types. In numerous cases, analysts had recorded ground stone artifacts as multiple types or had combined them into a single category, such as *abrader/anvil* or *mortar/pestle*; for this study these artifacts were designated as one type or the other. Recording of material types appeared to be even less consistent than that of artifact types, so this study did not attempt to evaluate this variable.

Some early reports presented data only for flaked stone implements and not for debitage. Other reports presented contradictory information on the same stone attributes in different sections of the report, so that raw counts were unreliable. Some data were presented differently within the same report, so that comparable raw counts were impossible. Some of the reports described unflaked cobble hammerstones in the ground stone analysis section although analysts most commonly include this artifact class with flaked stone, as was done in this analysis.

After recording of artifact types, the next step of the analysis was to combine some categories, for example, merging noncomparable artifact classes, such as wedges or spokeshaves, into general categories described as *other flaked stone tools* or *retouched tools*. The intent was not to reinterpret categories but to place them into the most appropriate class based on the original analyst's artifact descriptions.

Finally, grouping sites by similarities of function and regional location and characterization of these groups according to artifact counts, ratios, diversity measures, and presence/absence data allowed the project analyst to make comparisons among groups. Appendix A contains group data by site or project, which is summarized within each appropriate section of this report.

In spite of differences in sampling method and extent, each sample should be representative enough of each site to provide information on function, structure, and other research topics. State, tribal, and federal agencies review the adequacy of archaeological data recovery techniques, procedures, and design to ensure that information contained in sites eligible for the National Register of Historic Places is properly retrieved and available for future studies. Variability caused by the idiosyncrasies of any individual project cannot be measured or accounted for; therefore, this study is based on the assumption that data from each site is a representative sample of the range of variation within that universe. The relative frequency (the ratio of the number of times a particular variable occurs to the total number of observations) of an artifact attribute within a site describes how that variable compares to other similar attributes. The distribution of these frequencies enables the archaeologist to visualize patterns or trends within the target population (Thomas 1976:43). When the same attributes, or variables, are examined from similar sites with representative samples, frequency distributions should be comparable.

FUNCTIONAL COMPARISONS

Functional assessments in past lithic studies have incorporated both assemblage comparison and evaluations based on use-wear analyses. Use-wear studies of flaked stone assemblages have utilized many different attributes, including edge angle, wear pattern, edge shape, and a combination of these, to determine patterns of use on tools. Most analyses have evaluated wear based on a low-power approach and have concentrated on describing the functional variability within artifact categories. To date, formal use-wear analysis of ground stone has been confined to tabular tools, such as tabular knives and hoes. Researchers have mentioned wear patterns, when visible, in artifact descriptions or definitions. For example, a trough metate may have longitudinal, bidirectional striae indicating grinding motion that are incorporated as part of the definition of the metate type. In addition, traditional categories of ground stone artifacts in the Southwest are laden with functional implications.

Assemblage comparison examines artifact types from different site categories or intrasite feature categories. Researchers have taken either a descriptive approach (Elstien 1989:731–735; Hoffman 1985:591) or an exploratory approach that seeks to explain variability through type differences (Bernard-Shaw 1983:415–425; Hoffman and Doyel 1985:639–644; Howard 1989:761–765; Rozen 1984:582–602; Swidler 1989; Volume 2:Chapter 13; Volume 3:Chapter 6). Others have tested expectations based on site type or activity area definitions (Greenwald 1988b:200–206, 1988c:264–275). Because of these efforts, we now have a corpus of data to draw on that can provide the beginnings of a model of Hohokam lithic profiles for specific site types, particularly field house, farmstead, hamlet, and village. What follows is the introduction of such a model and an evaluation of how well the Pueblo Salado and Dutch Canal Ruin assemblages adhered to it.

Field House Assemblages

Archaeologists have fairly consistently used the designation *field house* to identify small structures that were located away from permanent habitations (Crown 1985:79) and were occupied seasonally for agricultural and other food procurement activities (Doyel 1985:682). They served as protection from the weather and as habitations during times of limited occupation. Periods in which they were probably occupied include the agricultural growing season, particularly harvest time, and seasonal excursions to procure wild food (Chapter 4).

Investigators find few lithic artifacts at these sites, due to the restricted period of occupation and the limited nature of the activities that residents performed. Even though field houses may represent structures that were revisited every year, perhaps several times annually, due to long periods of disuse occupational debris would have accumulated in limited amounts.

Because the Hohokam did not use field houses for the full range of year-round activities that they conducted at residence locations, the diversity of lithic artifacts at these sites will also be small (Cable and Doyel 1984b:259; Greenwald 1988a:266). Artifacts that should be part of the assemblage include those needed for short-term residential subsistence, such as food preparation articles and maintenance tools. Nonutilitarian artifacts would not be essential, nor would it be efficient to transport them to the site, unless they were related to ritual enhancement of the subsistence-related functions of the site. For example, the Hohokam might have considered artifacts related to the ceremonial spreading of corn meal on agricultural crops to be essential to ensuring a successful harvest.

The variety of material types should also be limited. Short-term occupation would have restricted the number of resources necessary for subsistence needs. However, the more often residents used a field house, the greater the variety of resources that should have been introduced. Availability of resources would also be a factor.

According to Crown (1985:84), field house artifacts may have been curated after use. Used in this context, the term *curation* would imply that the artifacts were manufactured and maintained in anticipation of future use and possibly recycled for other tasks after they had outlived their primary function. This is a logical extension of the assumption that field houses were far from residential locations, had a short-term function, and were used repeatedly. In anticipation of curation behavior, analysts might expect multipurpose tools; evidence of tool maintenance, reuse, and recycling; and the presence of technologically formal tools (Binford 1973, 1977, 1979). According to models developed for mobile hunting and gathering groups, which assume that these peoples would have anticipated not being able to find appropriate resources during trips, the biface would have been used both as a multipurpose tool and as a core, increasing the flexibility and decreasing the weight of the mobile assemblage. However, because the Hohokam technological strategy included the use of mediocre-quality lithic material, especially in the production of flakes, the necessity of anticipating a lack of adequate resources was probably minimal. In addition, resource availability at some field house locations was fairly high, as they were situated near convenient washes or streambeds and, in some cases, primary outcrops. Therefore, as Bamforth (1986:40) pointed out, why carry tools when raw material is available locally? It would be more efficient to produce flakes as needed from a core, following the traditional technological strategy of core reduction, than to resharpen or maintain a formal tool, as long as the material was easily acquired. According to Bamforth (1986:40), "tools made of locally available materials have shorter use lives, are maintained less frequently for a smaller time investment, and require a smaller time investment overall for procurement and subsequent maintenance than do tools made of non-local materials." The probability of finding evidence for curation behavior with the artifacts from field houses, therefore, is unlikely, given the availability of material resources in most locations. However, the assemblage should contain a prevalence of general-purpose tools that would reflect the short-term nature of the site. Multiple-purpose tools would be more efficient and less costly to produce or import than a full tool ensemble when occupation was only going to be short-term. In keeping with the scenario described above, general-purpose tools would include unretouched flake tools, as well as handstones and grinding slabs. Artifacts that are specific to the function of the site, such as agricultural production and food procurement items, may be present if the users did not return such tools to the residential site. Specialized tools may have been too expensive, in terms of labor and materials invested, to be left behind at periodic and distant subsistence activity stations.

Eight archaeological projects, including the Phoenix Sky Harbor Center Project, produced field house data that could be evaluated and compared to the model described above. Sites included Pueblo Patricio, represented by investigations from Block 28-North (Allen and Cable 1983), and Murphy's Addition (Allen and Cable 1984), part of the Central Phoenix Redevelopment Project; La Cuenca del Sedimento (Halbirt 1989a; Swidler 1989); Sites AZ T:4:13(ASM) and AZ T:4:22(ASM) from the Central New River Drainage Project (Hoffman 1985; Hoffman and Doyel 1985); and those from the Salt-Gila Aqueduct Project (Bernard-Shaw 1983), the Waddell Project (Elstien 1989; Howard 1989), and the Tucson Aqueduct Project (TAP), Phase A, Brady Wash sites (Ciolek-Torrello, Callahan, and Greenwald 1988). Flaked and ground stone data profiles and artifact totals, including ceramics, for all projects are available in Appendix A. Table 12.1 presents concise data, such as the means and ranges of artifact samples from field house sites at these projects, including Dutch Canal Ruin, as well as the same kinds of data from farmstead, hamlet, and village sites. Total artifacts varied greatly, mostly due to the variation in site type representation within various projects. However, field houses did display a smaller amount of flaked and ground stone than did other site types. The mean for total flaked stone (616) was approximately one-fourth of the next lowest total mean (2681), and the mean for total ground stone (28) was approximately one-sixth of the next lowest total mean (196).

This study measured artifact diversity in two ways. The first was by the number of artifact categories represented in the flaked and ground stone profiles. This approach is a crude measure that is influenced by the total sample and the evenness of the distribution. For example, although all artifact categories may be represented, one of these categories may be heavily weighted compared to the others and be distributed unevenly. Thus, out of a sample of 100, one category may contain 91, with only 1 item in each of the other nine artifact classes. To compensate for this type of deviation, in the current study a diversity index measured variability among site types in ground stone assemblages. A diversity index is a quantitative method based on information theory that ecologists and biologists often employ to measure the distribution of observations among categories. The index ranges between 0 and 1, with higher values indicating greater tool diversity. The Brillouin diversity formula, used for this study, is most appropriate for nonrandom samples (Pielou 1975:10; Zar 1984:34). It is calculated from the following equation (Zar 1984:34):

$$H = \frac{(\log n! - \sum \log f_i!)}{n}$$

where n = the sample size and f_i = the number of observations in type i. Flaked stone assemblages were not compared among site types by means of this index because of their usually large sample sizes, with totals exceeding 1000. The logs of factorials greater than a sample of 1000 were not available for this study. In addition, ground stone assemblages were more consistently divided into more numerous categories than flaked stone and therefore yielded a more detailed picture of their internal variability. Diversity indexes for field houses were the smallest of all site types (Table

Table 12.1. Site Type Comparisons by Flaked and Ground Stone Morphological Types and Assemblage Characteristics

Morphological Type/Assemblage Characteristic	Field House[1]		Farmstead[2]		Hamlet[1]		Village[3]	
	Mean	Range	Mean	Range	Mean	Range	Mean	Range
Debitage	81.1	68.2 - 93.7	83.6	71.8 - 91.9	79.8	70.1 - 90.1	86.4	78.7 - 94.0
Used Flake	5.8	0.1 - 17.9	2.8	0.4 - 7.2	7.3	1.1 - 14.2	3.6	0.1 - 8.6
Core	4.4	0.0 - 14.0	4.5	0.4 - 15.8	4.1	1.2 - 13.6	2.6	1.3 - 4.2
Core Tool	2.6	0.0 - 12.7	1.2	0.0 - 6.7	0.7	tr - 1.9	1.3	0.0 - 7.3
Core/Hammerstone	0.7	0.0 - 2.2	0.9	0.0 - 2.6	1.3	0.0 - 4.9	1.4	0.0 - 5.8
Hammerstone	2.9	0.0 - 9.4	4.0	0.0 - 18.6	2.3	1.1 - 4.2	0.9	tr - 4.2
Retouched Tool	2.4	0.0 - 8.3	3.1	0.0 - 6.9	3.9	0.1 - 7.1	3.8	0.2 - 8.3
Unretouched to Retouched Tool Ratio	1.8	0.0 - 10.0	0.8	0.0 - 1.6	3.2	0.6 - 8.9	1.1	0.1 - 2.7
Debitage to Core Ratio	30.3	0.0 - 133.0	23.3	3.9 - 77.0	16.4	4.2 - 28.8	24.0	7.5 - 59.6
Ground Stone Diversity Index	0.4	0.2 - 0.7	0.6	0.4 - 0.8	0.7	0.5 - 0.9	0.7	0.5 - 0.9
Flaked Stone Artifact Total	616.1	36 - 2045	4158.8	112 - 15,869	2681.4	1700 - 4771	6207.4	2837 - 10,960
Flaked Stone Category Total	6.8	4 - 9	7.6	5 - 10	8.6	7 - 10	9.1	6 - 11
Ground Stone Artifact Total	28.4	6 - 53	196.0	35 - 864	352.0	82 - 886	710.0	165 - 1754
Ground Stone Category Total	4.8	2 - 8	8.5	6 - 16	11.5	6 - 16	12.7	11 - 16

[1]Both flaked and ground stone sample size = 8
[2]Both flaked and ground stone sample size = 10
[3]Flaked stone sample size = 8, ground stone sample size = 10
tr = trace

12.1), although ranges overlapped. The smallest indexes were from the New River (0.17) and La Cuenca del Sedimento (0.20) field houses. Field house sites with the most diversity were from Murphy's Addition (0.67) and Dutch Canal Ruin (0.57).

The ground stone assemblage also provided data on the general frequency of activities, represented by artifact groupings (Figure 12.1). The statistic indicated for each artifact group in Figure 12.1 (food-processing activities, nonutilitarian functions, etc.) was calculated as the relative frequency of particular artifacts within the total sample of field house sites. Manos, metates, mortars, pestles, and tabular tools represented food-processing activities. Artifacts representing nonutilitarian functions were palettes, stone bowls, medicine stones or plummets, ornaments, and minerals. Some of these objects—such as plummets, stone bowls, and medicine stones (Sayles 1965:102), objects of vesicular material shaped like molar teeth—may have utilitarian functions but generally are very restricted in distribution, are specific in design, and require a large investment of labor for their manufacture, attributes not often associated with objects used for everyday household functions. Special-function tools are those that are generally useful for limited types of tasks. Although they may be reused in other capacities, their morphology and use-wear patterns are similar across the Hohokam region and beyond, denoting a similar, primary purpose. Axes and mauls, polishing stones, abraders, and perforated discs, or spindle whorls, were designated as special-function tools for this study. General-purpose tools, on the other hand, are artifacts whose morphology is not standardized and that may be useful for a variety of functions. Handstones and grinding slabs are in this category. Other ground stone includes all artifacts that were not designated by any of the above categories or whose function is unknown. Examples include raw material, indeterminate ground stone, and stone rings.

The majority of ground stone from field houses (65.2%) (Figure 12.1) was food-processing equipment, consisting mostly of manos and metates. Special-function tools were a large portion of the assemblage (11.8%). This category, however, was dominated by abraders, which formed a significant proportion of the assemblages from the Block 24-North (28.6%), Murphy's Addition (32.1%), and La Cuenca del Sedimento (33.3%) field houses. Abraders were 10.1% of all field house ground stone artifacts, indicating that tool maintenance had been a practical function at these site types. Nonutilitarian functions were not well represented at field houses (1.3%). Minerals, the only artifact type from this category, were present at Block 24-North and Murphy's Addition. General-purpose tools were most common at field houses, reflecting the short-term nature of that site type. In the flaked stone assemblages (Table 12.1), used flakes, another general-purpose artifact type, were more prevalent at field houses than were retouched tools, which are usually more function-specific in nature.

The field houses at Dutch Canal Ruin fit within the general range of variability of field house sites. They had a higher amount of debitage than the others but had a comparable debitage-to-core ratio. A high percentage of ground stone was indeterminate, but the remainder of the collection followed the general field house pattern.

Farmstead Assemblages

Farmsteads are seasonal residential settlements associated with agriculture and related activities (Gregory 1991:163). They may consist of multiple, contemporaneous field house structures (Volume 1:Chapter 1) and are usually occupied by a single social group (Gregory 1991:163). They differ from field houses in that they have a larger number of extramural features, may contain more structures, and were used as short-term residences for a social group.

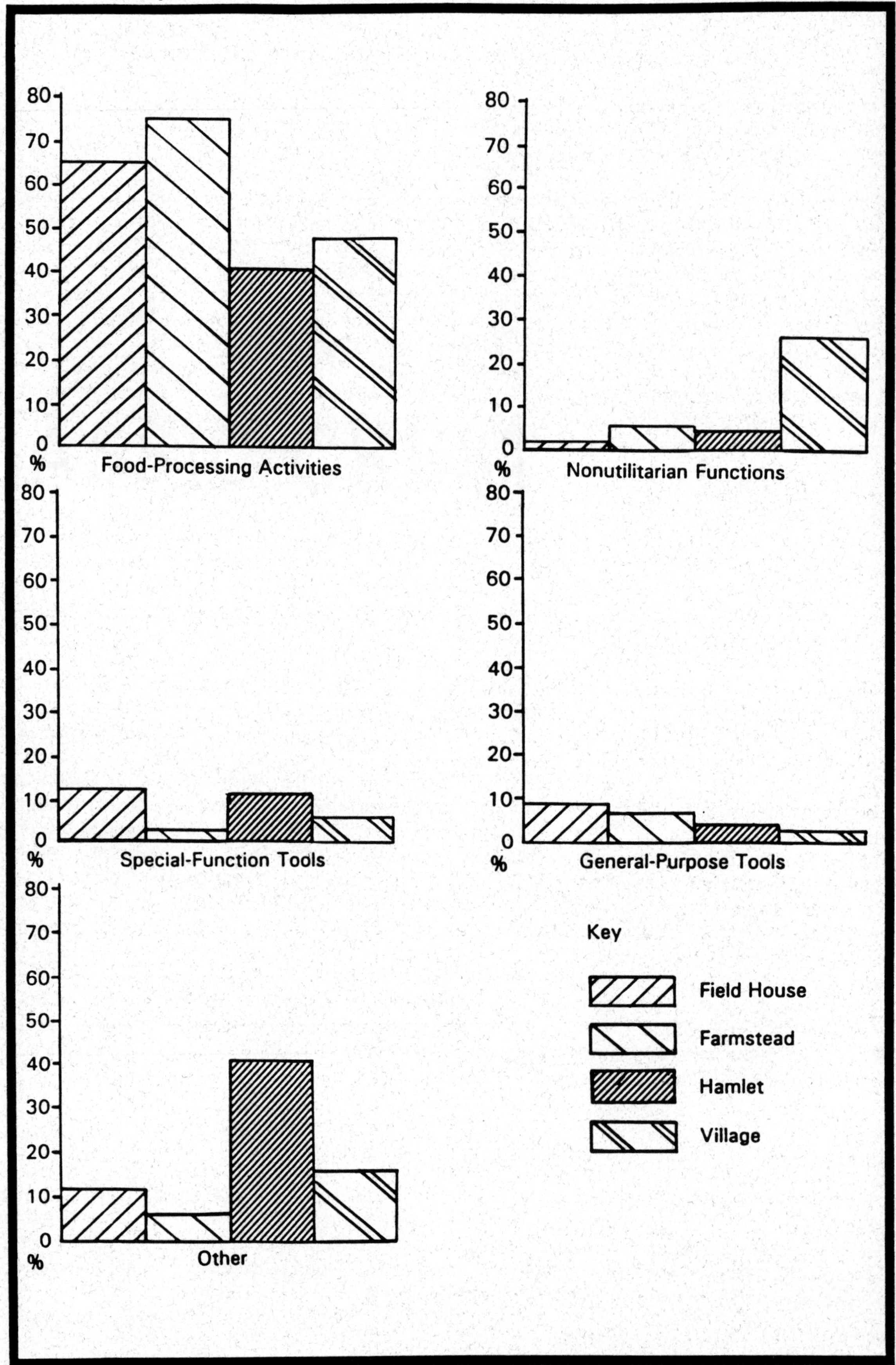

Figure 12.1. Relative frequencies of ground stone categories by site type within the Hohokam region.

Since farmsteads, like field houses, were not occupied year-round, they do not reflect the full range of activities associated with a year-round permanent residence. Farmstead assemblages will not be represented by the number and variety of tools that are found at hamlets or villages (Doyel 1985:686), but they will contain more and a larger diversity of tools than field houses because of the greater number of activities expected in a residential situation and because of the larger social groups occupying them. The larger social group implies greater social interaction, the probability of communal activities, and the possibility of some on-site activities based on a division of labor. Farmstead social units may have split up tasks, particularly gender-based activities, to support traditional residential social organization during the seasonal stay. Communal activities also would have contributed to the continuation of group identity and cohesiveness. Evidence of group activities that may appear in the archaeological record includes ritual paraphernalia.

Projects that contained farmstead data included the Central New River Drainage (Hoffman 1985; Hoffman and Doyel 1985); TAP, Phase A (Ciolek-Torrello, Callahan, and Greenwald 1988; Greenwald 1988a, 1988b) and Phase B, represented by the Hawk's Nest and Fastimes sites (Halbirt 1988, 1989a; Huckell 1988, 1989a); the Salt-Gila Aqueduct (Bernard-Shaw 1983); Adobe Dam (Bruder and Gasser 1983); La Cuenca del Sedimento (Halbirt 1989a; Swidler 1989); ANAMAX-Rosemont (Ferg 1984a; Rozen 1984; Tagg 1984); the Waddell Project (Elstien 1989; Howard 1989); and the Phoenix Sky Harbor Center Project (Volume 2). In some projects, only particular sites or site components represented farmsteads, such as Sites AZ T:8:16(ASM) and AZ T:4:12(ASM) from the Central New River Drainage; other projects, such as ANAMAX-Rosemont, were composed almost entirely of farmsteads (Ferg 1984b:821). This disparity among projects in the amount of data from farmsteads accounted for much of the variability in flaked and ground stone artifact totals presented in Appendix A and the high mean of total artifacts for farmstead sites listed in Table 12.1. In fact, the 4159 figure for total flaked stone was larger than the total mean for hamlets. When large projects that were weighted heavily by farmsteads, such as ANAMAX-Rosemont and Waddell, are removed from the sample, then the figure drops dramatically, to 1442, comparable to the farmstead pattern described above. Although the mean of total ground stone artifacts originally fell within the expected range for farmsteads, when ANAMAX-Rosemont and Waddell are dropped out the resulting figure of 72 suggests a closer affiliation to short-term field house sites than to year-round hamlets or villages.

An increase in the variety of activities from field houses to farmsteads was indicated by the average number of flaked stone (7.6) and ground stone (8.5) artifact categories, as well as by the Brillouin diversity index (Table 12.1). Although the Waddell sites generated a large amount of ground stone artifacts, the variety of types that were recorded was low, and the diversity index was only 0.45. The lowest index, 0.39, was from the Central New River Drainage sites, and the highest (0.75) was represented by the ANAMAX-Rosemont farmsteads.

The ratio of unretouched to retouched tools (Table 12.1) was low for farmsteads (0.8), and the amount of unused debitage to cores (23.3) decreased from the field house figure (30.3). Although field houses contained a substantial number of cores and the largest mean of core tools of all site types, farmsteads contained the largest mean for hammerstones (4.0). Hammerstones are associated with the production of flaked stone tools (Huckell 1986:15) and with food processing (Dodd 1979). Perhaps the occupation of farmsteads during the food production and procurement season involved intensive plant processing. The ground stone assemblages (Figure 12.1) supported this interpretation. Food-processing equipment was present in greater frequency (76.0%) at farmsteads than at any other site type. Other farmstead ground stone artifact categories were similar to each other in frequency.

The Dutch Canal Ruin farmsteads followed typical farmstead patterns. Food-processing equipment represented the majority of ground stone artifact categories, and hammerstones were a major component (8.9%) of the flaked stone assemblage. The total number of flaked and ground stone artifacts was among the lowest in this site class.

Hamlets and Villages

Sites that represent continuous, year-round, residential occupation are either hamlets or villages. Hamlets are small settlements composed of two to eight contemporaneous households (Volume 1:Chapter 1). Villages are aggregated settlements ranging from groups of six to eight households to large settlements of several hundred individuals. The village system may incorporate communal architecture, such as ballcourts or platform mounds (Gregory 1991:162; Masse 1991:199). The development of hamlets into large villages involved an increase in socioeconomic complexity, indicated in some cases by the specialization of production, a well-defined intravillage structure, and involvement in a regional social and economic system. The difference between large hamlets and small villages may not be obvious, however, so the distinction in stone assemblages may reflect merely a difference in degree or no difference at all. For this reason, the discussion on stone assemblages below will focus first on attributes typical of permanent habitation sites in general and then on the differences that analysts of stone tool assemblages can observe between hamlets and large villages.

Year-round occupation brought about the diversification of activities. Instead of a focus on seasonal subsistence practices, activities at permanent habitation sites represented all aspects of cultural organization, including social, political, economic, and religious practices. The nongrowing season would have afforded more time for nonsubsistence functions or for preparations for the next growing season, such as replenishing resources for stone tool production, tool manufacture and repair, ceremonies, production of nonsubsistence items such as ornaments, and leisure. Evidence in the archaeological record is the presence of both utilitarian and nonutilitarian items, with nonutilitarian artifacts, such as ritual paraphernalia, ornaments, and stone balls, representing a larger percentage of the assemblage at permanent habitation sites than at farmsteads. A concomitant decrease in the percentage of subsistence tools such as manos and metates also occurs at year-round habitations. This pattern can be translated to the flaked stone assemblage as an increase in special-function tools, such as projectile points, notches, and drills.

Storage facilities are more common at permanent habitations, which would have allowed residents to reserve food items, material resources, and implements until needed. Storage of resources at these sites means that more material types should be present than in locations that provide less variety of lithic sources.

As the population increased and became more aggregated into village-level sites, residents would have left more evidence for intersite relations, such as the presence of exotics, and a higher frequency of intrasite community activities, evidenced by ritual articles. Nonlocal material, such as obsidian and turquoise, will be higher in frequency at villages than at hamlets. Palettes, censers, and figurines will be more prevalent at villages as well. Concurrently, a larger percentage of village assemblages will be represented by nonutilitarian artifacts. The frequency of manos and metates, for example, will be lower at villages, with more of the assemblage made up of ornaments, stone balls, palettes, and other objects not related to subsistence.

Villages yield a greater diversity of artifact types and a greater number of artifacts than do hamlets. As site structure and intersite relations became more complex, residents would have conducted a greater variety of activities and special-function tasks as well. Division of labor may have diversified from gender-based tasks to individual specialization based on talent, status, or kinship ties. Researchers have documented specialization in artifact manufacture at some village sites, such as Snaketown, Las Colinas, and Cashion. The organization of production at these sites was differentiated, but the mechanism by which production and distribution were coordinated can only be hypothesized (Teague 1985; Wilcox and Shenk 1977:187–197).

For this study, lithics data were available on hamlets from the Salt-Gila Aqueduct (Bernard-Shaw 1983); TAP, Phase A (Ciolek-Torrello, Callahan, and Greenwald 1988; Greenwald 1988a, 1988b; Greenwald and Ciolek-Torrello 1987); El Caserío (Landis 1989a); Block 24-East (Allen 1985); La Lomita Pequeña (Mitchell 1988); and the Phoenix Sky Harbor Center (Volume 3). The statistics from the TAP, Phase A, hamlets were heavily weighted by data from NA18,003 (Locus I) and NA18,030, so these sites were treated separately from the rest of the assemblage (Appendix A). The Phoenix Sky Harbor Center sample was represented by Area 14 of Pueblo Salado. Village sites that were used for comparison included Water World from TAP, Phase B (Halbirt 1989c; Huckell 1989b); Eastwing (Rodgers 1987); Grand Canal Ruins (Landis 1989b; Mitchell 1989c); Cashion (Antieau and Greenwald 1981); AZ T:14:16(ASM) from the Central New River Drainage Project (Hoffman 1985; Hoffman and Doyel 1985); NA18,031 from TAP, Phase A (Heron and Ciolek-Torrello 1987); AZ EE:2:105(ASM) from the ANAMAX-Rosemont Project (Ferg 1984a; Rozen 1984; Tagg 1984); and the Civano phase features from Area 8/9 of Pueblo Salado.

Hamlet and village stone assemblages had the largest numbers of artifacts (Table 12.1), and although the mean of total flaked stone artifacts for hamlets was smaller than that for farmsteads, as discussed above, the large variation in assemblage size in the farmstead sample caused the farmstead total to be unusually elevated. If the outliers are taken out of the farmstead sample, the means fall satisfactorily within the predicted pattern.

Hamlets and villages exhibited greater artifact diversity, reflecting more diversification of activities, than field houses or farmsteads, and the diversity of stone assemblages between hamlets and villages followed expected patterns. Hamlets were represented by fewer flaked stone categories than villages. As occupational space had become more structured and production specialization more likely, activities would have become more diversified and the number of artifact types would increase.

Evidence of production occurs in the form of artifacts in various stages of manufacture, as large amounts of raw material, usually schist, and as an uneven distribution of artifact types. When ground stone tool manufacture was indicated in this study, usually only one artifact type or material type showed evidence of production. For example, the Grand Canal Ruins assemblage exhibited evidence for possible ornament manufacture (Mitchell 1989c:457), and Pueblo Salado, both hamlet and village components, contained evidence for artifact production based on elevated amounts of tabular metamorphic material (Volume 3:Chapter 7). El Caserío contained both unworked and worked pieces of schist (Landis 1989a:133), and La Lomita had many small fragments of schist (unworked?) and worked schist pieces that suggested the production of schist tools (Mitchell 1988:215). Evidence from Las Colinas, however, suggested that two different artifact types and material types had been processed for production, probably during different periods of time. Tabular schist material was present in large quantity during the Mound 8 excavations, along with perforated discs of the same material in various stages of production (Teague 1981:222–223, 228). In addition, the 1982–1984 excavations produced a large number of stone rings, some in various stages of manufacture, as well as others in complete and fragmentary form (Euler and Gregory 1988:308–309). Evidence for other types of production specialization was present at the TAP, Phase A, hamlet and village sites, whose ground stone profiles showed unusually large frequencies of polishing stones, used in pottery manufacture. Site NA18,031 exhibited an exceptionally large percentage of tabular tools as well, suggesting the intensive processing of a plant resource, such as agave, and thus the intensive production of that same resource.

The frequency of nonutilitarian ground stone was markedly different among site types (Figure 12.1). Villages had a high frequency (26.8%) of this artifact category compared to other sites (1.3–6.4). Ceremonial objects at hamlets and villages occurred at similar frequencies, except for palettes, which occurred at a frequency four times higher in villages than in hamlets. Ornaments accounted for the largest difference between hamlets and villages in the nonutilitarian category. These figures, however, are somewhat deceiving, as much of the high ornament frequencies can be attributed to beads that were elements of a composite ornament, although counted individually. For example, one cluster of 78 beads and another of 199 beads from Las Colinas probably represented a single ornament each (Euler and Gregory 1988:313). Beads often are recovered in quantity from burial contexts. The Las Colinas cluster of 78 beads came from a cremation, and 1143 of the 1147 beads from the Grand Canal Ruins were recovered from burials (Mitchell 1989c:Table 10.8).

The hamlet and village components of Pueblo Salado conformed to patterns noted for these site types in general. The Civano phase of Area 8/9, however, had the lowest frequency of cores of the village sites. In the ground stone assemblages, both hamlet and village components had a large frequency of tabular metamorphic material (Volume 3:Chapter 7), including schist and phyllite, probably used for tool production. In fact, flakes removed from one tabular specimen could be refitted, indicating that the material had been reduced on-site. Area 14 had a low frequency of tabular tools and no polishing stones, and Area 8/9 contained a low percentage of metates, with 2.5% of the assemblage composed of polishing stones. The presence of polishing stones in the village component of Pueblo Salado agreed with other evidence that ceramics had been locally produced (Chapter 11). Although polishing stones are indirect evidence of ceramic production, their distribution falls into distinctive patterns when viewed from a core/periphery perspective (see below).

SPATIAL COMPARISONS

The core-periphery model of the Hohokam system is based on notions of cultural dominance and geographic expansion. A variety of traits, such as red-on-buff pottery, ritual paraphernalia, complex irrigation systems, death ritual, and ballcourts, are more frequent in the core area and less frequent as distance from the core increases. Frequency is apparently linked to degree of involvement with the core system. This simplistic model lacks sophistication in describing the nature of core-periphery relationships, and although some archaeologists have attempted to analyze and interpret basic core cultural system structures, such as economic relations, the association between core and peripheral sites is still poorly understood. The tremendous increase in data from both core and peripheral areas in the past five years has afforded the opportunity to investigate the relationship between these spatially distinct units. By comparing and contrasting flaked and ground stone assemblages from core and peripheral sites, Hohokam scholars may be able to gain new insights into this regional structure; such a comparison in this study may reveal elements of the relationship of Dutch Canal Ruin and Pueblo Salado to the regional system.

Geographically, the core is "the triangular area along the Gila and Salt rivers from a little west of their juncture on the west to the Bartlett Dam on the lower Verde and Saguaro Dam on the Salt, and a little east of the South Butte area east of Florence" (Wilcox and Shenk 1977:183). Models of the structure of core-periphery relations have been posited focusing on different aspects of the cultural system: the technological, the economic, or the sociopolitical. Haury (1950) and Gasser (1980) have distinguished between riverine and desert Hohokam, emphasizing differences in farming techniques and general subsistence tactics. Other archaeologists (Teague and Crown 1980; Wilcox and Sternberg 1983) have analyzed regional social interactions using models of multiple peripheral zones. Wilcox and Sternberg (1983:219–222) delineated peripheral zones on the basis of access of major sites to ballcourt sites. The concept of peripheral zones developed by Teague and Crown (1980) emphasizes access to the Gila and Salt Basin. The first zone borders the basin via ephemeral streams, and the second zone is farther from the basin geographically. For the purposes of this study, the core area definition is based on Wilcox and Shenk's (1977) geographical structure, although other factors, including subsistence strategies and access to major waterways and the core area, are interrelated. Because the Salt-Gila Aqueduct Project included sites that may be argued to belong to both core and peripheral areas, they were not used in the following evaluations. Sites considered to be in the core area were from the following projects: Block 28-North, Murphy's Addition, La Cuenca del Sedimento, Dutch Canal Ruin, Pueblo Salado, El Caserío, Block 24-East, La Lomita Pequeña, Grand Canal Ruins, Cashion, and Las Colinas. Peripheral sites were from New River, Waddell, TAP Phase A, Hawk's Nest, Adobe Dam, Fastimes, ANAMAX-Rosemont, Water World, Baccharis, and Eastwing.

The approach taken with this study was exploratory. Due to the lack of agreement regarding the nature of the core-periphery structure, the SWCA analyst evaluated core and periphery groups for attributes in which they were similar and those in which they differed and also combined field house and farmstead flaked and ground stone assemblages and compared these separately from combined hamlet and village collections. This process allowed the inclusion of sites designated by various principal investigators as field house/farmstead (rather than either field house or farmstead) and hamlet/village (rather than hamlet or village), increasing the study sample for the two most comparable groups, one representing short-term, subsistence-related occupation and the other representing year-round residential sites. Tables 1-8 in Appendix A do not contain all of the site data used for the periphery and core comparisons because the volume of statistics was extremely large. When the mean statistic was used, it was calculated from sites or projects that yielded 25 or more artifacts, a consideration only when computing ground stone variables from the field house/farmstead group. In addition, the number of field house or farmstead sites that were located within the Hohokam core area (7) was only slightly over half the number that were located in the periphery (13), somewhat underrepresenting the core sample.

Field House and Farmstead Sites

Grouping field house/farmstead sites into core and peripheral areas revealed that a number of flaked stone types showed consistent differences between the groups (Table 12.2). Unifaces and bifaces both had higher means for periphery sites than core sites. Although site or project reports did not always consistently report these categories, the retouched tools category represents a consistently applied variable. It usually had to be calculated from the various data that were supplied in analytical reports, but it was also usually documented one way or another so that the total count could be retrieved. The retouched tools category followed the pattern noted for unifaces and bifaces, but to a greater degree: peripheral sites had three times more retouched tools than did core sites. In addition, the ratio of unretouched to retouched tools indicated that core sites had over twice as many unretouched tools, or used flakes, as retouched tools, and the peripheral sites had slightly more retouched than unretouched tools.

Core sites exhibited a higher mean of hammerstones and a smaller core tool mean than peripheral sites (Table 12.2). Additionally, the debitage-to-core ratio was larger at core sites,

Table 12.2. Core and Periphery Comparisons of Field House/Farmstead and Hamlet/Village Flaked Stone Morphological Types and Assemblage Characteristics

Morphological Type/ Assemblage Characteristic	Field House/Farmstead				Hamlet/Village			
	Periphery[1]		Core[2]		Periphery[3]		Core[2]	
	Mean	*Range*	*Mean*	*Range*	*Mean*	*Range*	*Mean*	*Range*
Debitage	83.0	72.5 – 91.2	84.3	68.9 – 93.7	85.3	74.8 – 94.0	83.3	70.1 – 93.0
Used Flake	4.1	0.1 – 11.5	4.6	0.3 – 17.9	5.4	0.1 – 12.9	5.7	1.1 – 14.2
Core	2.8	0.0 – 5.2	3.3	0.0 – 9.4	2.2	1.2 – 4.2	3.1	1.4 – 4.7
Core Tool	1.8	0.0 – 12.7	0.8	0.0 – 3.6	1.2	0.0 – 7.3	0.7	0.0 – 2.3
Core/Hammerstone	1.2	0.0 – 6.1	1.0	0.0 – 3.5	0.3	0.0 – 1.4	2.5	0.0 – 5.8
Hammerstone	2.7	0.0 – 18.6	4.3	0.2 – 9.4	1.5	tr – 4.2	1.5	0.3 – 4.2
Uniface	2.0	0.0 – 5.6	tr	0.0 – 0.3	1.8	0.0 – 3.5	0.6	0.0 – 2.7
Biface	1.1	0.0 – 2.9	0.5	0.0 – 2.7	1.5	0.0 – 3.0	0.2	0.0 – 0.6
Drill	tr	0.0 – 0.4	0.0	0.0 – 0.0	tr	0.0 – 0.1	tr	0.0 – 0.1
Projectile Point	0.2	0.0 – 0.6	tr	0.0 – 0.2	0.4	0.0 – 0.8	0.5	0.0 – 1.7
Other Tool	0.7	0.0 – 3.1	1.1	0.0 – 2.7	0.4	0.0 – 0.7	1.7	0.0 – 5.8
All Core Types	5.7	0.0 – 12.7	5.1	0.0 – 9.4	3.7	0.0 – 7.3	6.3	0.0 – 5.8
All Hammerstone Types	3.8	0.0 – 18.6	5.3	0.0 – 9.4	1.8	0.0 – 4.2	4.0	0.0 – 5.8
Retouched Tool	4.5	0.0 – 8.3	1.5	0.0 – 4.5	5.0	0.2 – 8.3	2.5	0.1 – 5.3
Unretouched to Retouched Tool Ratio	0.7	0.0 – 1.5	1.8	0.0 – 10.0	1.0	0.1 – 2.0	3.1	1.3 – 7.5
Debitage to Core Ratio	20.4	0.0 – 77.0	34.9	7.0 – 133.0	24.7	7.5 – 59.6	20.0	9.7 – 39.5
Flaked Stone Artifact Total	3638.3	36 – 15,869	493.7	86 – 1938	4424.4	2026 – 9485	4936.1	1700 – 11,825
Flaked Stone Category Total	7.9	4 – 11	6.1	4 – 8	9.0	5 – 10	8.9	7 – 11

[1]Sample size = 13
[2]Sample size = 7
[3]Sample size = 8
tr = trace

perhaps reflecting the greater use of unretouched flakes as tools. The number of flaked stone (7.9) and ground stone (8.6) artifact types for peripheral sites was greater than for core sites (6.1 and 6.7, respectively). The core site sample, though, comprised mostly field houses, which usually have a smaller number of artifacts. Brillouin diversity indexes for both were almost identical: 0.49 for core sites and 0.50 for peripheral sites. The statistics from this analysis thus indicate no variability in regional location for field houses and farmsteads with regard to assemblage diversity, a relative indicator of the variety of activities that were performed at a site.

Peripheral sites contained more ground stone food-processing tools than did core sites (Table 12.3). In contrast, the core sites contained more special-function tools, particularly abraders and axes or mauls. In food-processing equipment, all types except mortars were present more frequently in the periphery than in the core (Table 12.4). Tabular tools were especially uneven in their distribution, with no cases recovered from the core site sample and a mean of 5.7% from the sample of periphery sites. Palettes were present only in the peripheral site sample, and ornaments had higher frequencies in the periphery. However, higher frequencies in the periphery sample may be due to the small sample size for core sites, especially since palette and ornament means were so low.

Table 12.3. Core and Periphery Comparisons of Field House/Farmstead and Hamlet/Village Ground Stone Categories and Statistics

Ground Stone Category	**Field House/Farmstead**		**Hamlet/Village**	
	Periphery	Core	Periphery	Core
Food Processing Tool	*70.1	45.7	65.4	37.9
Nonutilitarian Item	3.9	3.2	5.7	16.8
Special-Function Tool	4.4	20.5	9.6	6.6
General-Purpose Tool	3.9	12.6	2.5	3.4
Other Item	17.8	18.3	15.8	35.2
Artifact Total (Mean)	225.5	29.0	319.4	739.7
Diversity Index	0.5	0.5	0.7	0.8
Artifact Category Total (Mean)	8.6	6.7	12.0	13.0

*Mean of sample column frequencies, Table 12.4

Hamlet and Village Sites

The core and periphery groups that comprised hamlet and village sites were more comparable in sample size than were the field house and farmstead sites: core sites were represented by seven flaked stone profiles and nine ground stone profiles, while peripheral sites were represented by eight flaked stone and ground stone assemblages each. The mean total number of artifacts for each group was comparable for flaked stone (Table 12.2), but the mean total of ground stone for core sites was more than twice that for peripheral sites (Table 12.3).

Table 12.4. Core and Periphery Comparisons of Field House/Farmstead and Hamlet/Village Ground Stone Morphological Types

Morphological Type	**Field House/Farmstead**		**Hamlet/Village**	
	Periphery	Core	Periphery	Core
Mano	*43.7	22.8	32.5	18.1
Handstone	3.7	11.1	2.5	3.0
Metate	19.3	22.4	17.7	15.9
Grinding Slab	0.2	1.5	—	0.4
Abrader	0.2	16.2	1.5	2.9
Anvil	0.1	—	0.2	0.3
Mortar	0.4	0.5	0.1	0.1
Passive Ground Stone NFS	0.1	0.4	0.5	0.4
Pestle	1.0	—	1.7	0.5
Axe/Maul	0.8	2.3	0.8	1.5
Tabular Tool	5.7	—	13.4	3.3
Polishing Stone	3.4	2.0	8.5	0.4
Grinder NFS	—	4.9	—	2.2
Palette	0.9	—	0.8	0.5
Ornament	2.2	0.5	3.9	14.0
Raw Material	1.6	1.1	2.7	16.8
Perforated/Unperforated Disc	tr	—	0.1	1.8
Stone Bowl	0.6	—	0.6	0.5
Stone Ring	0.2	—	0.2	0.8
Medicine Stone/Plummet	0.2	—	0.1	0.4
Mineral	—	2.7	0.3	1.4
Other Ground Stone	2.4	3.4	1.9	3.2
Indeterminate Ground Stone	13.4	8.5	10.3	11.5

*Mean of sample column frequencies
tr = trace
NFS = not further specified

Some of the flaked stone patterns noted between the core and the periphery for field house and farmstead sites were also present at hamlet and village sites. Unifaces and bifaces were present in higher frequencies at peripheral sites, supported by a relatively high frequency of retouched tools (Table 12.2). The ratio of unretouched to retouched tools was also comparable, with almost three times as many used flakes as retouched tools at the core sites compared to the peripheral sites.

The ratio of debitage to cores was similar for regional groups, although field houses differed from farmstead groups. The frequency of cores was slightly higher in the core area for hamlet and village sites, except for core tools. The large apparent difference in core/hammerstone mean frequencies between groups was probably due to the sample, in which some site data did not distinguish between core/hammerstones and hammerstones. Comparisons between regional groups for this category should be based on the mean frequencies for all hammerstone types. The pattern was similar to that of the field house/farmstead groups: the hammerstone mean was higher for core sites than for peripheral sites.

Peripheral sites contained both more food-processing tools and more special-function tools (Table 12.3), most notably tabular tools and polishing stones (Table 12.4). Although most core sites had small amounts of tabular tools, only two (Las Colinas and Pueblo Salado) had any polishing stones (Appendix A). In comparison, most peripheral sites had polishing stones, constituting a mean of 8.5% of the total assemblage. Core sites, by contrast, contained the only grinding slabs; they also had a high frequency of ornaments and raw material compared to periphery assemblages. Other artifact types more frequent from core sites were perforated discs, stone rings, and minerals. These items may account for the slightly higher ground stone diversity index calculated for core sites (0.8) than for peripheral sites (0.7) (Table 12.3).

Discussion

Detecting variability between core and peripheral sites using ground and flaked stone data emphasizes technological and economic aspects more than any other factors. Taking this bias into consideration, variability does indeed seem to reflect both of these cultural subsystems. Technological differences in the emphasis on tool production is notable in the ratios of unretouched to retouched tools and the relative frequencies of marginally and facially worked items. The preference for retouched tools over used flakes in the periphery could have been due to a number of reasons, including cultural affiliation, raw material applicability, or task applicability. One can only posit at this point what the relationship may have been between core and peripheral populations. If sociopolitical relationships were close, then the technological differences may have been simply a preference that was emphasized by geographical distance from the populated core area. If sociopolitical relationships were not so close, then the technological differences in the periphery could have reflected a separate cultural group affiliation, either as a hinterland of the core or as a colony (Schroeder 1980) of the core.

The use of local material is an attribute of Hohokam stone technology, and the farther sites are from the Salt and Gila rivers, the greater the use of nonriverine stone materials by site residents (Masse 1980; Rozen 1984:587). Local variation in raw materials does not, for the most part, change the composition of Hohokam flaked or ground stone assemblages. However, Greenwald (1985) previously noted a preference for riverine cobbles. Sites excavated along Brady Wash, as part of TAP, Phase A, had a larger amount of riverine cobble flaked stone, which was available at least 20 km (12 miles) away, than flaked stone made of a finer-grained felsite material that was available in outcrops about 10 km (6 miles) from the sites. However, on the opposite side of the Picacho Mountains, farther from the riverine source, the felsite material was more commonly used (Greenwald 1988a:262). In addition, the frequencies of retouched and unretouched tools also differed. The ratio of unretouched to retouched tools for the Brady Wash sites was 1.83, and the ratio for the sites at which toolmakers apparently preferred the locally available felsite was 1.09 (Greenwald 1985:Table 80). Although only a slight variation, it demonstrates that the pattern of high frequency of retouched tools in the Hohokam periphery may be not just fortuitous but related to the different material resources available outside of the Phoenix Basin. The ubiquity of riverine cobbles of dense, medium- to coarse-grained material along the Salt and Gila rivers may have promoted the use of unretouched tools in that area. Perhaps flakes from this resource were more durable (Beck and Jones 1990:284; Bostwick and Shackley 1987:21) and, therefore, more useful without retouch than materials found outside of the Phoenix Basin. Conversely, material sources in the periphery that were fine grained and had less durable edges than typical igneous and metamorphic types found in the Phoenix Basin would probably have been more useful as tools if their edges were shaped and strengthened by marginal retouch. Alternatives to basalts and other common materials in the core area, which usually are finer grained and less dense, are commonly present in the periphery. Some of these materials compose either the majority or a large proportion of flaked stone collections at peripheral sites, for example, dacites in the Waddell project area (Elstien 1989:Table 20.2), felsites and intermediate igneous types from the Picacho Mountains in the Picacho area sites of TAP, Phase A (Greenwald 1988a:262), silicified sediment at Water World (Huckell 1989b:Table 5.2) and Hawk's Nest (Huckell 1989a:Table 13.6), metasediment and silicified limestone at ANAMAX-Rosemont (Rozen 1984:Table 5.5), and chalcedony at the Adobe Dam project (Bruder and Gasser 1983:Table 17).

Variability in the relative frequencies of flaked stone artifact types may also be a reflection of varying emphases on subsistence resources. For example, the contrast in the availability and diversity of core and periphery resources for sedentary populations is striking (Gasser 1980:72). The differences among farming techniques and general subsistence tactics between riverine and desert Hohokam noted by Haury (1950) and Gasser (1980) may have affected other aspects of the technological system as well.

Ground stone assemblages consistently exhibited greater frequencies of food-processing tools at sites in the periphery (Table 12.3). In contrast, field houses and farmsteads in the core sites showed a higher frequency of both special-function and general-purpose tools. At hamlets and villages, core sites contained more nonutilitarian artifacts, such as ornaments, raw materials, and minerals, and also had the only grinding slabs recorded for these site types. Core area populations apparently were engaged more often than those in the periphery in tasks other than food production. Some of these tasks involved the production of specialized artifacts, such as ornaments, perforated discs, and stone rings. Field houses and farmsteads in the core area also yielded more evidence for the use of abraders and axes/mauls, probably for the production of tools and timber and for plant processing.

Inhabitants of peripheral sites more often used tabular tools for food processing. This is not surprising, given the common co-occurrence of tabular knives and rock piles, especially in the northern Tucson Basin (Fish, Fish, and Madsen 1985), that are widespread in bajada environments. Although the function of rock piles is controversial (Crown 1984b:25), archaeologists most often interpret them as a method of retaining soil moisture for the cultivation of agave (Fish, Fish, and Madsen 1985:11; Fish et al. 1985). The association of tabular knives with agave processing also has been derived from the ethnographic record (Castetter, Bell, and Grove 1938) and from their contextual association with roasting pits. Fish and coauthors (1985:107) have indicated that tabular knives constituted 19.2% of the entire flaked stone collection from their survey of the northern Tucson Basin. At NA18,017, a rock pile site located on the upper bajada at the southeast end of the Picacho Mountains, tabular tools accounted for 56.0% of the ground stone assemblage. The distribution of this particular tool type within the Hohokam region indicates more production of agave outside of the core area, although some production did take place within the core. The unique function and form of tabular tools, the types of material used, the differential access to natural or cultivated stands of the agave plant, and the particular features associated with processing of these plants suggest that agave exploitation was a specialized subsistence practice.

Another ground stone artifact that was much more common in peripheral sites was the polishing stone. Associated with the final production stages of pottery, both ethnographically (Colton 1931; Colton 1952; Hill 1937) and archaeologically (Geib and Callahan 1988; Sullivan 1988), this artifact is only one indication of pottery making. Little direct evidence for ceramic manufacture, such as kilns and sherd wasters, exists in the Hohokam region (Sullivan 1988). Core sites that have produced evidence of ceramic manufacture include Snaketown (Haury 1976:194–197) and Las Colinas (Abbott

1988; Crown 1988). The results of the Pueblo Salado temper study (Chapter 11) were inconclusive, but the presence of plainware of mixed-sand temper may be indicative of local production, although analysts have not yet tested this hypothesis. Temper studies by Abbott (1992) indicated that ceramic manufacture also may have occurred at the sites of La Ciudad and Pueblo Grande. Las Colinas and Pueblo Salado were represented in the core area sample of this study and were the only sites in the hamlet/village group where excavators found polishing stones; by contrast, most of the peripheral sites had polishing stones in their assemblages (Table 12.4). Hamlet and village sites from TAP, Phase A, had multiple types of indirect evidence of ceramic manufacture, including similarities between temper materials of unfired prehistoric clay samples and the constituents of local sand samples and the presence of a ceramic anvil, a wooden paddle, and a variety of unfired clay objects (Heacock and Callahan 1988:110, Table 29). In addition, because archaeologists for that study did not wash polishing stones before examination, the analyst found clay adhering to the surfaces of 14.8% of the examples (Greenwald 1988b:130). Other indirect evidence for local ceramic manufacture was present at Water World, including a large percentage of red-on-brown pottery, temper materials, and polishing stones (Deaver 1989:180–186). Local pottery production was suspected within the project area of the Central New River Drainage (Doyel and Elson 1985:517) but was not documented at the ANAMAX-Rosemont Project (Deaver 1984).

The Phoenix Sky Harbor Center sites varied in their flaked and ground stone patterns. They usually exhibited some of the highest debitage frequencies within the core groups, except for the Dutch Canal Ruin farmsteads, which had less debitage than the mean for field houses/farmsteads (Appendix A). Area 8/9 of Pueblo Salado also had a high debitage-to-core ratio, more similar to hamlet and village sites in the periphery. A relatively low frequency of used flakes and overall low percentages of cores and hammerstones at Pueblo Salado also were similar to some of the peripheral site attributes, although the frequencies fell within the wide range of variability for core sites. Pueblo Salado had a relatively small amount of ceremonial ground stone artifacts but relatively high frequencies of axes or mauls, polishing stones, and raw material. Area 14 was more comparable than Area 8/9 to the core means of flaked and ground stone types and attributes. The only significant difference exhibited by the Area 14 ground stone assemblage was a high frequency of such general-purpose tools as handstones and grinding slabs.

Dutch Canal Ruin generally had more retouched than unretouched tools, with a ratio similar to that for peripheral sites. Used flakes were less frequent than is typical at core sites, and the farmsteads had a high percentage of retouched tools. Farmsteads at Dutch Canal Ruin also had a low debitage-to-core ratio, more similar to that of sites in the periphery. Although the flaked stone from Dutch Canal Ruin varied from the core group norm, the ground stone followed the typical core pattern.

TECHNOLOGICAL COMPARISONS

Many analysts do not find or even attempt to explore the possibility of technological variation among Hohokam flaked stone assemblages. Those who do rely on ratios of debitage to cores, the frequency of shatter, and the amount of dorsal cortex retained on flakes. More recently, some analysts have attempted to compare debitage types to those recommended by Sullivan and Rozen (1985) for unbiased classifications. These debitage classes (complete flakes, broken flakes, flake fragments, and debris) are part of a model that Sullivan and Rozen developed to describe different technological systems. They derived their typology from the technological processes of flaked stone artifact manufacture and based it on attributes discernible in the debitage, taking data from the Tucson Electric Power (TEP) St. Johns Project (Rozen 1981). A hierarchical cluster analysis produced four groups that reflected different methods and intensities of flaked stone production.

One group represents tool manufacture, one a combination of tool manufacture and core reduction, one unintensive core reduction, and one a technology of intensive core reduction.

When analysts have applied Sullivan and Rozen's system to Hohokam artifact assemblages (Volume 2:Chapter 13; Volume 3:Chapter 6; Mitchell 1988), large proportions of debitage have been interpreted as by-products of tool manufacture. The Pueblo Salado assemblage was comparable to Sullivan and Rozen's tool manufacture group, and the Dutch Canal Ruin collection resembled a combination of unintensive core reduction and tool manufacture. Particularly incongruent with the classification of the debitage into these groups was the small amount of retouched tools, whether cobble or flake forms. The assemblage from La Lomita Pequeña reflected a combination of intensive core reduction and tool manufacture, according to Sullivan and Rozen's system. Although the frequency of retouched tools from that site was also low, Mitchell (1988:187) suggested that the evidence for tool manufacturing actually reflected maintenance of cobble tools.

The data that Sullivan and Rozen (1985) used to produce these technological groups were from east-central Arizona, from sites with a wide temporal range (Paleoindian to Pueblo III), and chert was the only material type used in the analysis (Sullivan and Rozen 1985:760–761). The difference in regional technology, including raw material factors, is probably large enough to reinterpret the debitage patterns that occur in the Hohokam region. Although both culture groups (the one from east-central Arizona, the other from south-central Arizona) exhibited core reduction technologies dominated by the hard-hammer percussion technique and the utilization of local materials, cobble tools and dense riverine cobbles of basalt and other materials of medium-grained texture predominate at Hohokam sites. According to Phagan (1980:253), "flake attributes resulting from similar stone-working techniques may be quite different in different raw materials." Some of the material characteristics that would be expected to contribute to differences in flake attributes are mass, hardness, and granularity. Core stability is easier to maintain with larger masses of material (Phagan 1980:246–247), and harder, denser materials will tend to have a lower incidence of crushing or destruction of platforms than will softer, more brittle materials. In addition, granularity can affect the fracturing quality of a core. Coarser materials will produce a higher proportion of sheared flakes than finer-grained lithics, which would be recorded as flake fragments under Sullivan and Rozen's (1985) debitage classification system.

Another factor that may influence the interpretation of results from Dutch Canal Ruin, Pueblo Salado, and La Lomita Pequeña is the homogeneity of the Hohokam technological system. Since reduction technology was similar throughout the region, perhaps other conditions were responsible for the debitage patterns that are being observed. After evaluating debitage groupings from the ANAMAX-Rosemont Project, Sullivan and Rozen (1989:171) suggested that occupational factors, rather than technological factors, had influenced the composition of the assemblage, that is, differences between debitage patterns may be due to variation in the number of features and the artifact density between sites. They posited that low percentages of whole flakes and high percentages of flake fragments were due to artifact breakage after core reduction by "the cumulative effects of artifact trampling, disposal, and scavenging" (Sullivan and Rozen 1989:171). Thus, sites that had more features and a greater density of artifacts might have been occupied longer and experienced more postreduction artifact breakage.

To determine how and whether debitage patterns were varying within the Hohokam region, the SWCA analyst compared site assemblages having debitage categories similar to Sullivan and Rozen's (1985) classification system (Table 12.5). Some analyses did not separate broken (i.e., platform-remnant-bearing [PRB]) flakes from flake fragments, and these are indicated in the table by a horizontal line in the PRB flake category. The sample exhibited a large range of variability in debitage frequencies, from a low of 17.7% to a high of 69.9% of whole flakes and from 12.7% to

Table 12.5. Comparison of Projects or Sites by Debitage Categories

Project/Site	Debitage Category				
	Whole Flake	PRB[1] Flake	Flake Fragment	Angular Debris	Whole/Fragment Ratio[2]
Waddell	69.5	—	29.9	0.6	2.32
TAP[3], Phase A	61.0	—	27.5	11.5	2.22
ANAMAX-Rosemont	48.0	15.3	31.3	5.5	1.03
Las Colinas[4]	17.7	—	50.6	31.7	0.35
El Caserío	43.6	4.8	28.6	23.0	1.31
Grand Canal Ruins	41.9	5.5	33.3	19.3	1.08
La Lomita Pequeña	34.8	18.8	12.7	33.7	1.10
Baccharis	35.0	—	49.4	15.6	0.71
Dutch Canal Ruin	25.2	8.9	60.1	5.7	0.37
Pueblo Salado	37.3	20.6	37.6	4.5	0.64

[1]Platform-remnant-bearing
[2]Ratio of whole flakes to flake fragments and broken flakes
[3]Tucson Aqueduct Project
[4]1982–1984 excavations

60.1% of flake fragments. The highest frequencies of whole flakes came from peripheral sites, such as those in the Waddell Project; TAP, Phase A; and the Santa Rita Mountains of the ANAMAX-Rosemont Project. The percentage of whole flakes was not very high for the ANAMAX-Rosemont sites compared to the other two periphery collections, but it was larger than in core site assemblages. Lower percentages of whole flakes were present at core sites, in addition to either high frequencies of flake fragments (Baccharis, Dutch Canal Ruin, and Pueblo Salado), high frequencies of debris (El Caserío, Grand Canal Ruins, and La Lomita), or both (Las Colinas). Given previous core/periphery data indicating that occupants of peripheral sites produced more retouched tools and fewer unretouched tools than had core populations, analysts would expect to see evidence of more tool production at peripheral sites. According to the model provided by Sullivan and Rozen (1985), this would include high relative frequencies of broken flakes and flake fragments. Instead, the opposite pattern occurred in this study, with relatively low percentages of these debitage categories. A ratio of whole flakes to flake fragments and broken flakes also was used to compare assemblages. (Flake fragments and broken flakes were combined for this study because some of the prior analyses did not distinguish broken flakes.) In this comparison, the ANAMAX-Rosemont sites were thrown into the core range of variability, but the Waddell and TAP, Phase A, assemblages still were markedly different.

SUMMARY AND CONCLUSIONS

This flaked and ground stone analysis compared assemblages from numerous sites within the Hohokam region for functional, spatial, and technological distributions, using data obtained from published analytical and site reports. Most data were usable; where information was absent, incomplete, contradictory, or not well enough defined for use by another analyst, it was not included.

The model described in this chapter of the lithic composition of various site type assemblages (field houses, farmsteads, hamlets, and villages) was based on previous archaeological data and on expectations formed from site type definitions. Hohokam flaked and ground stone assemblages, separated by site type, exhibited lithic patterns that supported those postulated by the model.

Field house assemblages represented short-term occupation and a limited number of activities. The amount and diversity of artifacts were low, and almost all artifacts had only utilitarian function. General-purpose tools were most common at field house sites, and the ground stone assemblage was dominated by food-processing equipment.

Farmstead assemblages were larger and more diverse than field houses but smaller and less diverse than year-round residential site collections. The farmstead sample represented a variety of activities, similar to those expected for short-term residential occupation, and included a high incidence of food-processing artifacts.

Hamlets and villages had the largest amount and greatest diversity of lithic artifacts. The greatest variety of activities was represented at these site types, with a concomitant decrease in the emphasis on food processing. Nonutilitarian artifacts were common at villages, and evidence of specialization of production was exhibited at both site types, although more often at villages. Villages contained more flaked stone artifact types than did hamlets, although both site types had similar ground stone diversity indexes. Due to the temporal distribution of sites between hamlet and village groups, detection of variability due to temporal differences was not possible.

The Dutch Canal Ruin and Pueblo Salado lithic assemblages exhibited patterns that were well within the range of variability for each site type. Field houses and farmsteads were represented at Dutch Canal Ruin, and although the total number of artifacts at these components was small compared with similar site-type collections, these assemblages were indicative of short-term occupation with an emphasis on food-processing activities. Hamlet and village lithic assemblages at Pueblo Salado exhibited larger amounts and greater diversity than those at Dutch Canal Ruin field houses and farmsteads. Food-processing equipment constituted a smaller percentage of the ground stone collection, and nonutilitarian artifacts, such as ritual paraphernalia, ornaments, and minerals, were more frequent. Pueblo Salado also contained evidence of ground stone artifact production.

Site assemblages were also evaluated based on their regional distribution. The premise of a core/periphery structure within the Hohokam region provided an analytical tool for assemblage comparisons. Field house and farmstead sites were evaluated as a unit, and hamlet and village sites were examined as a unit. Both analytical units exhibited some similarities, such as more retouched tools compared to unretouched flake tools at peripheral sites and the opposite trend at core sites; peripheral sites also exhibited a larger percentage of ground stone food-processing tools, including tabular tools.

Pueblo Salado was one of the few core area sites with polishing stones in its ground stone assemblage. This artifact type, in concert with ceramic evidence (Chapter 11), raises the possibility that the Area 8/9 Civano phase component of Pueblo Salado was a locus of ceramic manufacture. Other core sites with polishing stones documented as part of their ground stone assemblage were Las Colinas, both the field house and farmstead components of Dutch Canal Ruin, and the farmstead component of La Cuenca del Sedimento.

Flaked and ground stone from Area 14 of Pueblo Salado were most similar to each other in core site patterns compared to the rest of the assemblage from Pueblo Salado and Dutch Canal Ruin. Area 8/9 of Pueblo Salado and field houses and farmsteads at Dutch Canal Ruin exhibited some flaked stone attributes that were typical of peripheral sites, such as low frequencies of used flakes and similar debitage-to-core ratios. Debitage classes, however, fell into a typical core pattern of a high percentage of flake fragments or angular debris.

Sullivan and Rozen's (1985) interpretation-free classification of debitage was discussed for its utility in the Hohokam region. Although few analysts have employed this system, its limited application has indicated that large proportions of the flaked stone assemblage were the result of tool manufacture. The small amount and limited facial or marginal reduction of retouched tools in Hohokam flaked stone assemblages would argue against this interpretation. Two factors may affect the interpretation of the debitage patterns. One is the data base from which Sullivan and Rozen's model was structured and its applicability to the Hohokam region. Chert was the raw material used in the original analysis on which the model was based, whereas the dominant materials used in the Hohokam area were basalt and metamorphic materials. Another factor that may affect the interpretation of the debitage categories is the homogeneity of Hohokam technology. Patterns noted in debitage groupings may represent variability based on conditions other than technology. This is Sullivan and Rozen's (1989:171) suggestion for variability noted in the ANAMAX-Rosemont assemblage. In the current study, the comparison of debitage classes from 10 project or site assemblages revealed that differences in category frequencies were patterned on a core/periphery structure. Because this is a preliminary finding, based on only a small sample of core assemblages, more documentation of debitage attributes in the future by other analysts would help clarify the distributions noted here.

CHAPTER 13

CLASSIC AND POST-CLASSIC PERIOD SHELL INDUSTRY

Arthur W. Vokes

The shell material recovered from the excavations during the Phoenix Sky Harbor Center Project provided an opportunity to examine assemblages that dated largely from the Classic and post-Classic periods. This chapter presents the analysis of the project assemblage and a regional comparison of Hohokam Classic and post-Classic shell assemblages. The analyst examined the shell material from a number of different perspectives to characterize the nature of the post-Classic period Polvorón phase assemblage and to attempt to distinguish factors that may have influenced it, particularly looking at the diversity of artifact forms present and the genera represented. From this base and by comparing the assemblage to others, the analyst attempted to characterize the nature of the Classic and post-Classic period shell industry.

ANALYTICAL METHODS

The analyst compiled a descriptive record, including written documentation; in many cases, a scale drawing, accompanied by a set of linear measurements; and notes on condition, shape, decorative motifs, and technological aspects. A digital vernier caliper provided measurements to the nearest tenth of a millimeter, and each artifact was weighed on a triple beam balance. For purposes of analysis, fragments that could be refitted were considered to be single occurrences (with the number of pieces recorded in the field notes). In some instances (e.g., with *Laevicardium* and *Anodonta* valves) not all the fragments could be joined, but it was logical to assume that the fragments were from the same valve; such examples also were recorded as single objects. Specimens generally were considered complete if a full set of linear measurements could be obtained.

The analyst employed a classification structure based on that developed by Haury for the material from Snaketown (1937, 1976) and Los Muertos (1945); additions or modifications generally are noted in the text. The biological nomenclature and identifications follow Keen 1971. Abbott 1974 and Rehder 1981 are additional sources.

GENERA AND SPECIES

Sources

The Hohokam obtained shell from local freshwater rivers and lakes and the two provinces (Panamic and Californian) of the Pacific Ocean. Haury (1937:136) claimed that some material may have originated from the Gulf of Mexico; however, subsequent research (Nelson 1981) has indicated that these identifications are erroneous. In the Hohokam region, researchers have confirmed no occurrences of material from the eastern Gulf.

Archaeologists working in the Southwest benefit from a natural biotic division of the Pacific Ocean that occurs off the western coast of the Baja Peninsula in the area near Magdalena Bay, with the Panamic province to the south and the colder Californian province to the north. As a result of the temperature differential, many species of shell occur in only one of these zones or have limited distributions and frequency in the other. Although the Hohokam obtained shell from both biotic communities, the principle source appears to be the Gulf of California, a northern finger of the tropical Panamic province.

Identified Shell

Table 13.1 lists the shell species identified in the assemblage recovered from Phoenix Sky Harbor Center. Thirteen marine genera and five freshwater or terrestrial genera were present.

Marine Shell

Laevicardium and *Glycymeris* were the most common marine genera in the assemblage. The latter's popularity is clearly related to its use in the production of shell bracelets; of the 40 artifacts made of *Glycymeris*, 38 were bracelets, and the other 2 were rings. The Hohokam most likely obtained raw *Glycymeris* valves from the Gulf of California. The species of *Glycymeris* present in the Californian province are quite small and would not have been suitable for bracelet production. However, the valves employed in production of the rings may have come from the Californian source.

Similarly, *Olivella* is a genus represented in both provinces, although the species are distinct. One distinguishing characteristic that appears to be diagnostic of the regional separation is the shape of the callus (Silsbee 1958). In all cases where this feature could be observed, it indicated a southern, or gulf, origin. Additionally, in 23 specimens residual coloration suggested the species was *O. dama*, which is endemic to the Gulf of California.

Laevicardium elatum is one of the few species represented within both communities, with the northern limits of its range extending to the area near San Pedro, California (Abbott 1974:486). However, archaeologists have discovered no evidence of its extensive use by the aboriginal populations of southern California (Gifford 1947). Therefore, the *Laevicardium* recovered from Hohokam sites was most likely derived from the Gulf, where it is quite abundant.

In some cases genera are represented by species in both provinces, and the probable source region cannot be determined. For instance, both *Chione* and *Trivia* have species that are very similar to each other in size and shape and cannot easily be distinguished without the complete valve.

Freshwater/Terrestrial Shell

The nonmarine shells, with the exception of *Anodonta*, appear to have been inadvertently introduced into the site. Researchers in the Phoenix Basin have recovered both *Pisidium* and *Sphaerium* from sediments derived from prehistoric canals (Masse 1988:349–351; Vokes and Miksicek 1987:181–182). No evidence to date has indicated that either genus was ever economically important to the prehistoric inhabitants of the region.

Similarly, the occurrences of the gastropod genera *Succinea* and *Helisoma* also appeared to be fortuitous. *Helisoma* is an aquatic snail that is frequently recovered from drift assemblages and was present in the canal sediments of the Hohokam Expressway (Masse 1986:350–351) and the Las Acequias–Los Muertos systems (Vokes and Miksicek 1987:180–181). Although *Succinea* is a

Table 13.1. Shell Genera and Species Identified in the Phoenix Sky Harbor Center Assemblage

Genus/Species	Source		Frequency	
	Gulf of California	Pacific Ocean	Count	MNI
Marine				
Pelecyopods				
Glycymeris sp.	X	*	44	40
Laevicardium elatum	X	X	146	99
Argopecten circularis	X	*	1	1
Ostrea (cf. *O. corteziensis*)	X	*	1	1
Chione	X	X	1	1
Unidentified Marine Bivalve			8	8
Gastropods				
Olivella sp.	X	X	44	44
O. dama	X		(23)	(23)
Turritella leucostoma	X		2	2
Conus	X		27	27
C. regularis	X		(1)	(1)
cf. *C. perplexus*	X		(4)	(4)
Trivia	X	X	1	1
Nassarius moestus	X		4	4
Vermicularia	X	X	3	3
Kelletia kelleti		X	1	1
Haliotis sp.		X	2	2
Unidentified Marine Univalve			5	5
Unidentified Marine Shell			17	17
Freshwater and Terrestrial				
Pelecyopods				
Anodonta californiensis			183	86
Pisidium (cf. *P. casertanum*)			7	7
Sphaerium (cf. *S. striatinum*)			2	2
Gastropods				
Succinea sp.			2	2
Helisoma sp.			1	1
Unidentified Freshwater/Terrestrial Univalve			6	4

MNI = minimum number of individuals
*Other possible sources
() = included in genus total

terrestrial snail, the genus favors moist, well-vegetated areas along the periphery of marshes and watercourses, including canals.

Anodonta californiensis is a comparatively large although gracile bivalve that was endemic to most of the permanent watercourses in Arizona prior to the development and impoundment of rivers earlier this century (Bequeart and Miller 1973:220–223). Today, this species is listed as endangered within Arizona, with a remnant population restricted to the upper reaches of the Black River (Landye 1981:26). Investigators commonly recover it in considerable quantities in some prehistoric sites along the major Arizona rivers, suggesting that it may have been a food resource as well as a raw material for local artisans (Vokes 1988:373). Because it was one of the most common shells encountered during the current project, the inhabitants of the project area also may have used it for both purposes. However, of the 183 *Anodonta* shells and fragments found in the Phoenix Sky Harbor Center, representing 86 individual specimens (Table 13.1), the only identifiable artifacts were two pendants.

ARTIFACT FORMS

Most of the major artifact forms generally associated with the Classic period were represented in the assemblage. Several forms, such as the perforated shell, that were popular during the pre-Classic period were either not well represented or entirely absent. The following discussion summarizes the material in the current assemblages without reference to the temporal or spatial aspects, which are discussed in the subsequent section.

Beads

Beads are common among all of the cultures in the Southwest. In sheer quantity, they may outnumber all other artifact forms. However, this apparent abundance is deceptive, as they represent a compound artifact form of which numerous individual specimens may have formed a single strand. Excavators recovered three distinct forms of beads: (1) whole valve beads; (2) cylindrical, choker, or tubular beads; and (3) disc beads (Table 13.2).

Table 13.2. Bead Forms by Genus

Artifact Form	**Genus**					**Total**
	Olivella	*Conus*	*Nassarius*	Vermicularia	Unidentified	
Whole Shell Beads	22(15)	4(1)	3(3)	–	–	**29(19)**
Cylindrical Beads	6(5)	7(5)	–	2(2)	–	**15(12)**
Disc Beads	–	1(1)	–	–	22(11)	**22(11)**
Total	**28(20)**	**12(7)**	**3(3)**	**2(2)**	**21(10)**	**66**

#(#) = frequency(number of contexts)

Whole Shell Beads

Whole shell beads were the most common bead form, with 29 individual beads representing 19 distinct contexts. Two methods of stringing these beads were represented. For spire-lopped beads, the bead makers had removed the spire by grinding and had then passed the cord through this perforation and out the natural aperture. They formed the other type by punching or drilling a hole through the back of the body whorl and passing the cord through the natural aperture and out this hole. The former style was the most prevalent, with 26 individual beads from 16 different contexts. This type of bead was made from *Olivella* (Figure 13.1a) and *Conus* (Figure 13.1b), with the former more common in the sample (n=22). All of the *Conus* whole shell beads were present in one deposit on the floor of Feature 8/9-100.

Nassarius valves were used for the body-perforated bead form (Figures 13.1c, d). Three of these beads were present in the floor fill of separate structures in Areas 8/9 and 14. Each was unmodified except for the placement of the perforation. The specimen from Feature 8/9-44 was the largest, over 13 mm in length (Figure 13.1c). The other two were just under 7 mm in length. In Hohokam assemblages the use of the genus *Nassarius* for beads is not common and seems to have been largely restricted to the Classic period (Nelson 1991:56).

Cylindrical or Choker Beads

The Hohokam bead maker manufactured a cylindrical or choker bead by grinding away the anterior segment of the body whorl and the spire, leaving only the upper or posterior portion of the body whorl and the penultimate whorl. The Phoenix Sky Harbor Center assemblage included *Olivella*, *Conus*, and Vermicularia valves, most from separate deposits.

The cylindrical bead made from an *Olivella* valve (Figure 13.1e) first became popular in the Sedentary period. Haury (1937:Plate cxiii; 1976:309) reported four occurrences at Snaketown during the Sedentary period. The form became very popular in the ensuing Classic period, as indicated by the recovery of large numbers of cylindrical beads from mortuary deposits, particularly from the Tucson Basin (Hemmings 1969:200) and the Gila Bend region (Nelson 1991:58). Investigators recovered six beads representing five occurrences during the excavations at the Phoenix Sky Harbor Center.

The use of *Conus* valves for this type of bead was first identified by DiPeso (1956:94–95) from the work he undertook at Paloparado. Approximately half of the *Conus* bands DiPeso recovered were elements of a "neck choker," with the remainder worn either as finger rings or in contexts that did not indicate function. This dual function has resulted in typological confusion for the archaeologist. Analysts have attempted to develop criteria for distinguishing between the two uses (Vokes 1984:478): the presence of a segment of the internal structure in examples that are otherwise finished, and the presence of well-polished, beveled margins. Bands with small diameters (less than 16 mm) would be likely candidates for choker beads, as they would not have fit most people's fingers. These criteria are subjective and do not exclude larger bands from use as beads or preclude the possibility that the smaller bands might have served as finger rings on young children or infants. For this reason, all of the *Conus* bands are discussed here, with attempts to infer the function of specific specimens.

Seven *Conus* band fragments exhibited traits that suggested they may have served as choker beads (Figure 13.1f). Two retained substantial portions of the interior shelf that is present along

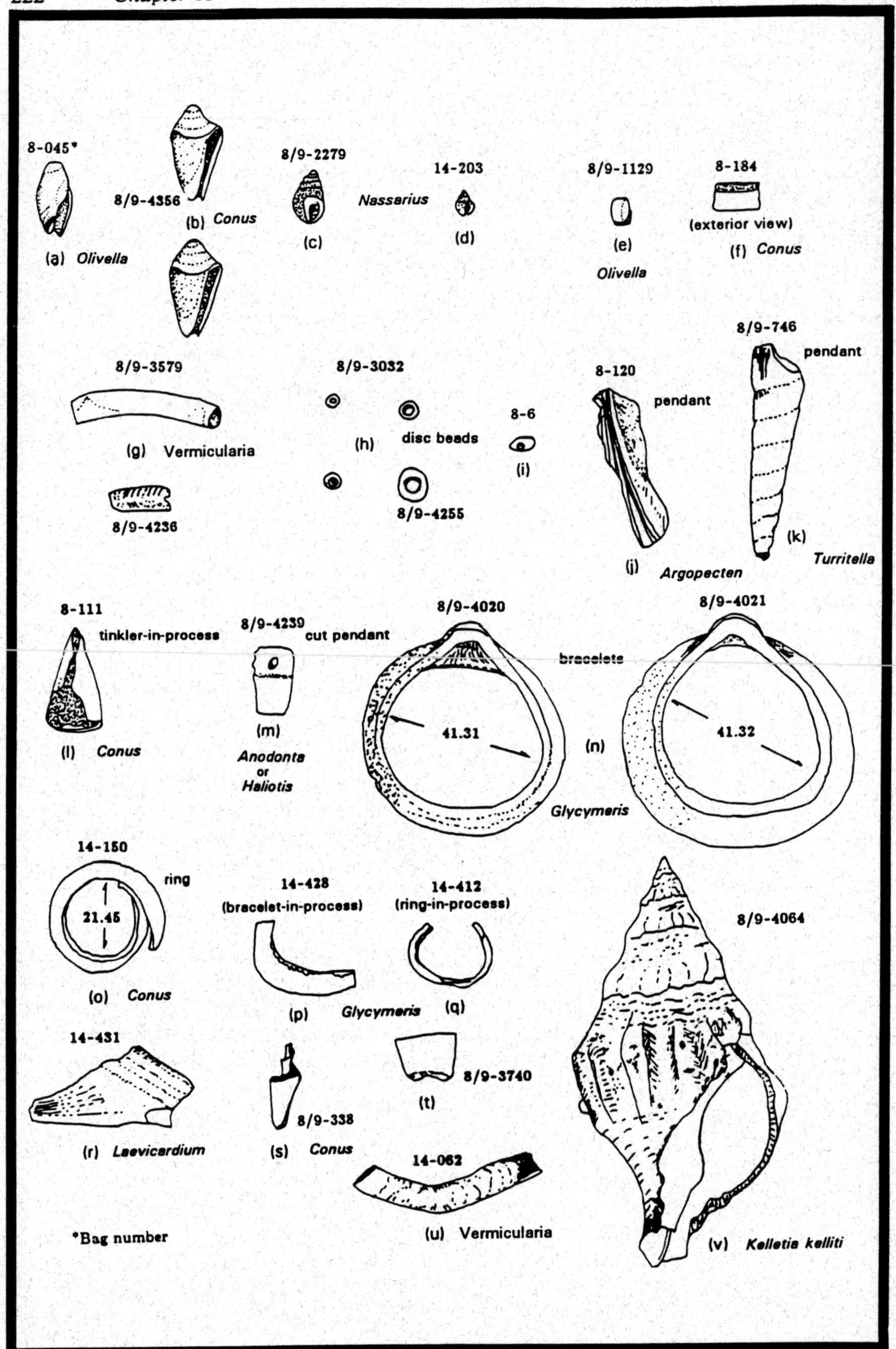

Figure 13.1. Shell artifacts from the Dutch Canal Ruin and Pueblo Salado assemblages.

the suture between the body and the penultimate whorls. The internal diameter of one of these was estimated at 10 mm. None of the bands appeared to have had diameters greater than 14 mm.

An additional four band segments, including one complete band, may have been finger rings. These *Conus* specimens were judged to be possible rings rather than choker beads on the basis of their shape, size, and the finish of the band's edge. All had projected diameters in excess of 18 mm. The complete specimen (Figure 13.1o) was unusually large, with an interior diameter of 21.45 mm and a maximum band width of 10.21 mm. In this specimen, a pronounced ridge encircled the upper interior margin; however, the band was large enough that it could easily have been worn as a finger ring. A fifth specimen appeared to be a ring in process.

One small segment, recovered from the general fill of Area 8 at Dutch Canal Ruin, had been decorated with a series of parallel lines incised into the surface. The piece was too small, however, to permit identification of the design. The interior margin of the specimen was smoothed and somewhat beveled in a manner that may have indicated its use as a finger ring. The analyst could not determine the original size of the band.

The remaining cylindrical or tubular beads had been made from segments of worm tubes from *Vermetus* or similar shells (Figure 13.1g). Although the initial study (Volume 3) identified these shells as *Vermetus*, further inspection revealed that distinguishing the genus was not possible, as all diagnostic features were missing. Investigators recovered two examples: one from the general fill in Area 8/9 and one from the level of roof fall in Feature 8/9-64. The latter was an unusually long segment (39.16 mm), with a maximum diameter of 6.83 mm. The maker had shaped both ends by cutting progressive grooves around the perimeter of the tube, giving the ends a somewhat faceted finish. The other was considerably narrower, 5.34 mm in diameter; it was too fragmentary for the analyst to estimate its original length.

Disc Beads

Excavators recovered 22 disc beads (Figure 13.1h), most of them as single occurrences. However, 10 were from two multiple deposits; 4 beads were from the courtyard in Area 8/9 (Feature 8/9-61), and 6 were from the subfloor deposits of Feature 8/9-25. The diameters of these beads generally were between 2.79 mm and 5.06 mm, with a few as large as 8.07 mm. Two beads were unusual in appearance. One had a shallow groove cut across one of its faces, bisecting the perforation. The function of this groove is at present unknown. The other was not a true disc; it was quite thin (1.52 mm) and had a pronounced lenticular shape (Figure 13.1i). It was 6.26 mm in length and 2.65 mm in width. This bead, recovered from the general fill of Area 8 at Dutch Canal Ruin, probably dates to the Polvorón phase.

Pendants

Excavators recovered three pendant forms: whole shell pendants, tinklers, and carved forms. Three whole shell pendants were made from two distinct genera. Five tinklers were present, one of which was still in the final stage of being finished. The carved forms were represented by four examples, all geometric in shape.

Whole Shell Pendants

Whole shell pendants are the simplest of the pendant forms. The valve is essentially unmodified beyond what is required to prepare it for suspension. Two of the pendants were *Turritella* valves that the maker had perforated by punching a hole through the back of the body whorl opposite the natural aperture (Figure 13.1k). In one case, the maker had ground the perforated surface to form a nearly flat platform prior punching the perforation through. Investigators recovered both pendants from Area 8/9 at Pueblo Salado. The third pendant was represented by a fragment of *Argopecten* recovered from the lower fill of a pit house in Area 8 of Dutch Canal Ruin (Figure 13.1j). Enough of the dorsal margin remained for the analyst to observe that the umbo had been ground to achieve a relatively small perforation for suspension. Segments of the side and ventral margins also appeared to have been ground to smooth and shape the edges.

Tinklers

Tinklers, like whole shell pendants, are relatively simple in form compared to carved pendants. The maker grinds away the spire of a moderate-sized univalve, creating a cone of shell, and then cuts or drills a small perforation at the apex of the cone for suspension. In the Southwest, most of the tinklers are made from *Conus* valves; the natural shape of this genus's shell is readily adapted to the artifact's form. Excavators recovered one piece that appeared to be still in the final stages of production from the general fill of Area 8 at Dutch Canal Ruin (Figure 13.1l). The spire had been ground away, but the resulting edge was somewhat uneven, and the cone was unperforated, although a portion of the outer lip where a hole could have been placed was missing. The shell tinkler, although known from Sedentary deposits (Bradley 1980:45; Vokes 1986:317, 1988:337), is primarily a phenomenon of the Classic and post-Classic periods. It has a wide distribution throughout the Southwest and is very common in Mesoamerica (DiPeso, Rinaldo, and Fenner 1974:398–399, 467–487), where the form may have originated (Kidder, Jennings, and Shook 1946:147–149).

Cut/Carved Pendants

The shape of three of the four carved pendants was clearly geometric, while that of the fourth was unknown. All were manufactured from nacreous shell materials, two from *Anodonta* and two from *Haliotis*. The three geometrics were all variations of rectangular pendants (Figure 13.1m). One was subrectangular with a slightly tapered form. No perforation was present; however, the narrow end where it might have been was missing.

The fourth pendant was made from the back of an *Anodonta* valve. The remaining portion was a small corner near the perforation, part of which was present along a broken edge. The finished edge formed a rounded arc that suggested it had been part of an ovate or other curvilinear form.

Bracelets

Excavators recovered 38 bracelets or bracelet fragments, all of *Glycymeris*. All but one were finished, plain bracelets (Figure 13.1n). The exception, a specimen in the final stages of production, is discussed below.

Two of the plain bracelets were complete bands that were found encircling the upper arm of a young child. The rest of the sample was fragmentary, with the dorsal margin present in nearly 46% of the pieces, the side areas present in 37%, and the ventral margin in 40%.

The umbo was present in 11 cases. In all but 2, it was unmodified. However, on 4 specimens the band to either side had been ground back and smoothed. Four of the umbones had been left unperforated, while the remaining seven had been perforated by the makers, who had ground the beak of the umbo until they penetrated it. In four of these specimens, the hole appeared to have been widened with a small reaming tool. Researchers have suggested several reasons for perforating the umbo. Officer (1978:117) speculated that the perforations were for sewing the bracelet onto a backing cloth. Haury (1976:313) suggested that the hole could have served as a means for attaching small objects such as feathers to the bracelets as tassels.

The exterior surfaces of most of the bands were ground and polished to some degree. Seven specimens had been left relatively unmodified or had been ground along the margin only. Of the remaining specimens, the surface treatment of two others could not be ascertained because of weathering of the exterior surface. The rest were ground across the entire surface. In part, this reflects efforts to create a nearly flat appearance on the exterior face.

The vast majority of these bands measured less than 6 mm in width and thickness. Haury (1976:313) suggested that the Snaketown samples comprised three distinct classes of bracelets, with the more massive bands becoming more common in the Sedentary period. Figure 13.2 plots the measurements of band width and thickness for the plain bracelets in the current sample that included nondorsal segments, with Haury's suggested divisions indicated. Although his typology incorporated other features, such as treatment of the umbo and the shape of the band's cross section, the measurements of the band's width and thickness were a major factor. The current sample had only one nondorsal segment that would fall within the Type 3 (broad band) category. The measurements were limited to the nondorsal segments because of the bias introduced by the more massive area of the taxadontic plate. The assemblage included a number of Type 1 (narrow band) examples, which Haury (1976:313) characterized as being more prevalent in the Pioneer period. Thus, band width appears not to be a temporally sensitive index, leaving the typology useful only descriptively.

Rings

Excavators recovered one finished ring segment made from a *Glycymeris* valve and one unfinished specimen. Another four fragments of *Conus* shell may have been either rings or cylindrical beads, as discussed above. The specimen still in the final stages of manufacturing is discussed below. The single finished *Glycymeris* band fragment was present in the floor fill of Feature 14-15 at Pueblo Salado. The band was 2.7 mm wide, but the diameter could not be determined.

Geometric Forms

Investigators recovered two geometrically shaped fragments of cut shell from the excavations at Pueblo Salado. Both had been cut from the middle of the valve's back. One was rectangular, while the other appeared to have been part of a disc. These were probably intended to be pendants, although other uses are possible. Haury (1937:146, 1976:317) suggested that such items may also have served as gaming pieces or as mosaic elements. The current examples appeared too large for

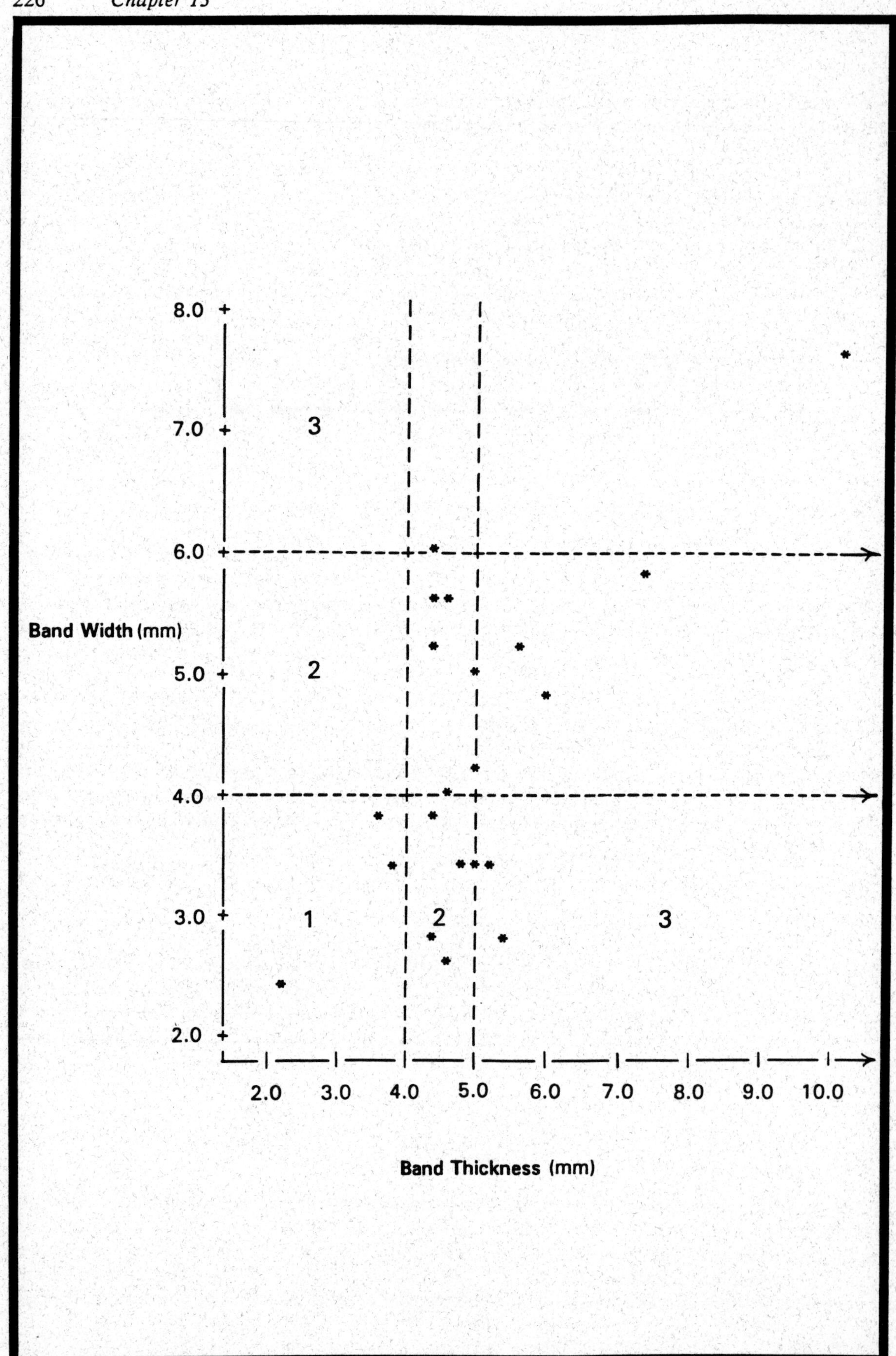

Figure 13.2. Scatter plot of nondorsal segments of plain bracelets with Haury's types indicated: Type 1, width 2.5-4.0 mm, thickness 2.0-4.0 mm; Type 2, width 4.0-6.0 mm, thickness 4.0-5.0mm; Type 3, width 6.0-10.0 mm, thickness 5.0-17.0 mm (Haury 1976:313).

use in mosaics and did not exhibit beveled edges, a trait often associated with tesserae. Neither exhibited evidence of perforations; however, the disc fragment was small enough that a perforation may have been present elsewhere on the original artifact. Although a perforation might have been present on the rectangular form as well, the missing portion would have been relatively small.

MANUFACTURING EVIDENCE

Unfinished artifacts and waste material provided clear evidence of ongoing manufacturing in a number of areas at Dutch Canal Ruin and Pueblo Salado, all associated with Classic and post-Classic period loci. The manufacturing activities appeared to have been largely focused on the carving of shell and the cutting of choker beads or rings. The presence of an unfinished bracelet and a *Glycymeris* ring at Area 14 indicated that the local inhabitants had been familiar with the procedures for the production, or at least the finishing, of these artifacts.

Both the bracelet (Figure 13.1p) and the ring (Figure 13.1q) were in the final stages of production when they were discarded. The lack of debris chips and other items common in early phases of production suggested the possibility that these bands had been roughed out at another location and imported into the community in the final stages of manufacturing. However, the production of bracelets or rings appears to have been a minor activity at the Phoenix Sky Harbor Center sites.

By contrast, the production of cut shell forms such as pendants, geometrics, and *Conus* bands and tinklers appears to have been relatively common. Four recovered fragments of *Laevicardium* exhibited clear indications of a carving technique commonly referred to as "groove and snap" (Figure 13.1r) (Vokes 1984:524). This technique involved the cutting of a groove into the surface of the shell to control the direction and shape of a break, which was initiated by twisting the shell. All of these pieces included portions of the valve's back, and all had one edge that had been modified, with the other edges exhibiting simple breaks. A fifth piece of *Laevicardium* appeared to be a rectangular segment that was being roughly shaped by grinding. Three of the edges had some areas with high-point grinding, while a fourth edge was a raw break. This appeared to be a geometric form or pendant in the early stages of production.

Four fragments of the body whorl of a *Conus* valve appeared to be the waste by-products from the manufacturing of choker beads or rings (Figure 13.1s, t). The fragments were lower body whorls including the siphonal canal and the internal columellar structure. The edges of these fragments exhibited the characteristic cuts and grooves employed by makers when separating the band from the anterior portion of the valve. A fifth fragment may have been an example of a valve being converted to a tinkler. In this instance the cut marks were somewhat higher up the body whorl, closer to the shoulder.

The final example of on-site production was a large segment of Vermicularia tube with one end extensively ground and beveled (Figure 13.1u). The opposing end was a raw break. This tube segment measured over 48 mm in length. Although this specimen was present in a trash deposit in Area 14 at Pueblo Salado, the two beads made from Vermicularia were both from Area 8/9, suggesting a multiple acquisition of this material over the course of the occupation.

FRAGMENTARY MATERIAL

All assemblages that contain a quantity of shell artifacts will have some fragmentary material that lacks either the diagnostic features that might lead to identification of the artifact's form or all evidence of having been worked. The current sample is no exception, with 14 worked fragments and 260 unworked pieces.

Worked Fragments of Unknown Form

The fragmentary materials that exhibited some evidence of having been worked but whose original form could not be defined were largely pieces of *Laevicardium elatum* (n=5) or segments of *Anodonta californiensis* (n=5). The shells of both species were often employed as a carving medium in these communities. Therefore, some of the fragmentary material may have been by-products of this activity. Many of these fragments probably were pieces of finished artifacts as well; both species are comparatively gracile in nature and can fragment easily.

One worked *Chione* fragment was also in the sample. The piece may have been part of a ring or pendant, although the analyst could not be certain of the form. The portion recovered was from the dorsal margin and the beak. Examples of whole shell pendants made of this material were recovered at Paloparado by DiPeso (1956:92, Plate 15b5) but were mistakenly identified as *Glycymeris* in the text.

Excavators in the Phoenix Sky Harbor project area also recovered three fragments of univalves; one was a piece of *Conus*, another appeared to be a very small segment of *Haliotis*, and the third could not be identified with any confidence. All fragments had been ground along one or more edges.

Unworked Fragments

As with the worked material, the majority of the unworked fragments were either *Laevicardium* (47%) or *Anodonta* (50%). Of the former, 122 fragments represented approximately 86 individual pieces. *Anodonta* fragments numbered 130 and represented as few as 78 valves. This shell tends to shatter and split apart along its nacreous layers, which makes estimating the number of individuals represented very difficult. The unworked material estimate represents approximately 95% of this shellfish in the sample. The presence of such quantities of unworked *Anodonta* valves suggested that the local inhabitants may have used it as a food resource.

The sample included four unworked fragments of marine univalves. Three could be identified as to genus: *Olivella*, *Conus*, and *Trivia*. The final piece appeared to be a small segment of the siphonal canal region from another *Olivella* valve; however, it was too small for positive identification. Both the *Olivella* and the *Trivia* may have been fragmentary pieces of beads or may have been intended for such use. The fragment of *Conus* incorporated much of one side of the body whorl. It may have been a part of a tinkler, or it could have been from an unmodified valve.

WHOLE SHELLS

The sheer number of whole unmodified valves in the assemblage was unusual. Discounting the terrestrial and small freshwater bivalves and univalves other than *Anodonta*, the assemblage consisted of 19 potential artifact valves, including 18 marine shells and a single *Anodonta*. The great majority of the marine specimens were *Olivella* valves (n=15) and were concentrated in the trash deposit (Feature 14-1) at Area 14, Pueblo Salado, with a scattering in other features. Of the remaining three marine shells, a *Nassarius* valve was recovered from the general fill of Area 8 at Dutch Canal Ruin, an *Ostrea* valve was recovered from the general fill of Area 14 at Pueblo Salado, and a large univalve, *Kelletia kelleti* (Figure 13.1v), was recovered from Feature 8/9-81, an inhumation at Pueblo Salado. This is the first known occurrence of this univalve in the Southwest. The valve was relatively large, 103 mm in length, and exhibited no evidence of any modification. The shell had apparently been picked up as beach drift, as the sculpture was rounded and pitted from worm boring. Over 40 barnacles still adhered to the exterior surface, indicating that the shell had not been handled much after collection. This species has a limited natural distribution along the California coast. How the local Hohokam acquired and buried it in such an apparently short period of time is unknown.

It is unclear if the complete *Ostrea* valve recovered from the general fill of Area 14 was associated with the prehistoric occupation. Historically, oyster shells were occasionally ground up and included in chicken feed as a food supplement. However, this valve was similar in shape to *O. corteziensis*, a species endemic to the Gulf of California, and could therefore have been acquired by the prehistoric population.

DISCUSSION OF THE PROJECT ASSEMBLAGE

The shell assemblage from the Phoenix Sky Harbor Center excavations was mainly representative of the later components of the Hohokam tenure in the Phoenix Basin. The areas of Dutch Canal Ruin that dated to the Colonial and Sedentary periods produced only a few shell artifacts. In part, this probably reflects the transient nature of these occupations. The more substantial character of the Classic period communities at Pueblo Salado was reflected in the quantity and diversity of its assemblages, similar to those recovered from other Classic period sites in the basin. Nearly all of the principal artifact forms were present, and evidence indicated that some of the material had been produced on-site. The discussion below focuses on the spatial and temporal distribution of the shell material within the settlements. Following this is an attempt to characterize the assemblages generally associated with the Classic and post-Classic periods.

Genera

Tables 13.3 and 13.4 summarize by minimum number of individuals the genera present in the assemblage by site area and temporal association. Little can be said regarding the early components; however, beginning with the Soho phase, the sample was substantial enough to permit observations of the assemblage's composition.

Four marine genera stand out in the collection in terms of the frequency of specimens: *Laevicardium*, *Olivella*, *Glycymeris*, and *Conus*. *Laevicardium* was the most common marine genus throughout the sequence. Although some valves may have been employed as containers, evidence indicated that some on-site production of artifacts from this genus had taken place; carved shell artifacts were present but did not appear to be as frequent as might be expected given the number of fragments in the sample.

The relative frequency of *Glycymeris* was related to the presence of the shell bracelets. It was the sole genus used for this artifact form in the current assemblage, and bracelets were the primary

Table 13.3. Shell Genera by Intrasite Location, Minimum Number of Individuals

Site (AZ)	Area/ Locus	Phase	Genus																	Total
			Marine Bivalve						Marine Univalve									Unidentified Marine Shell	Freshwater Bivalve	
			Glycymeris	*Laevicardium*	*Argopecten*	*Chione*	*Ostrea*	Unidentified	*Olivella*	*Conus*	*Turritella*	*Nassarius*	*Trivia*	*Kelletia*	Vermicularia	*Haliotis*	Unidentified		*Anodonta*	
T:12:62[1]	2	Gila Butte	—	1	—	—	—	—	—	—	—	—	—	—	—	—	—	—	—	1
T:12:62	3	Sacaton	—	—	—	—	—	—	—	—	—	—	—	—	—	—	—	—	1	1
T:12:47[2]	14	Soho	15	20	—	—	1	5	21	4	—	1	—	—	1	—	1	—	12	81
T:12:47	8/9-A	Civano	7	5	—	—	—	—	6	1	1	2	1	1	—	1	1	9	11	46
	8/9-B	Civano	2	9	—	—	—	1	4	1	—	—	—	—	—	—	—	—	11	28
	8/9-C	Civano	2	10	—	—	—	—	3	1	—	—	—	—	—	—	—	—	3	19
	8/9	Civano	4	16	-	—	—	—	2	6	—	—	—	—	1	1	2	4	12	48
	8/9	Civano/ Polvorón	—	6	—	—	—	—	3	2	—	—	—	—	1	—	1	—	10	23
	8/9	Polvorón	2	4	—	—	—	—	—	1	1	—	—	—	—	—	—	—	10	18
T:12:62	8	Polvorón	8	27	1	1	—	2	5	11	—	1	—	—	—	—	—	4	12	72
T:12:62	4	Unknown	—	—	—	—	—	—	—	—	—	—	—	—	—	—	—	—	2	2
T:12:62	10	Unknown	—	1	—	—	—	—	—	—	—	—	—	—	—	—	—	—	2	3
Total			40	99	1	1	1	8	44	27	2	4	1	1	3	2	5	17	86	342

Note: Table does not include freshwater or terrestrial shell other than *Anodonta*.
[1]Dutch Canal Ruin
[2]Pueblo Salado

Table 13.4. Shell Genera by Phase, Minimum Number of Individuals

Phase	Genus																Total
	Marine Bivalve						Marine Univalve									Unidentified Marine Shell	Freshwater Bivalve	
	Glycymeris	*Laevicardium*	*Argopecten*	*Chione*	*Ostrea*	Uniden-tified	*Olivella*	*Conus*	*Turritella*	*Nassarius*	*Trivia*	*Kelletia*	Vermic u-laria	*Haliotis*	Uniden-tified		*Anodonta*	
Gila Butte	—	1	—	—	—	—	—	—	—	—	—	—	—	—	—	—	—	1
Sacaton	—	—	—	—	—	—	—	—	—	—	—	—	—	—	—	—	1	1
Soho	15	20	—	—	1	5	21	4	—	1	—	—	1	—	1	—	12	81
Civano	15	40	—	—	—	1	15	9	1	2	1	1	1	2	3	13	37	141
Civano/ Polvorón	—	6	—	—	—	—	3	2	—	—	—	—	1	—	1	—	10	23
Polvorón	10	31	1	1	—	2	5	12	1	1	—	—	—	—	—	4	22	90
Unknown	—	1	—	—	—	—	—	—	—	—	—	—	—	—	—	—	4	5
Total	40	99	1	1	1	8	44	27	2	4	1	1	3	2	5	17	86	342

Note: Table does not include freshwater or terrestrial shell other than *Anodonta*.

type of artifact made from the shell. In other Classic period assemblages, *Glycymeris* has been associated with other artifact types, such as whole shell beads, pendants, and rings/pendants (Nelson 1981:181; Vokes 1984:473, 1988:326). Of these forms, only rings were present in the current assemblage, and these were limited to the Soho phase. The reason for this is unclear, as they have been recovered from other Civano phase sites.

The high incidence of *Olivella* reflected its use as a major medium in the production of two types of beads: whole shell and cylindrical forms. The number of unworked valves recovered from trash and other features in Area 14 at Pueblo Salado suggested that the local inhabitants had produced some of their own beads. That *Olivella* was most frequent in the Soho phase in this sample indirectly supported this possibility. During the Civano and Polvorón phases, its use appeared to have declined.

Conus was part of the assemblage throughout the Classic period. However, this genus increased in popularity over time, from 4.9% of the Soho phase assemblage to a high of 13.3% in the Polvorón phase. It was employed in three types of artifacts: whole shell beads, choker beads or rings, and tinklers. The whole shell beads were present only in a single deposit in the Civano phase, while the choker beads or rings were present throughout the Classic and post-Classic period components. Tinklers were from both Civano and Polvorón phase deposits. In general, use of this genus appeared to have gradually increased as the residents of the project area adopted and popularized new forms.

One genus that was present consistently although not in any great quantity in the Classic and post-Classic period remains was *Nassarius*. Within the Hohokam culture area, this genus appears to have been largely restricted to the Classic period (Nelson 1981:125). Although it was never as popular with the Hohokam as were some other univalves, it was present in vast quantities at the site of Casas Grandes in Chihuahua, Mexico (DiPeso, Rinaldo, and Fenner 1974), which was contemporary with the Classic and post-Classic period Hohokam (Lekson 1984; Ravesloot, Dean, and Foster 1986). At present, the relationships that existed between these cultures are unclear.

The freshwater shellfish *Anodonta* was recovered from deposits from virtually all of the temporal periods represented and was always present in significant numbers. Although it was employed in the manufacturing of cut pendants, its primary role may have been as a dietary resource. This shellfish thrived in the nearby Salt River and also grew in the muddy bottoms of the canals, where it would have been relatively easy to gather.

Artifact Distributions

Some general patterns appeared in the relative use of certain shell artifacts over the tenure of the Phoenix Sky Harbor Center sites. The distribution of artifact forms by minimum number of individuals is presented in Table 13.5 and summarized in Table 13.6.

In Area 14 of Pueblo Salado, which represented the Soho phase in the sample, most of the shell was concentrated in the western portion of the occupation area, particularly in Features 14-7 and 14-8 (structures), along with the associated trash deposit, Feature 14-1, which together produced approximately half of the sample. None of the artifacts appeared to be from well-defined, intact floor assemblages. Although the material in Feature 14-7 was largely concentrated on the floor and in the floor fill, reasons existed that led the analyst to suspect that this material had been deposited after the abandonment of the structure. This structure may have burned during its occupation; however, none of the shell material exhibited charring or other indications of having been exposed to intense heat, suggesting introduction after the structure burned and was abandoned. The material

Table 13.5. Shell Artifacts by Intrasite Location, Minimum Number of Individuals

Site (AZ)	Area/ Locus	Period/ Phase	Bead			Pendant			Bracelet	Ring	Geometric/ Carved	Manufacturing			Fragment		Whole Shell	Total
			Whole Shell	*Choker*	*Cut/ Disc*	*Whole Shell*	*Tinkler*	*Cut/ Carved*				*Bracelet*	*Ring*	*Carved*	*Worked*	*Unworked*		
T:12:62[1]	2	Gila Butte	—	—	—	—	—	—	—	—	—	—	—	—	—	1	—	1
T:12:62	3	Sacaton	—	—	—	—	—	—	—	—	—	—	—	—	—	1	—	1
T:12:47[2]	14	Soho	4(3)	5(4)	5(4)	—	—	1	12	1	1	1	1	4	3	29	13	80
T:12:47	8/9-A	Civano	8(5)	—	6(1)	1	—	1	7	—	1	—	—	1	1	16	1	43
	8/9-B	Civano	1(1)	1(1)	4(1)	—	1	—	2	—	—	—	—	—	—	21	2	32
	8/9-C	Civano	3(2)	1(1)	—	—	—	—	2	—	—	—	—	—	2	11	—	19
	8/9	Civano	6(3)	1(1)	3(3)	—	1	1	4	—	—	—	—	1	3	26	—	46
	8/9	Civano/ Polvorón	3(2)	2(2)	—	—	1	1	—	—	—	—	—	—	1	15	—	23
	8/9	Polvorón	—	—	—	1	—	—	2	—	—	—	—	1	1	13	—	18
T:12:62	8	Polvorón	4(3)	5(3)	4(2)	1	2	—	8	—	—	—	—	5	3	35	2	69
T:12:62	4	Unknown	—	—	—	—	—	—	—	—	—	—	—	—	—	2	—	2
T:12:62	10	Unknown	—	—	—	—	—	—	—	—	—	—	—	—	—	2	1	3
Total			29(19)	15(12)	22(11)	3	5	4	37	1	2	1	1	12	14	172	19	337

Note: Table does not include freshwater or terrestrial shell other than *Anodonta*.
[1]Dutch Canal Ruin
[2]Pueblo Salado
#(#) = frequency(number of contexts)

Table 13.6. Shell Artifacts by Phase, Minimum Number of Individuals

Phase	Artifact Type															Total
	Bead			Pendant			Brace -let	Ring	Geom/ Carved	Manufacturing			Fragment		Whole Shell	
	Whole Shell	*Choker*	*Cut/ Disc*	*Whole Shell*	*Tinkler*	*Cut/ Carved*				*Brace -let*	*Ring*	*Carved*	*Worked*	*Unworked*		
Gila Butte	—	—	—	—	—	—	—	—	—	—	—	—	—	1	—	1
Sacaton	—	—	—	—	—	—	—	—	—	—	—	—	—	1	—	1
Soho	4(3)	5(4)	5(4)	—	—	1	12	1	1	1	1	4	3	29	13	80
Civano	18(11)	3(3)	13(5)	1	2	2	15	—	1	—	—	2	6	74	3	140
Civano/ Polvorón	3(2)	2(2)	—	—	1	1	—	—	—	—	—	—	1	15	—	23
Polvorón	4(3)	5(3)	4(2)	2	2	—	10	—	—	—	—	6	4	48	2	87
Unknown	—	—	—	—	—	—	—	—	—	—	—	—	—	4	1	5
Total	29(19)	15(12)	22(11)	3	5	4	37	1	2	1	1	12	14	172	19	337

Note: Table does not include freshwater or terrestrial shell other than *Anodonta*.
#(#) = frequency(number of contexts)

in Feature 14-8 was from floor fill and fill contexts, also indicating that much of it had been deposited as trash subsequent to the abandonment of the structure. The presence of whole valves in the trash deposit, Feature 14-1, was perplexing.

Some evidence indicated that the ramada, Feature 14-9, might have been the locus of some of the on-site manufacturing activities at Area 14. A fragment of marine bivalve that appeared to be a piece of manufacturing discard was recovered from the floor fill along with a worked *Conus* fragment nearby on the floor. These are hardly conclusive indications of manufacturing, but the ramada would have been a logical area for such activities.

Why the shell was not more evenly distributed about Area 14 remained unclear. Although the lack of shell in the other structures may have reflected a later occupation, with residents depositing trash selectively in the earlier depressions, the near absence of shell in the other trash deposits was more perplexing.

In contrast, the distribution of shell materials in the Civano phase component of Area 8/9 appeared to have been relatively even across the compound. The exception to this generality was Feature 25, which had a total of 26 different pieces of shell representing 7 different artifact forms, including whole shell beads, bracelets, and pendants. The reason for this concentration was unclear. Investigators suggested that the presence of an infant's body on the floor near the southeast corner may have led to a ritual abandonment of the room. However, much of the shell material (n=17) lay in the fill below the floor, indicating it had been deposited prior to the interment of the infant. Furthermore, the material on the floor and in the floor fill was scattered across the room.

A relatively light scatter of shell material characterized the Polvorón phase occupation of this area of Pueblo Salado. Much of this material was unworked fragments of *Anodonta*, which may have represented dietary remains more than cultural artifacts.

The majority of Polvorón phase shell in the sample was recovered from Area 8 at Dutch Canal Ruin. In contrast to the relatively impoverished nature of the Polvorón phase material at Pueblo Salado, the features in Area 8 reflected a more affluent occupation. Cylindrical or choker beads made of *Conus* were comparatively common, with a few bracelets and pendants. The assemblage was concentrated in the floor fill of the two structures, Features 8-2 and 8-3, and the fill of the fire pit, Feature 8-10. Since investigators proposed that these structures had been abandoned in a leisurely fashion, these artifacts likely represented postabandonment trash from an unknown source.

Dividing the shell assemblage temporally resulted in relatively equivalent numbers representing the Soho and Polvorón phases, with the number from the Civano phase somewhat higher. With the exception of the material from Area 8/9, these collections were relatively discrete, and the material from Area 8/9 appeared to date largely to the Civano phase. It is possible to make some observations regarding the temporal patterns in the data.

Whole shell forms and the related, simpler artifact forms appeared to increase during the Civano phase at the expense of the more elaborate carved types of shell ornaments. Tinklers and whole shell beads and pendants appeared to increase, with cut pendant forms, *Glycymeris* rings, and even bracelets decreasing in popularity. These trends were also reflected in the nature of the manufacturing debris and unfinished artifacts.

REGIONAL COMPARISON

The composition of the shell assemblages associated with the Soho and Civano phases is reasonably well defined, especially in contrast to that of the ensuing Polvorón phase. Beginning with the Hemenway excavations at the end of the last century (Cushing 1890; Haury 1945), researchers have considered the shell industry to be one of the characteristic features setting the Hohokam culture apart from other populations in the prehistoric Southwest (Fewkes 1896). In recent years, a number of excavations at Classic period sites have helped to clarify the nature of Classic period assemblages (Debowski 1974; Urban 1981; Vokes 1984, 1988). However, during this same interval a new source of confusion arose: the question of late occupations or reoccupations of portions of these sites (Doyel 1974; Gregory et al. 1988; Teague 1989). Investigators noted the presence of this late occupation at a number of sites, but its nature remained poorly understood. Not until the excavation of El Polvorón, a single-component site, during the Salt-Gila Aqueduct Project was the Polvorón phase formally identified and characterized (Crown and Sires 1984; Sires 1984).

Attempting a diachronic study of Hohokam Classic and post-Classic period shell at this time may be premature. Because few of the past studies recognized post-Classic assemblages, the number of comparative data sets is limited. Furthermore, few past studies presented the shell assemblages according to phase. In reviewing the available literature, five studies provided a total of eight sites that could be used in this study (Tables 13.7a, b, c) (Debowski 1974; Gross 1989; Urban 1981; Vokes 1984, 1988). In some cases (Urban 1981; Vokes 1988), sources of data were limited to those features that investigators had confidently assigned to specific time periods. This analyst re-examined the early investigations at Las Colinas (Hammack and Sullivan 1981) regarding temporal assignment of specific architectural features based on morphological attributes, archaeomagnetic results, and, to a certain degree, ceramic associations. The assignment of these features to the Soho, Civano, or Polvorón phases corresponded to Chenault's (Chapter 8) assessment of these same features. These included Feature 100 in the Soho phase, Feature 26 in the Civano phase, and Features 34, 35, 39, 40, 41, 74, and 114 in the Polvorón phase.

Tables 13.7a, b, and c and 13.8a, b, and c and Figures 13.3, 13.4, and 13.5 summarize shell assemblages from each of the major phases represented in the Phoenix Sky Harbor Center for a selected sample of Phoenix Basin sites. The Classic period shell artifact assemblages (Tables 13.7a, b; Figures 13.3 and 13.4) were characterized by a relatively high incidence of *Glycymeris* bracelets, with lesser quantities of beads, pendants, and rings, and infrequent occurrences of other artifact forms such as trumpets. However, techniques such as etching and perforating, which were relatively common in the Sedentary period, were not being used in subsequent phases. Bracelets had been important to the Hohokam since early in their historical development. In the pre-Classic periods, bracelets often represented 60% or more of the formal artifacts in an assemblage (Vokes 1984, 1988:381). By the Classic period, the proportion of bracelets had declined somewhat, representing as little as 22% of one formal assemblage, excluding the worked and unworked fragments and whole shells (which constituted 59% of the Soho phase assemblages and 63% of the Civano phase assemblages [Tables 13.7a and b]). The decline in the relative popularity of bracelets probably reflects the emergence of other forms, including cylindrical or choker beads and tinklers. The earliest beads and tinklers thus identified are in assemblages of the Sedentary period (Bradley 1980:45; Urban 1981:321, 324; Vokes 1984:530, 1988:336–337). In the Classic period assemblages the quantities of whole shell beads and the related cylindrical forms were roughly similar to those of the cut shell beads. However, noticeable shifts in the forms of cut shell pendants were apparent in the Classic period assemblages. Many of the faunal design forms that had been prevalent in the pre-Classic periods were no longer represented, although other forms increased in popularity. The Cipactli (Haury 1976:Figure 17.3) and other quadrupeds had been abandoned in favor of lizards, frogs, and thunderbirds, often in highly abstract representations. The bulk of the cut pendants were

Table 13.7a. Shell Artifacts from Selected Soho Phase Sites

Site	Artifact Type																		Total
	Bead			Pendant			Bracelet	Ring	Geometric/ Carved	Perforated	Tool	Other Artifact	Manufacturing			Fragment		Whole Shell	
	Whole Shell	*Cylinder/ Choker*	*Cut/ Disc*	*Whole Shell*	*Tinkler*	*Cut/ Carved*							*Bracelet*	*Ring*	*Carved*	*Worked*	*Unworked*		
AZ T:12:47(ASM)	4	5	5	—	—	1	12	1	1	—	—	—	1	1	4	3	29	13	80
AZ U:15:32(ASM)	—	1	2	1	2	—	13	6	1	—	—	1	—	1	—	—	16	—	44
AZ U:15:27(ASM)	1	—	2	1	2	3	11	2	2	1	1	—	—	—	—	1	25	—	52
AZ T:12:14(ASU)	1	—	—	—	—	2	2	1	—	—	—	—	—	—	—	1	13	—	20
AZ T:12:16(ASU)	—	—	—	—	—	—	2	—	—	—	—	—	—	—	—	—	19	—	21
AZ T:12:10(ASM)[1]	5	3	—	—	1	1	8	—	—	—	—	—	—	—	—	—	2	1	21
AZ T:12:10(ASM)[2]	1	1	2	3	—	1	10	3	—	—	—	1	1	—	—	7	67	—	97
Total	12	10	11	5	5	8	58	13	4	1	1	2	2	2	4	12	171	14	335

AZ T:12:47(ASM) = Pueblo Salado; AZ U:15:32 and AZ U:15:27(ASM) = Escalante Ruin Group (Debowski 1974); AZ T:12:14 and AZ T:12:16(ASU) = Grand Canal Ruins (Gross 1989); AZ T:12:10(ASM)[1] = Las Colinas (Urban 1981); AZ T:12:10(ASM)[2] = Las Colinas (Vokes 1988)

Table 13.7b. Shell Artifacts from Selected Civano Phase Sites

Site	Bead			Pendant			Bracelet	Ring	Geometric/ Carved	Perforated	Tool	Other Artifact	Manufacturing			Fragment		Whole Shell	Total
	Whole Shell	*Cylinder/ Choker*	*Cut/ Disc*	*Whole Shell*	*Tinkler*	*Cut/ Carved*							*Bracelet*	*Ring*	*Carved*	*Worked*	*Unworked*		
AZ T:12:47(ASM)	18	4	13	1	2	2	15	—	1	—	—	—	—	—	2	6	74	3	141
AZ U:15:3(ASM)	5	4	14	2	1	—	7	—	1	—	2	1	—	—	—	12	205	4	258
AZ U:15:19(ASM)	11	9	5	2	3	8	82	9	2	1	3	1	1	5	4	30	110	2	288
AZ T:12:14(ASU)	3	—	—	1	—	1	8	3	—	—	—	—	—	—	—	3	45	—	64
AZ T:12:10(ASM)[1]	3	—	—	—	—	—	2	—	1	—	—	—	—	—	—	—	1	7	14
AZ T:12:10(ASM)[2]	5	4	1	1	1	3	48	1	—	—	—	2	1	2	—	13	47	1	130
Total	45	20	32	7	7	14	162	13	5	1	5	4	2	7	6	64	482	17	895

AZ T:12:47(ASM) = Pueblo Salado; AZ U:15:3(ASM) = Escalante Ruin (Debowski 1974); AZ U:15:19(ASM) = Las Fosas (Vokes 1984); AZ T:12:14(ASU) = Grand Canal Ruins (Gross 1989); AZ T:12:10(ASM)[1] = Las Colinas (Urban 1981); AZ T:12:10(ASM)[2] = Las Colinas (Vokes 1988)

Table 13.7c. Shell Artifacts from Selected Post-Classic Period Sites

Site	Artifact Type																		Total
	Bead			Pendant			Bracelet	Ring	Geometric/ Carved	Perforated	Tool	Other Artifact	Manufacturing			Fragment		Whole Shell	
	Whole Shell	*Cylinder/ Choker*	*Cut/ Disc*	*Whole Shell*	*Tinkler*	*Cut/ Carved*							*Bracelet*	*Ring*	*Carved*	*Worked*	*Unworked*		
AZ T:12:47(ASM)	—	—	—	1	—	—	2	—	—	—	—	—	—	—	1	1	13	—	18
AZ T:12:62(ASM)	4	5	4	1	2	—	8	—	—	—	—	—	—	—	5	3	35	2	69
AZ U:15:59(ASM)	10	12	—	8	3	3	11	—	2	—	—	—	—	—	3	20	28	1	101
AZ T:12:14(ASU)	—	—	—	1	—	—	2	1	—	—	—	—	—	—	—	1	14	—	19
AZ T:12:10(ASM)[1]	4	1	2	—	1	—	11	1	1	—	—	—	—	—	—	2	2	1	26
AZ T:12:10(ASM)[2]	—	1	1	—	—	—	1	—	—	—	—	—	—	—	—	—	—	3	6
Total	18	19	7	11	6	3	35	2	3						9	27	92	7	239

AZ T:12:47(ASM) = Pueblo Salado; AZ T:12:62(ASM) = Dutch Canal Ruin; AZ U:15:59(ASM) = El Polvorón (Vokes 1984); AZ T:12:14(ASU) = Grand Canal Ruins (Gross 1989); AZ T:12:10(ASM)[1] = Las Colinas (Urban 1981); AZ T:12:10(ASM)[2] = Las Colinas (Vokes 1988)

Table 13.8a. Comparison of Marine Shell from Selected Soho Phase Sites

Genus	Site					Total
	AZ T:12:47 (ASM)	AZ U:15:32 (ASM)	AZ U:15:27 (ASM)	AZ T:12:10 (ASM)[1]	AZ T:12:10 (ASM)[2]	
Bivalve						
Glycymeris	15	22	14	8	8	67
Laevicardium	20	13	23	1	39	96
Pecten	—	—	5	1	—	6
Argopecten	—	—	—	—	2	2
Spondylus	—	2	—	—	—	2
Ostrea	1	—	—	—	—	1
Chione	—	—	—	—	—	—
Unidentified	5	—	—	—	—	5
Univalve						
Olivella	21	—	—	6	2	29
Conus	4	3	2	—	—	9
Nassarius	1	1	1	—	—	3
Turritella	—	—	—	—	—	—
Vermicularia	1	—	—	—	—	1
Agaronia	—	—	—	—	—	—
Theodoxus/Nerita	—	—	—	—	—	—
Columbella	—	—	—	—	—	—
Strombus	—	—	—	—	—	—
Kelletia	—	—	—	—	—	—
Trivia	—	—	—	—	—	—
Crucibulum	—	—	—	—	—	—
Haliotis	—	2	—	—	—	2
Unidentified	1	—	—	—	—	1
Unidentified	—	1	6	5	3	15
Total	69	44	51	21	54	239

AZ T:12:47(ASM) = Pueblo Salado; AZ U:15:32 and AZ U:15:27(ASM) = Escalante Ruin Group (Debowski 1974); AZ T:12:10(ASM)[1] = Las Colinas (Urban 1981); AZ T:12:10(ASM)[2] = Las Colinas (Vokes 1988)

Table 13.8b. Comparison of Marine Shell from Selected Civano Phase Sites

Genus			**Site**			**Total**
	AZ T:12:47 (ASM)	**AZ U:15:3 (ASM)**	**AZ U:15:19 (ASM)**	**AZ T:12:10 (ASM)[1]**	**AZ T:12:10 (ASM)[2]**	
Bivalve						
Glycymeris	15	10	122	2	2	151
Laevicardium	40	193	126	1	28	388
Pecten	—	—	—	—	1	1
Argopecten	—	—	—	—	—	—
Spondylus	—	2	2	—	—	4
Ostrea	—	—	—	—	—	—
Chione	—	—	—	—	—	—
Unidentified	1	—	4	—	—	5
Univalve						
Olivella	15	4	8	9	7	43
Conus	9	4	24	—	—	37
Nassarius	2	14	1	—	—	17
Turritella	1	—	—	—	—	1
Vermicularia	1	2	—	—	—	3
Agaronia	—	—	2	—	—	2
Theodoxus/Nerita	—	—	2	—	—	2
Columbella	—	—	1	—	—	1
Strombus	—	2	—	1	—	3
Kelletia	1	—	—	—	—	1
Trivia	1	—	—	—	—	1
Crucibulum	—	—	—	—	1	1
Haliotis	2	—	2	—	—	4
Unidentified	3	—	2	—	—	5
Unidentified	13	53	3	1	5	75
Total	104	284	299	14	44	745

AZ T:12:47(ASM) = Pueblo Salado; AZ U:15:3(ASM) = Escalante Ruin (Debowski 1974); AZ U:15:19(ASM) = Las Fosas (Vokes 1984); AZ T:12:10(ASM)[1] = Las Colinas (Urban 1981); AZ T:12:10(ASM)[2] = Las Colinas (Vokes 1988)

Table 13.8c. Comparison of Marine Shell from Selected Polvorón Phase Sites

Genus	Site					Total
	AZ T:12:47 (ASM)	AZ T:12:62 (ASM)	AZ U:15:59 (ASM)	AZ T:12:10 (ASM)[1]	AZ T:12:10 (ASM)[2]	
Bivalve						
Glycymeris	2	8	23	12	—	45
Laevicardium	4	27	28	2	1	62
Pecten	—	—	—	—	—	—
Argopecten	—	1	1	—	—	2
Spondylus	—	—	—	—	1	1
Ostrea	—	—	—	—	—	—
Chione	—	1	—	—	—	1
Unidentified	—	2	1	—	—	3
Univalve						
Olivella	—	5	7	6	—	18
Conus	1	11	36	1	1	50
Nassarius	—	1	2	—	—	3
Turritella	1	—	2	—	—	3
Vermicularia	—	—	—	—	—	—
Agaronia	—	—	—	—	—	—
Theodoxus/Nerita	—	—	—	—	—	—
Columbella	—	—	—	—	—	—
Strombus	—	—	—	—	—	—
Kelletia	—	—	—	—	—	—
Trivia	—	—	—	—	—	—
Crucibulum	—	—	—	—	—	—
Haliotis	—	—	3	—	—	3
Unidentified	—	—	—	—	—	—
Unidentified	—	4	—	7	—	11
Total	8	60	103	28	3	202

AZ T:12:47(ASM) = Pueblo Salado; AZ T:12:62(ASM) = Dutch Canal Ruin; AZ U:15:59(ASM) = El Polvorón (Vokes 1984); AZ T:12:10(ASM)[1] = Las Colinas (Urban 1981); AZ T:12:10(ASM)[2] = Las Colinas (Vokes 1988)

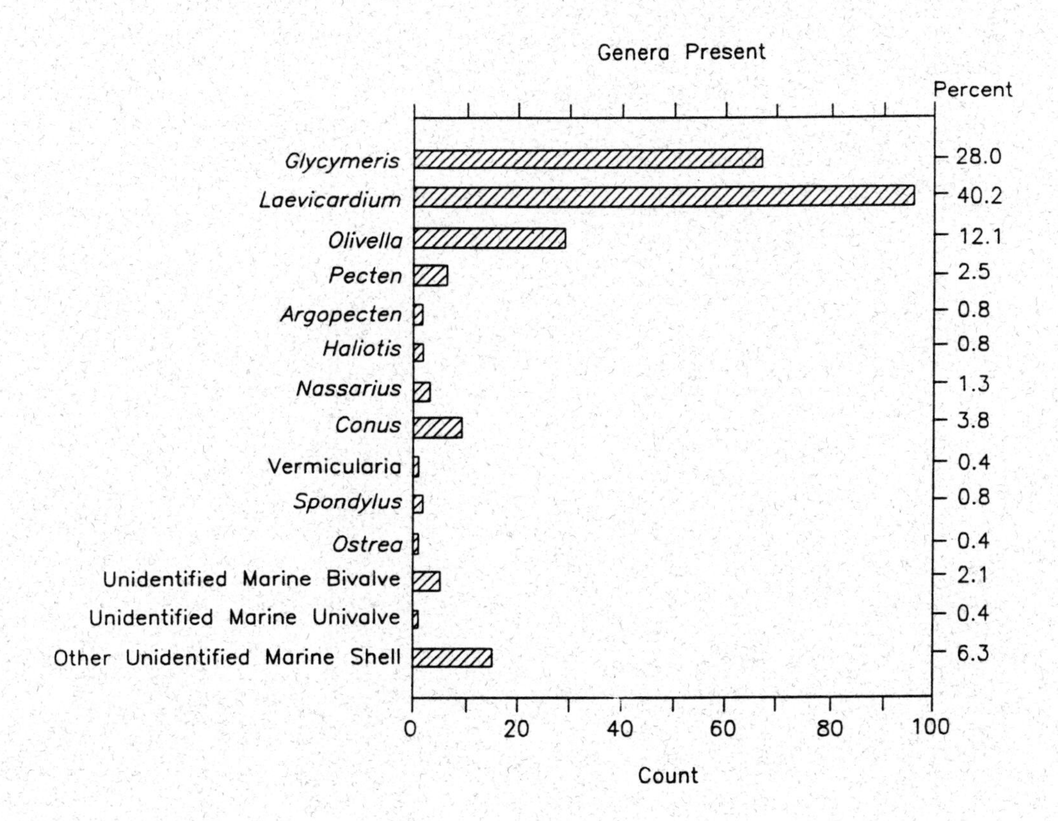

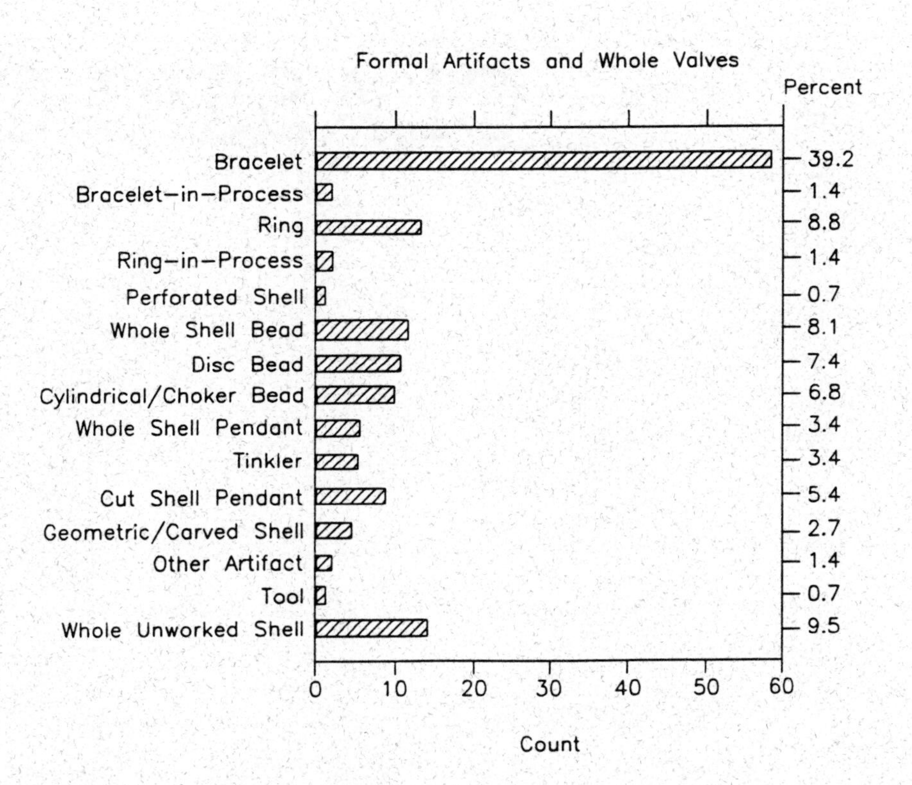

Figure 13.3. Soho phase marine shell assemblage, selected sites.

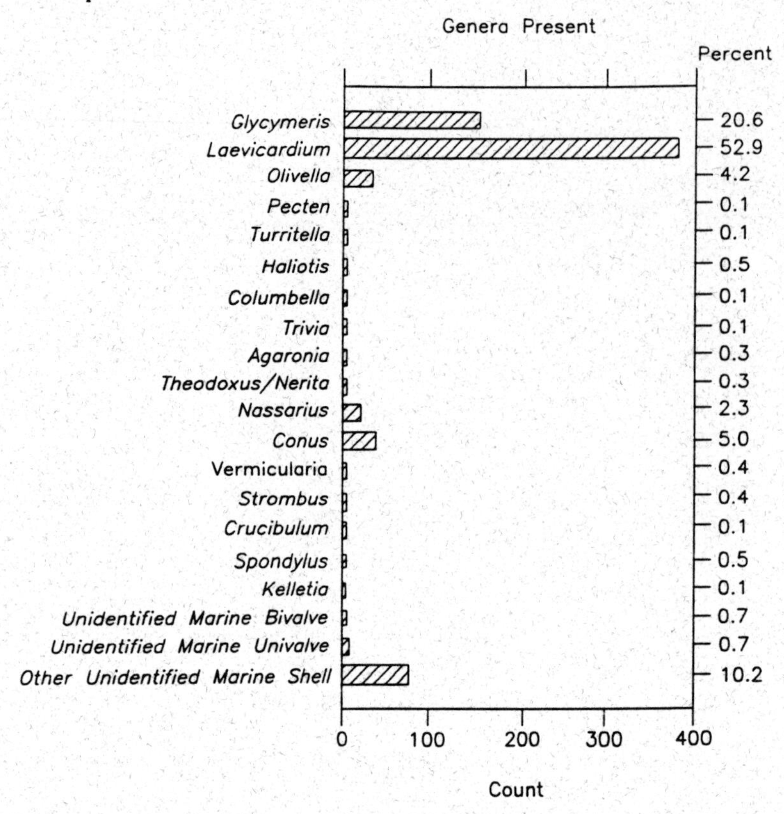

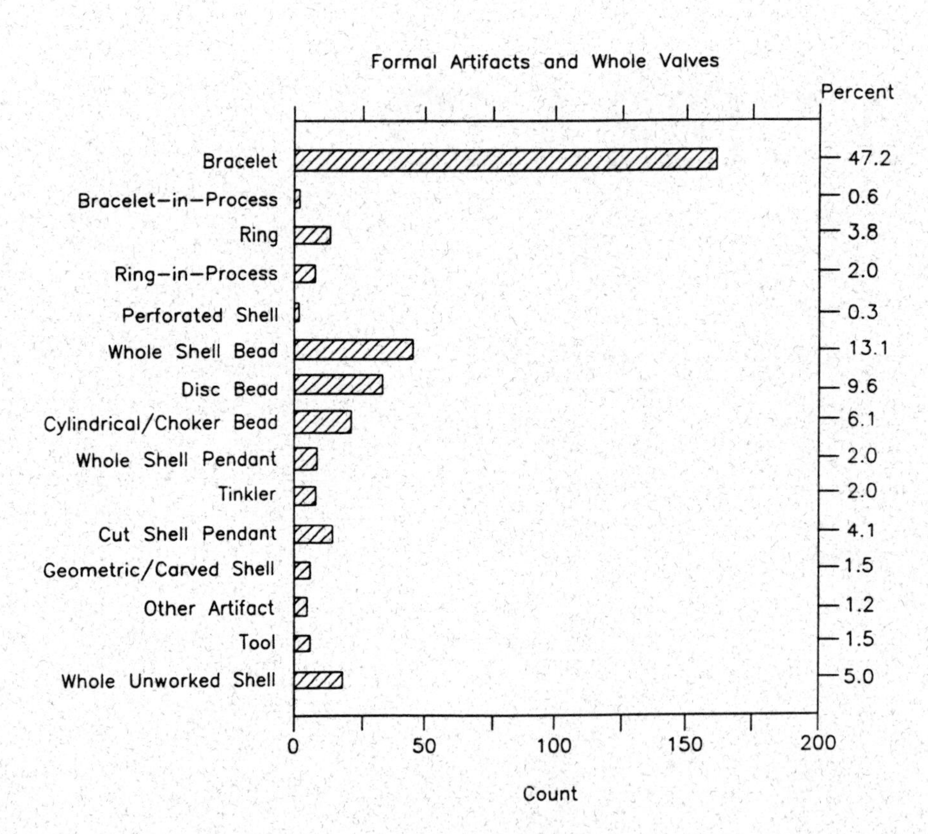

Figure 13.4. Civano phase marine shell assemblage, selected sites.

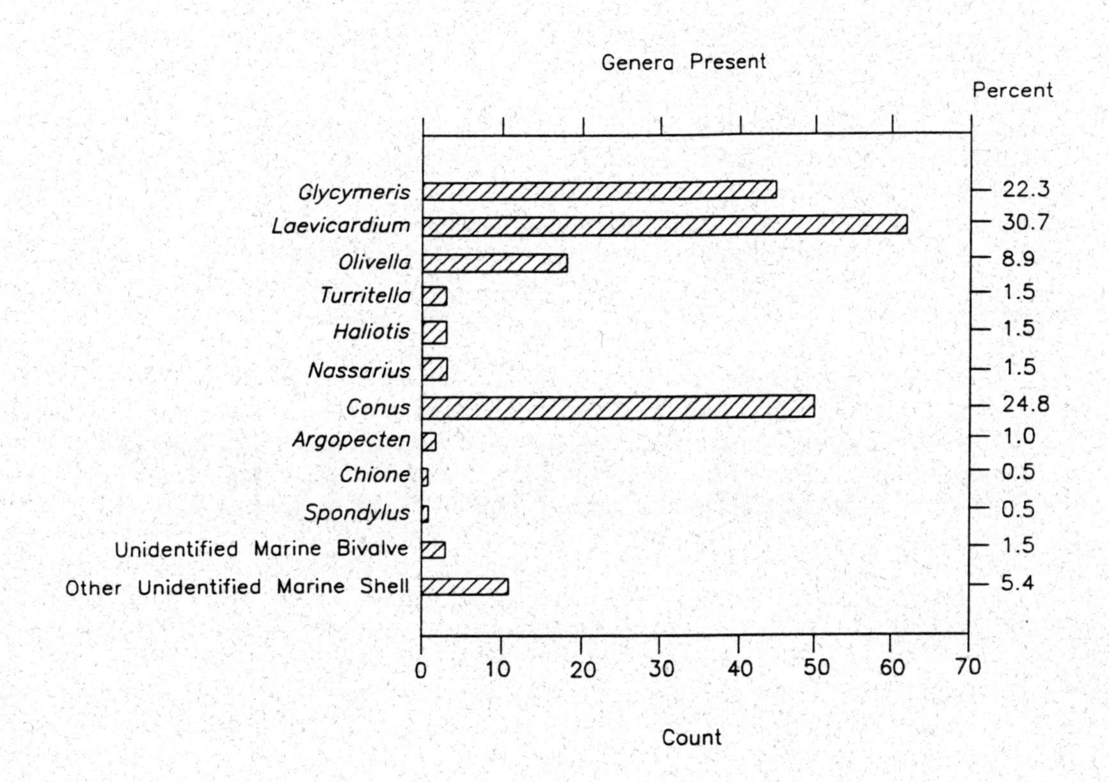

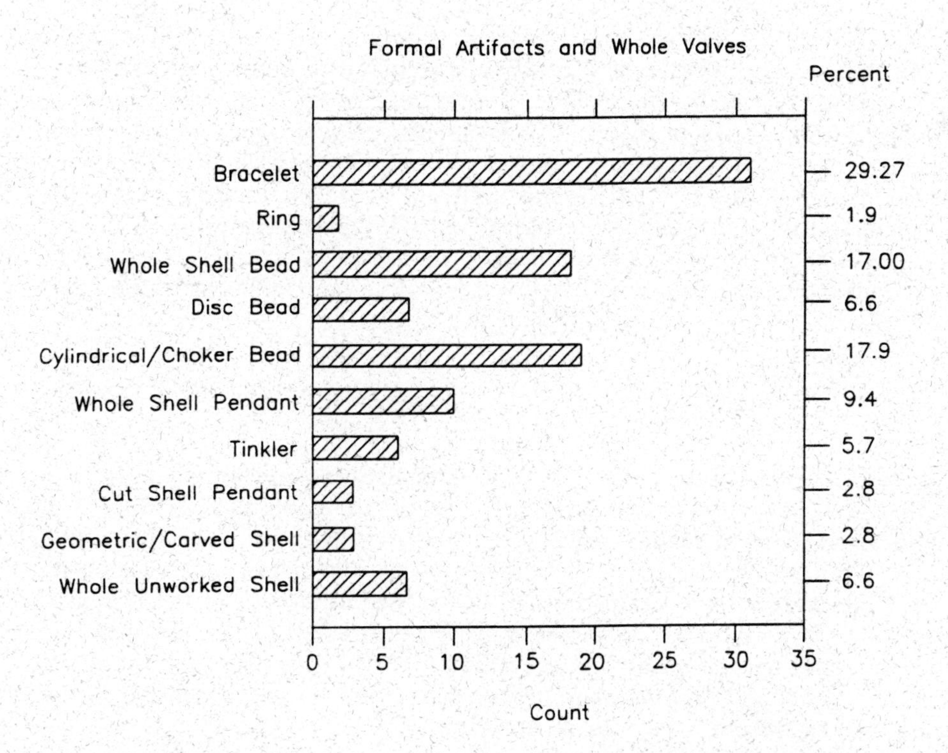

Figure 13.5. Polvorón phase marine shell assemblage, selected sites.

geometric forms and, as with the Phoenix Sky Harbor Center assemblage, were generally rectilinear shapes with rectangular and triangular forms being more common than other shapes.

The marine shell genera most often associated with the Classic period assemblages (Tables 13.8a and b and Figures 13.3 and 13.4) were *Glycymeris, Laevicardium, Olivella,* and *Conus*. The first three have a long history of use by the Hohokam. *Glycymeris* and *Laevicardium* were by far the most common. Bivalves, in general, often represent 75% of an assemblage. In addition to *Olivella* and *Conus*, numerous other univalves were present in low frequencies in the Classic period assemblages, especially during the Civano phase; most of the univalves had been employed as whole shell or cylindrical beads, forms for which many genera can be substituted.

The shell artifact assemblage from the Polvorón phase (Table 13.7c, Figure 13.5) was characterized by an increase in whole shell forms and related artifact types, with a decrease in bracelets to 31%. Whole shell forms appeared to have increased in relative popularity at the expense of cut shell beads, pendants, and bracelets. At the Phoenix Sky Harbor Center, bracelets represented 31% of the Polvorón phase assemblage (Table 13.6), quite a high percentage by comparison with the 22% reported for El Polvorón (Vokes 1984:542, 543). Investigators have not yet recovered any *Glycymeris* rings/pendants from Polvorón phase contexts in the region. The bead form that exhibited the greatest increase in its popularity was the cylindrical bead, with *Conus* choker beads or rings the principal variety. Disc beads, not present at El Polvorón, were a minor aspect of the Phoenix Sky Harbor Center assemblage. As with the beads, the pendants tended to be either whole forms or tinklers, both in the Sky Harbor Center assemblage and regionally. The cut pendants reported from these sites were geometric forms, principally rectilinear shapes.

Although *Laevicardium* was the most common genus in the combined Polvorón phase assemblages (Table 13.8c, Figure 13.5), it was not as well represented as the medium of choice for formal artifacts as in earlier phases. The abundance of unworked fragments may reflect the use of these valves as containers but more likely represents the residue from on-site manufacturing of artifacts. The major change in the relative frequency of genera present was the increase in *Conus* (24.7% in Polvorón phase assemblages versus 3.8% and 5.0% in Soho and Civano phase assemblages, respectively [Tables 13.8a, b, c]). This was clearly a reflection of the popularity of the associated bead and tinkler forms. The decline of *Glycymeris* to approximately 22% of the assemblage is a reflection of the decline of bracelets in the Polvorón phase assemblages. In general, marine bivalves declined to approximately 56% of the assemblage, while the marine univalves increased to roughly 38%.

SUMMARY

The excavations at Phoenix Sky Harbor Center provided an opportunity to examine material from both the Classic and post-Classic periods of the Hohokam tradition. The phases were represented by relatively discrete assemblages from specific areas of the sites investigated. This was particularly important for the Polvorón phase material, about which archaeologists know relatively little. The Polvorón phase was defined in 1984 with the publication of the results of the excavations along the Salt–Gila Aqueduct (Crown and Sires 1984; Sires 1984). The Phoenix Sky Harbor Center Project presented a second assemblage attributed to the phase and, through application of the attributes used to define Polvorón phase features and contexts, provided the opportunity to further assess shell materials from other sites in the Hohokam region.

In general, the results of the Phoenix Sky Harbor Center shell analysis supported the patterns reported from El Polvorón, with some modifications. One pattern that appears obvious is the increase in frequency of occurrence of shell from the Soho to the Civano phase, followed by a rapid decline during the Polvorón phase. The relative frequency of whole shell forms more than doubled during the post-Classic period, at least from the sites examined. The differences between the assemblages may reflect subtle temporal factors or other local factors, such as changes in exchange patterns or, with the Polvorón phase materials, a decline in the social organization of the Hohokam culture. As investigators gather more information on the post-Classic period, the portrait of the shell assemblage of the Polvorón phase will undoubtedly undergo further clarification.

CHAPTER 14

PROJECT REVIEW AND SUMMARY

David H. Greenwald
Dawn M. Greenwald
Richard V. N. Ahlstrom

The Phoenix Sky Harbor Center Development Project, sponsored by the City of Phoenix Community and Economic Development Department, consisted of four related archaeological projects. (1) In 1985, BRW, Inc. (1986) undertook a Class I overview (records search and review of relevant information) and Class III survey of the project area (Figure 14.1). That survey identified surface remains and located archival information concerning the spatial extent of Dutch Canal Ruin (AZ T:12:62, ASM) and Pueblo Salado (AZ T:12:47, ASM) within the Phoenix Sky Harbor Center. (2) During the fall of 1988, BRW conducted testing on approximately 24 ha (60 acres) within a portion of the Phoenix Sky Harbor Center bounded by Buckeye Road, Mohave Street, 18th Street, and the Squaw Peak Parkway (BRW 1989). Two loci, Areas 7 and 20, and portions of a third locus, Area 8/9, of Pueblo Salado were located within the southern half of that area. (3) In February of 1989, SWCA, Inc., initiated testing within the remaining areas of the Phoenix Sky Harbor Center. Volume 1 of this multivolume publication focuses on the results of those testing efforts and contains (4) the research design and work plan for the data recovery phase, which SWCA initiated in 1991. The results of the data recovery field and analytical efforts are presented in Volumes 2 and 3; this volume contains synthetic studies and project conclusions.

TESTING APPROACH AND SUMMARY OF RESULTS AT DUTCH CANAL RUIN

When SWCA began its testing program at Dutch Canal Ruin in February 1989, the site was known primarily through the archival maps produced by Frank Midvale in the 1930s (Wilcox 1987), although investigations in the Squaw Peak Parkway in 1986 had identified a pre-Classic period field house locus with associated canals in that area of the site (Greenwald and Ciolek-Torrello 1988a). Midvale's maps indicate that Dutch Canal Ruin extended for over 2.1 km (1 mile) east-west by approximately 0.8 km (0.5 mile) north-south and contained several features interpreted as trash mounds (Volume 1:Figure 4.13). On the basis of these mounds, researchers considered Dutch Canal Ruin a habitation site, and within the site, two areas (refer to Volume 1:Figure 1.6) appeared to have high potential for containing intact subsurface habitation features. Most of the area containing the mounds in the eastern portion of the site lay beyond the project boundaries, whereas the mounds in the western portion of the site appeared to be primarily within those boundaries. Systematic backhoe trenching based on this information achieved a 2% sampling level in the western portion of the site and a 1% level elsewhere within the site boundaries. In areas where field personnel encountered subsurface features, judgmental trenches placed at 5-m intervals to each side of the systematic trench determined if other features were present. This strategy was appropriate to the goals of the testing phase and resulted in the definition of 16 areas that contained habitation features (Volume 1:Figure 6.1).

The testing results indicated that at least five canal alignments partially or completely traversed the project area, with two of these alignments originally identified during testing and excavation in the Squaw Peak Parkway (Greenwald and Ciolek-Torrello 1988a). Investigators also identified field house loci throughout the project area, although at a higher concentration in the western half (Volume 1:Figure 6.1). With the exception of one concentration of features, all of the habitation

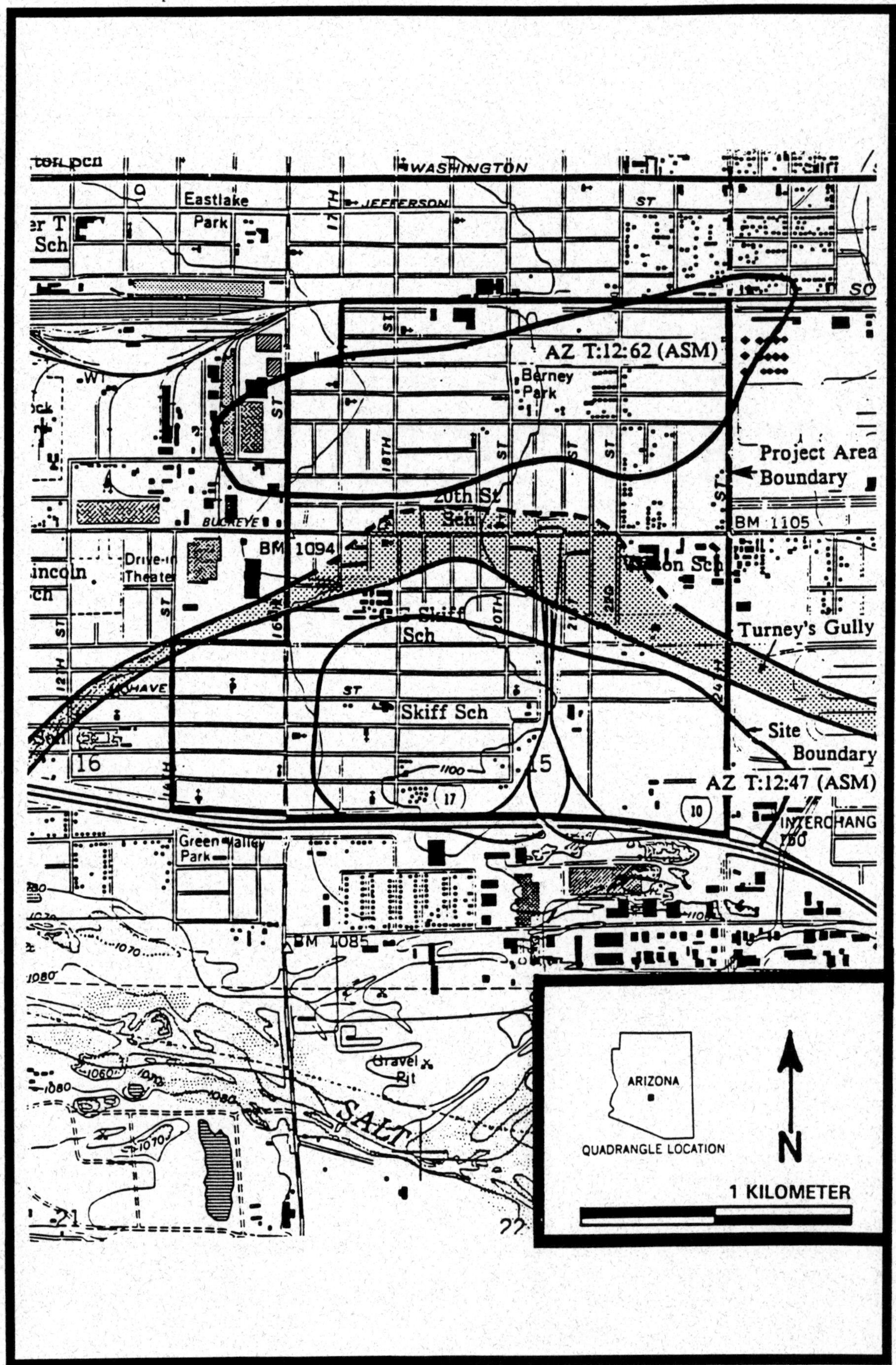

Figure 14.1. Map showing the Phoenix Sky Harbor Center and the site boundaries of Dutch Canal Ruin (AZ T:12:62, ASM) and Pueblo Salado (AZ T:12:47, ASM). Base map is Phoenix, Arizona, 7.5 minute USGS topographic map, photorevised 1982.

loci were assigned to the pre-Classic periods. Area 8 was markedly different from the other habitation loci, containing higher frequencies of artifacts at a stratigraphically higher position than those assigned to the pre-Classic period loci. Given the occurrence of Roosevelt Red Ware (Salado Polychromes), investigators assigned Area 8 to the Classic period. The Squaw Peak Parkway investigations (Greenwald and Ciolek-Torrello 1988a) had concluded that occupation of the freeway corridor area was associated with the Pioneer and Colonial periods. Preliminary interpretations of the Phoenix Sky Harbor Center testing phase data primarily supported a Colonial period occupation, although Pioneer and Sedentary period ceramics were present. Occupation during the pre-Classic periods appeared to have been of a temporary, seasonal nature, whereas occupation of Area 8 appeared to represent a more intensive use of the project area, perhaps permanent settlement. The subsequent data recovery phase supported this interpretation.

The results of the testing did not demonstrate a direct correlation between the features discovered and the Midvale maps, possibly due to the fact that the maps were not to scale or had never been intended as accurate representations of the site. The only correlation that project investigators could find was the increased number of subsurface features in the western portion of the project area, where Midvale's maps indicated mounds. The subsurface remains at Dutch Canal Ruin indicated a significantly different site function from that which archaeologists had inferred based on Midvale's archival records. Further review of the Midvale maps indicated that they had apparently been revised in ways that suggested the remains were more complex than they actually were. Project researchers could not determine if Midvale had made these revisions or if they were the product of later use of his maps by someone else. No notes have been discovered that provide explanations for his maps, and therefore researchers do not know precisely how Midvale perceived Dutch Canal Ruin.

To summarize, during testing at Dutch Canal Ruin, field personnel excavated over 11,275 m of backhoe trenches. The backhoe trenches revealed 65 habitation features, 5 cultural anomalies, 17 stratigraphic anomalies (unconfirmed cultural features), and 4 canal alignments (see Table 14.1 for a summary of testing results). Extensive recording and evaluation of the canals during the testing phase defined much of the canal network in the project area according to its spatial, temporal, and functional attributes. The field research also focused on the geomorphology of the project area, including identification of irrigation-impacted soils and definition of the paleo-geomorphology of the geologic floodplain. Project investigators considered these two topics important for understanding the physical characteristics of the area and the limitations that these characteristics may have imposed on prehistoric land-use strategies. Based on the testing results, a data recovery phase was recommended to mitigate potential impacts to the cultural resources at Dutch Canal Ruin by the proposed construction within the Phoenix Sky Harbor Center.

TESTING APPROACH AND SUMMARY OF RESULTS AT PUEBLO SALADO

Ahlstrom and Phillips (1983) initially recognized archaeological surface remains in the southern half of the Phoenix Sky Harbor Center (Site AZ T:12:47, ASM) during a survey of the area for the U.S. Army Corps of Engineers. Testing in the Squaw Peak Parkway for the Arizona Department of Transportation identified canal and habitation features, confirming at least in the freeway corridor the association of surface and subsurface remains (Euler and Sires 1984). Survey of the Phoenix Sky Harbor Center by BRW (1986) further identified the extent of surface artifacts associated with the archaeological site identified by Ahlstrom and Phillips (1983) within an area measuring approximately 0.83 km (0.75 mile) east-west by 0.65 km (0.40 mile) north-south. Testing within a portion of the Phoenix Sky Harbor Center by BRW (1989) also confirmed associations between surface artifacts and subsurface remains. Given the initial testing efforts at Pueblo Salado,

Table 14.1. Features at Dutch Canal Ruin (AZ T:12:62, ASM) Identified during Testing and Monitoring and Investigated during Data Recovery

Feature Type	Feature Number	Temporal Association	Area Investigated
Testing			
All Areas			384 trenches/11,275 m
Pit structure	15		
Possible pit structure	10		
Structure NFS	5		
Pit NFS	20		
Fire pit	4		
Pit with burning	1		
Pit with rocks	1		
Trash deposit	1		
Inhumation	3		
Cultural anomaly	5		
Canal	4		
Crosscut channel	3		
Monitoring			175,000 m^2
Areas 9 and 10			
Pit structure	1		
Pit NFS	2		
Horno	2		
Fire pit	1		
Trash deposit/pit	4		
Artifact concentration	3		
Canal	1		
Data Recovery			
Area 1			189 m^2
Pit structure	2	Colonial	
Pit with burning	2	Colonial	
Primary inhumation	1	Colonial	
Historic pit	1	Historic	
Area 2			134 m^2
Pit structure	1	Colonial	
Possible pit structure	1	Colonial	
Area 3			500 m^2
Pit structure	7	Colonial	
Fire pit	4	Colonial	
Pit with burning	6	Colonial	
Pit with rocks	1	Colonial	
Rock cluster	1	Colonial	
Area 4			146 m^2
Pit NFS	1	Unknown	
Area 5			26 m^2
Pit structure	1	Colonial/Sedentary	
Canal	2	Colonial	

Table 14.1. Features at Dutch Canal Ruin (AZ T:12:62, ASM) Identified during Testing and Monitoring and Investigated during Data Recovery, continued

Feature Type	Feature Number	Temporal Association	Area Investigated
Area 7			
Pit structure	1	Colonial	
Canal	2	Colonial/Classic	
Area 8			620 m²
Pit house	1	Polvorón	
Pit structure	1	Polvorón	
Pit NFS	10	Polvorón	
Wall segment	1	Polvorón	
Fire pit	6	Polvorón	
Trash pit	2	Polvorón	
Posthole	1	Polvorón	
Secondary cremation	2	Polvorón	
Area 9			8400 m²
Pit NFS	1	Unknown	
Trash deposit	1	Pioneer through Classic	
Area 10			5600 m²
Pit structure	1	Colonial	
Pit NFS	1	Classic	
Horno	2	Classic	
Fire pit	1	Classic	
Trash deposit/pit	3	Classic	
Artifact concentration	3	Classic	
Canal	1	Unknown	

NFS = not further specified

SWCA investigators designed a testing program that focused primarily on the surface artifact concentrations previously designated Loci 8, 9, and 14 (Ahlstrom and Phillips 1983; BRW 1986). Field personnel excavated trenches at the 2% level of coverage in the systematic format used at Dutch Canal Ruin. Judgmental trenches further delineated the distribution of subsurface features within the previously defined loci and tested other areas considered to have low potential for containing subsurface remains. Excavators also used trenches to examine the geomorphology of the area, which provided important information on the occurrence of intact subsurface remains. The extreme southern portion of the project area (south of Mohave Street) appeared to have been downcut or scoured by the Salt River, and investigators considered the potential for intact deposits in that area to be extremely remote. Therefore, they restricted testing to geomorphic and judgmental trenches between Mohave Street and the Interstate 10 freeway.

Testing efforts confirmed the presence of intact subsurface remains at Loci 8 and 9 (Area 8/9) and Locus 14 (Area 14) and traced the canal alignment initially identified by Euler and Sires (1984) and later followed by BRW (1989) from the Squaw Peak Parkway west to 18th Street. Habitation features were associated with the Classic period. Investigators considered features concentrated in Area 8/9 to be part of a compound that appeared to be temporally later than the remains in Area 14, based on the associated ceramics (Volume 1:Chapter 7). The presence of Roosevelt Red Ware sherds suggested a late Classic period affiliation, and SWCA investigators anticipated that a post-Classic period occupation might be represented. Geomorphic studies identified subsurface areas of oxidized soils that might be prehistoric agricultural fields buried by Salt River floods, possibly preserving evidence of field attributes. Since archaeologists have identified few agricultural fields in the Hohokam area, these features had the potential to be very significant.

During the testing phase of the project, field personnel excavated over 4525 m of backhoe trenches at Pueblo Salado. In addition to identifying features (Table 14.2), they traced the north branch of Canal Salado from 18th Street west to 16th Street, extending the confirmed alignment of this canal for over 0.3 km (0.5 mile). The testing efforts of both SWCA and BRW archaeologists revealed cultural deposits that contained important information, and project investigators made recommendations for a data recovery phase to recover important information concerning the prehistoric occupation and settlement of the geologic floodplain within the Phoenix Basin and to mitigate potential impacts to those resources from the proposed construction activities.

DATA RECOVERY APPROACH AND SUMMARY OF RESULTS AT DUTCH CANAL RUIN

The archaeological remains at Dutch Canal Ruin consisted of habitation features that had been occupied during the pre-Classic, Classic, and post-Classic periods and canal features that had been in use during the pre-Classic and Classic periods. The pre-Classic period occupation had been neither continuous nor permanent, and some phases were poorly represented or not represented at all. Classic and post-Classic period settlement, however, appeared to have been permanent or year-round, although settlements did not appear to have exceeded the size of small, extended family hamlets. The general focus of prehistoric land use was on farming, and investigators identified the agricultural zone at Dutch Canal Ruin as an area that extended in a linear fashion along the various canal alignments on the geologic floodplain.

Investigative Strategy

Investigators at Dutch Canal Ruin employed the general strategy of excavating samples of the feature categories and temporal periods represented; the excavations included approximately one-half of the habitation loci. Project investigators selected field house loci for their apparent degree of preservation, use of space, and potential for yielding information. Field personnel investigated Classic period loci primarily as a result of their discovery during construction monitoring. Initially, investigators selected the single post-Classic period locus (Area 8) because they thought it was a permanent habitation area dating to the Classic period; they did not recognize it as post-Classic until the data recovery phase. Canal mechanics and maintenance were a focus during the data recovery phase, as the strategy during the testing phase had been to gather as much information as possible regarding the paleohydraulics of canals. The investigating team specifically targeted two areas: the bifurcation of the canal systems in Area 6, and channel maintenance in Area 7. An opportunity to determine the temporal relationship between two canal alignments in Area 7 was incidental to the primary objectives. These efforts resulted, however, in the identification of the only Classic period canal known to exist in the portion of Dutch Canal Ruin located within the project area.

During the testing phase, investigators identified 16 habitation areas (Figure 14.2) in the Phoenix Sky Harbor Center at Dutch Canal Ruin. During data recovery they further investigated 9 of these areas, all dating to the Colonial period or later, per the techniques defined in the plan of work (Volume 1:Chapter 11). The only Pioneer period component was in the Squaw Peak Parkway and

Table 14.2. Features at Pueblo Salado (AZ T:12:47, ASM) Identified during Testing and Investigated during Data Recovery

Feature Type	Feature Number	Temporal Association	Area Investigated
Testing			
Area 8/9			65 trenches/1880 m
Adobe-walled structure	23		
Pit structure	13		
Possible structure	8		
Pit feature	47		
Burial	14		
Borrow pit	4		
Trash pit	5		
Miscellaneous feature	22		
Area 14			23 trenches/730 m
Pit structure	2		
Pit feature	3		
Burial	2		
Miscellaneous feature	8		
Data Recovery			
Area 8/9			5886 m^2
Semisubterranean adobe-walled structure	12	Civano	
Surface adobe-walled structure	3	Civano	
Pit house	4	Polvorón	
Adobe-walled structure NFS	14	Civano	
	*1		
Pit structure	2	Polvorón	
	*1		
Structure NFS	*1	Civano	
Courtyard area	5	Civano	
	*1		
Hearth	*1	Civano	
Pit NFS	4	Civano	
	*1		
Fire pit	4	Civano	
Pit with burning	2	Civano	
Ash pit	2	Civano	
Adobe-lined pit	7	Civano	
	*14		
Storage pit	1	Polvorón	
Borrow pit	1	Civano	
Pit with rock	*1	Civano	
Possible pit	*1	Civano	
Trash deposit/pit	7	Civano	
Cache/artifact concentration	2	Civano	
Primary inhumation	33	Civano	
Primary inhumation	3	Polvorón	
Inhumation NFS	1	Civano	
Human bone concentration	1	Civano	
Primary cremation	3	Civano	
Secondary cremation	2	Civano	
Canal/related feature	5	Civano	

Table 14.2. Features at Pueblo Salado (AZ T:12:47, ASM) Identified during Testing and Investigated during Data Recovery, continued

Feature Type	Feature Number	Temporal Association	Area Investigated
Area 14			335 m^2
Semisubterranean adobe-walled structure	2	Soho	
Pit house	1	Soho	
Ramada	1	Soho	
Adobe-walled structure NFS	1	Soho	
Semisubterranean post-reinforced adobe-walled structure	1	Soho	
Pit NFS	3	Soho	
Fire pit	1	Soho	
Pit with burning	1	Soho	
Adobe-lined pit	1	Soho	
Trash deposit	4	Soho	
Primary inhumation	3	Soho	

NFS = not further specified
*partially excavated

had been investigated in 1986 prior to construction of the freeway (Greenwald and Ciolek-Torrello 1988a). The excavations included four Colonial period loci (Areas 1, 2, 3, and 7), ranging from single-structure field house loci to multiple-structure farmsteads, a Colonial period field house component within Area 10, and a single Sedentary period locus (Area 5) that contained an isolated structure. Two primarily Classic period loci (Areas 9 and 10) investigated during construction monitoring of the remote parking facilities for Phoenix Sky Harbor International Airport in the eastern part of the site contained habitation features but no Classic period structures. Excavation of a third locus (Area 8) in the western portion of the site indicated, that Area 8 was a post-Classic period hamlet, the only post-Classic period locus identified at Dutch Canal Ruin. Area 4, originally thought to be a field house locus possibly associated with the pre-Classic period, could not be assigned to a specific period, as it contained few artifacts and had been disturbed by historic activity, but may have been pre-Classic.

In addition to Area 5, two other investigated areas held the potential to address questions related to canal mechanics, maintenance, and temporal associations. Canal studies successfully defined the function of the bifurcation in Area 6 and defined the temporal association of Alignments 8501 (Canal Viejo), a pre-Classic period canal, and 8537 (Canal Nuevo), a Classic period canal, both in Area 7.

Settlement Structure and Land-Use Patterns at Dutch Canal Ruin

Pre-Classic Period Field Houses and Farmsteads

Pre-Classic period settlement and land use at Dutch Canal Ruin consisted of seasonally occupied habitation areas associated with canal irrigation. Field house and farmstead loci varied from single structures to areas with multiple structures (e.g., Area 3 included at least seven structures), although

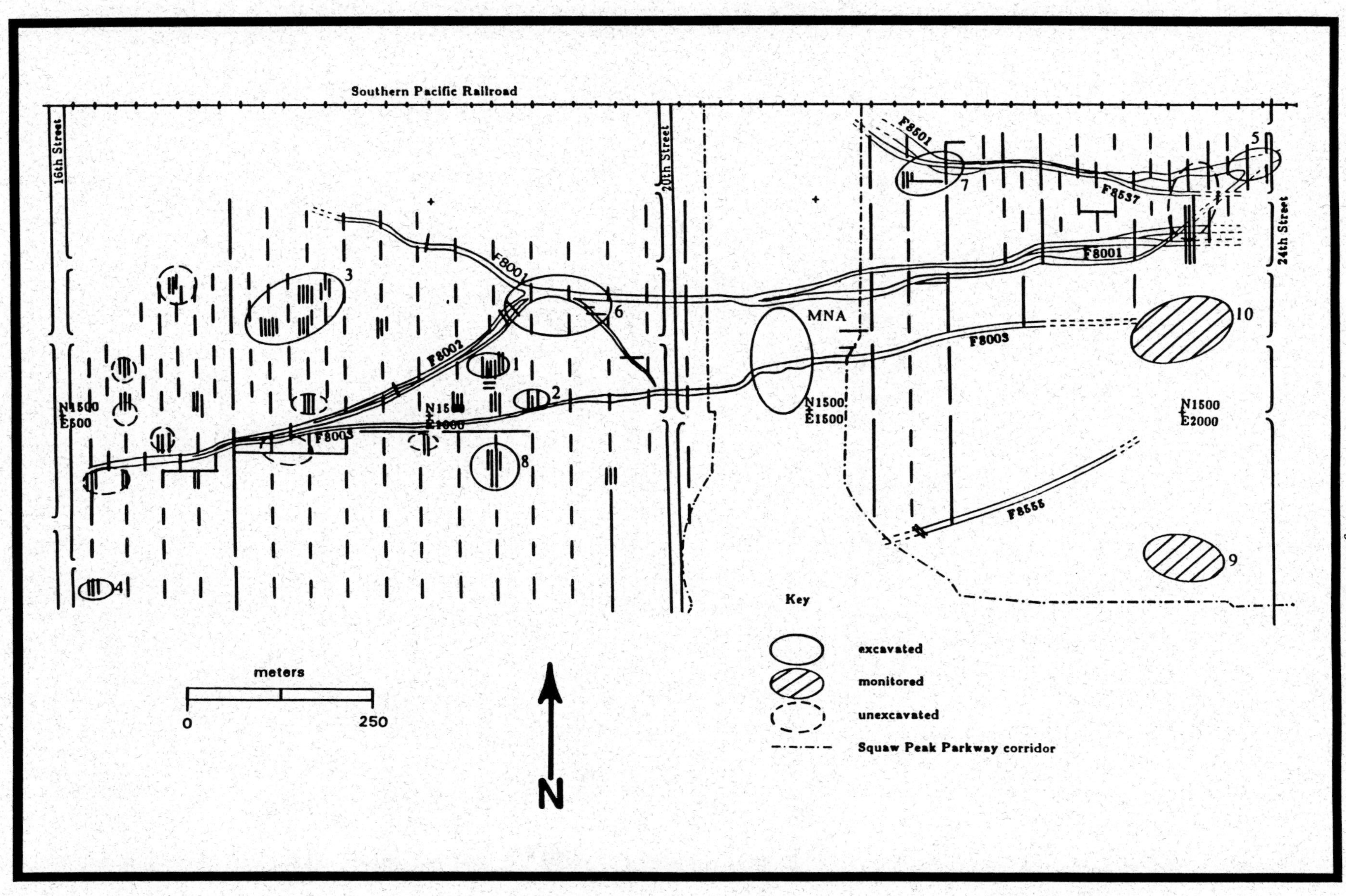

Figure 14.2. Plan view of Dutch Canal Ruin as defined within the Phoenix Sky Harbor Center, showing canal alignments and feature locales. Investigated areas are indicated by numbers; the feature locale excavated by the Museum of Northern Arizona (MNA) is indicated in the Squaw Peak Parkway corridor.

temporal data could not demonstrate contemporaneity for all of the structures. Excavations focused on 12 pre-Classic period pit structures (Table 14.1) of simple form that appeared to have been used on a limited basis and therefore were considered field house structures. Pioneer period occupation apparently had been restricted to an area that corresponds to the Squaw Peak Parkway (Greenwald and Ciolek-Torrello 1988a), although several of the features excavated in that area were also affiliated with the Colonial period. The Colonial period appeared to have been the primary period of field house and farmstead occupation, with the greatest number of features having been occupied during the Santa Cruz phase. The Sedentary period occupation of Dutch Canal Ruin had taken place during the early Sacaton phase. This occupation was not well represented, which may be related to discontinued use of the canals after about A.D. 900 during the late Colonial period, when discharges of the Salt River appear to have reached the highest of their levels during the Hohokam occupation (Nials, Gregory, and Graybill 1989). Following this period, the Hohokam concentrated canal irrigation on the lower bajada and abandoned the Dutch Canal Ruin agricultural zone in favor of the broad bajada that lies west of the Papago Buttes.

Pre-Classic Period Canals

Canals were an important aspect of the land-use and subsistence strategies practiced by the Hohokam farmers at Dutch Canal Ruin during the pre-Classic periods. This conclusion is based on the number of canals present and the use episodes observed within each canal alignment. The Hohokam did not design the canals simply to deliver water to the Dutch Canal Ruin agricultural zone but to service an area that extended along most of the length of each canal. The dispersed pattern of field house loci along these canals may be repeated beyond the project boundaries and the boundaries of Dutch Canal Ruin as currently defined. In summary, the Hohokam may have used the portion of the geologic floodplain extending downstream from the Park of Four Waters to the area south of Las Colinas (inclusive of Dutch Canal Ruin) primarily for agriculture rather than as a residential area, because this area was more susceptible to flooding than the lower bajada and because of its deep alluvial soils that were especially good for irrigation agriculture.

This investigation assigned three canal alignments to the pre-Classic periods based on the data recovered at Dutch Canal Ruin. Each alignment exhibited multiple use episodes, suggesting operation for extended periods of time. The earliest was the South Main Canal (Alignment 8003), initially used during the late Pioneer period. Reorientation to the southwest of a late channel within the North Main Canal (Alignment 8001) allowed the canal builders to reuse the South Main Canal alignment during the Colonial period. (This was the last documented use of pre-Classic period canals at Dutch Canal Ruin.)

The North Main Canal (Alignment 8001) consisted of three distinct and temporally separated canals (Chapter 6; see also Greenwald 1988). The earliest, North Main Canal 1, followed a westerly trajectory through most of the project area, then angled to the northwest (Figure 14.2; Volume 2:Figure 7.1), following an orientation that would eventually extend onto the lower bajada in the vicinity of Pueblo Patricio. This canal may have been constructed during the late Pioneer period, but it appeared to have been replaced by North Main Canal 2 by the early Colonial period. North Main Canal 2 followed a trajectory similar to that of North Main Canal 1 to the extent that it often lay within the earlier canal. Late in its use life, however, it was reoriented to the southwest, eventually intersecting the South Main Canal alignment. This reorientation may account for the bifurcation illustrated by Turney (1924, 1929) and identified in Area 6 during the testing phase of this project. Data recovery in Area 6 determined that the earliest channel of this system branched to the north, whereas the latest branched to the south. The third canal in Alignment 8001, North Main Canal 3, was constructed during the Santa Cruz phase and possibly abandoned by A.D. 900, that period when Salt River discharges may have created major flood events and damaged much of the canal systems in the Phoenix Basin (Nials, Gregory, and Graybill 1989). North Main Canal 3 followed a trajectory similar to that of North Main Canal 2, although it was generally constructed adjacent to the earlier canal rather than within it. It also followed a southwesterly trajectory after reaching the area of the bifurcation, eventually entering and following the South Main Canal alignment in the same channel as North Main Canal 2.

The third pre-Classic period canal at Dutch Canal Ruin, Canal Viejo (Alignment 8501), was located in the northeastern portion of the site. According to their stratigraphic relationships, this alignment predated North Main Canal 3. Investigators recognized at least six rechanneling episodes in this canal, further supporting the interpretation that the canal had a lengthy use history. The orientation of this canal, as with North Main Canals 1 and 2, indicated that in addition to supplying the Dutch Canal Ruin agricultural zone, it also had transferred water onto the lower bajada in the vicinity of Pueblo Patricio and La Ciudad (Turney 1924).

Lateral canals or field ditches associated with North Main Canals 2 and 3 (Greenwald 1988) moved irrigation water from the main canals to field areas. These field ditches contained fine sediments that were high in organic matter, suggesting restricted water movement within these ditches and that they had not been designed to transfer water over long distances. Excavators could trace the laterals only short distances, suggesting that fields had been close to the main canals, reducing or eliminating the need for distribution canals of intermediate size.

Classic Period Loci

Two locations (Areas 9 and 10) in the eastern portion of the project area of Dutch Canal Ruin contained Classic period remains. Area 9 included intermixed pre-Classic period materials, whereas Area 10 consisted almost entirely of Civano phase remains aside from a single Colonial period pit structure. Both areas probably were related to agricultural production, as charred corn kernels were plentiful. Area 9 was poorly preserved and yielded only two features, a pit and a trash deposit, although artifact frequencies were higher than in pre-Classic period loci elsewhere at Dutch Canal Ruin.

Area 10 contained 11 features. A pit structure to the southwest of the main concentration of features probably had been used during the pre-Classic period occupation. The remaining features probably were associated with the Classic period. The feature assemblage also included two hornos, the only two discovered at Dutch Canal Ruin. The number and type of features present suggested permanent occupation, although investigators discovered no Classic period structures. Architecture during this period probably would have consisted of surface adobe structures, which may not have survived truncation by historic and modern activities. The basal portions of pits, however, lay below the level of disturbance.

The presence of charred corn remains at these loci suggested that corn had been part of the diet and that Hohokam farmers may have cultivated corn during the Classic period at Dutch Canal Ruin. At least one canal may have been used in association with these occupation loci (see below).

Classic Period Canals

Canal Nuevo (Alignment 8537), which extended from the eastern project boundary in a west-northwesterly orientation and exited the project area east of the Squaw Peak Parkway, was the only canal dated to the Classic period. Although the fine, laminated deposits within the channel indicated numerous flow episodes and possibly a lengthy use history, investigators found none of the evidence of the successive rechanneling that they observed in the pre-Classic period canals. This suggests that flow velocity (discharge) was regulated at a nonerosive level and that the canal operated at equilibrium, with a tendency toward accumulation of fine sediments. The canal also retained a parabolic shape, suggesting little or no bank erosion and channel widening. In many regards, Canal Nuevo was similar morphologically to North Main Canal 3, the latest pre-Classic period canal at Dutch Canal Ruin. Both canals exhibited reduced bank erosion, increased water regulation, and decreased discharge levels.

Investigators could not assign Canal Barranca (Alignment 8555), which extended from the eastern project boundary in a southwesterly direction, to any temporal period. The portion of the canal within the Phoenix Sky Harbor Center was its terminal section and contained coarse sands rather than fine sediments such as clay that could be archaeomagnetically sampled. Stratigraphic relationships with other canals could not be demonstrated due to the isolated nature of this canal. It may be possible to demonstrate relative associations in the near future by examining the relationship of Canal Barranca to other dated canals in the adjacent parcels of the Phoenix Sky Harbor International Airport.

Post-Classic Period Hamlet

Area 8 at Dutch Canal Ruin had been occupied on a year-round, permanent basis, and investigators assigned it to the Polvorón phase on the basis of architectural morphology and material culture attributes and frequencies (Chapter 8). Support for this assignment came from radiocarbon dates that fit within the range of A.D. 1350–1450 defined for the Polvorón phase (see Chapter 9; Sires 1984). The investigating team could not demonstrate absolute contemporaneity for the two pit structures discovered in Area 8, and the structures may have been from sequential occupations; however, such paired structures may have been a characterizing feature of the Polvorón phase. The small size of these structures supported the notion that the occupants had been nuclear families. If the occupation of the structures was contemporaneous or overlapped, an extended family may have been present during some of this time. Area 8 produced more than 80% of the artifacts from Dutch Canal Ruin but only these two structures. Despite the lack of physical evidence, it is likely that the Dutch Canal Ruin site originally included the same kind of surface structures observed at other Polvorón phase sites (Greenwald and Ciolek-Torrello 1988b, 1988c), but that because of the shallow depth of the surface deposits, historic activities such as plowing destroyed these remains. The quantity of artifacts recovered from Area 8 indicated intensive use, which suggested that structures in addition to those discovered had once existed at the site.

DATA RECOVERY APPROACH AND SUMMARY OF RESULTS AT PUEBLO SALADO

Pueblo Salado archaeological remains dated primarily to the Classic period; a Polvorón phase component discovered at Area 8/9 appeared to have been a continuation of occupation of the compound into the post-Classic period, although the number of individuals during this later occupation was much reduced. Investigators assigned no habitation features to the pre-Classic period, and few diagnostic ceramics represented that period (Volume 3:Chapter 5; BRW 1989). The north branch of Canal Salado, found west of the Squaw Peak Parkway, was in use during the Classic and possibly the post-Classic periods. The generally coarse sediments, similar to those in Canal Barranca (Alignment 8555) at Dutch Canal Ruin, probably represented in-filling of the terminal section of the canal when little or no regulation of discharge was taking place.

Investigative Strategy

The investigating team at Pueblo Salado planned to define the full extent of two habitation loci, Areas 8/9 and 14, and to investigate a sample of all types of features in these occupation areas. Ceramic analysis conducted during the testing phase indicated that the two areas were temporally different, with Area 14 exhibiting an earlier occupation range than Area 8/9. On the basis of the available chronometric data, Area 14 was assigned to the Soho phase and Area 8/9 to the Civano and Polvorón phases. A major goal of the project was to recover all known burials, including the two reported by BRW (1989) in Area 7. The efforts to relocate the BRW burials uncovered a concentration of habitation features that SWCA investigators designated Area 20. The excavation recovered the burials in Area 20 and documented the habitation features in the backhoe trench profiles. The features at Area 20 shared several similarities with those in Area 14, and ceramic data gathered by BRW indicated that Area 20 dated to the Soho phase, an interpretation that was supported by the small sample of sherds recovered by SWCA during the data recovery phase. Area 20 thus appeared to be contemporaneous with Area 14.

Definition of Pueblo Salado within the Phoenix Sky Harbor Center came about initially from surface recording that identified 13 loci (Ahlstrom and Phillips 1983; BRW 1986). Testing by the Arizona State Museum (Euler and Sires 1984) confirmed subsurface remains at 4 loci in the project area. Researchers have reported three additional loci east of the Phoenix Sky Harbor Center: Area 15 and Area 16 in Phoenix Sky Harbor International Airport (Greenwald, Ballagh, and Zyniecki 1996; Greenwald and Zyniecki 1993) and Area 11 east of 24th Street and south of the airport (Cable 1985) (Figure 14.3). The distribution of these remains indicated that Pueblo Salado was aligned east-west, a pattern conforming to the orientation of Canal Salado and influenced by the braided channels of the Salt River.

Settlement Structure and Land-Use Patterns at Pueblo Salado

Soho Phase Settlement

The Soho phase loci at Pueblo Salado formed a dispersed settlement pattern. Areas 6, 14, 15, 16, and 20 were all located immediately south of the north branch of Canal Salado. This relationship may have been influenced by the proximity of the canal to Turney's Gully, a braided channel of the Salt River that has been used to define the north boundary of Pueblo Salado. In many places, the distance between the south edge of Turney's Gully and the canal was less than 100 m. The canal orientation may have been related to optimal use of the floodplain that avoided erosional channels associated with Turney's Gully.

The similar location of Areas 6, 14, 15, 16, and 20 relative to the north branch of Canal Salado suggested that the Soho phase occupation at Pueblo Salado had consisted of scattered, evenly spaced farming loci permanently occupied and dispersed along the alignments of Canal Salado. This pattern was repeated over a distance of approximately 1.4 km, most of the length of the north branch of Canal Salado. The occupations at these four areas each may have consisted of two households, probably extended families. Following the Soho phase, perhaps during the early part of the Civano phase, population aggregation resulted in the establishment of courtyard groups within the compound at Area 8/9 (see below) and also the compound at Area 6 (Greenwald et al. 1996) (Figure 14.3). Recent investigations in Areas 6, 15, and 16 for the expansion of Phoenix Sky Harbor International Airport (Greenwald, Ballagh, and Zyniecki 1996; Greenwald et al. 1996; Greenwald and Zyniecki 1993) have provided further information on the settlement patterns of the Soho, Civano, and Polvorón phases at Pueblo Salado.

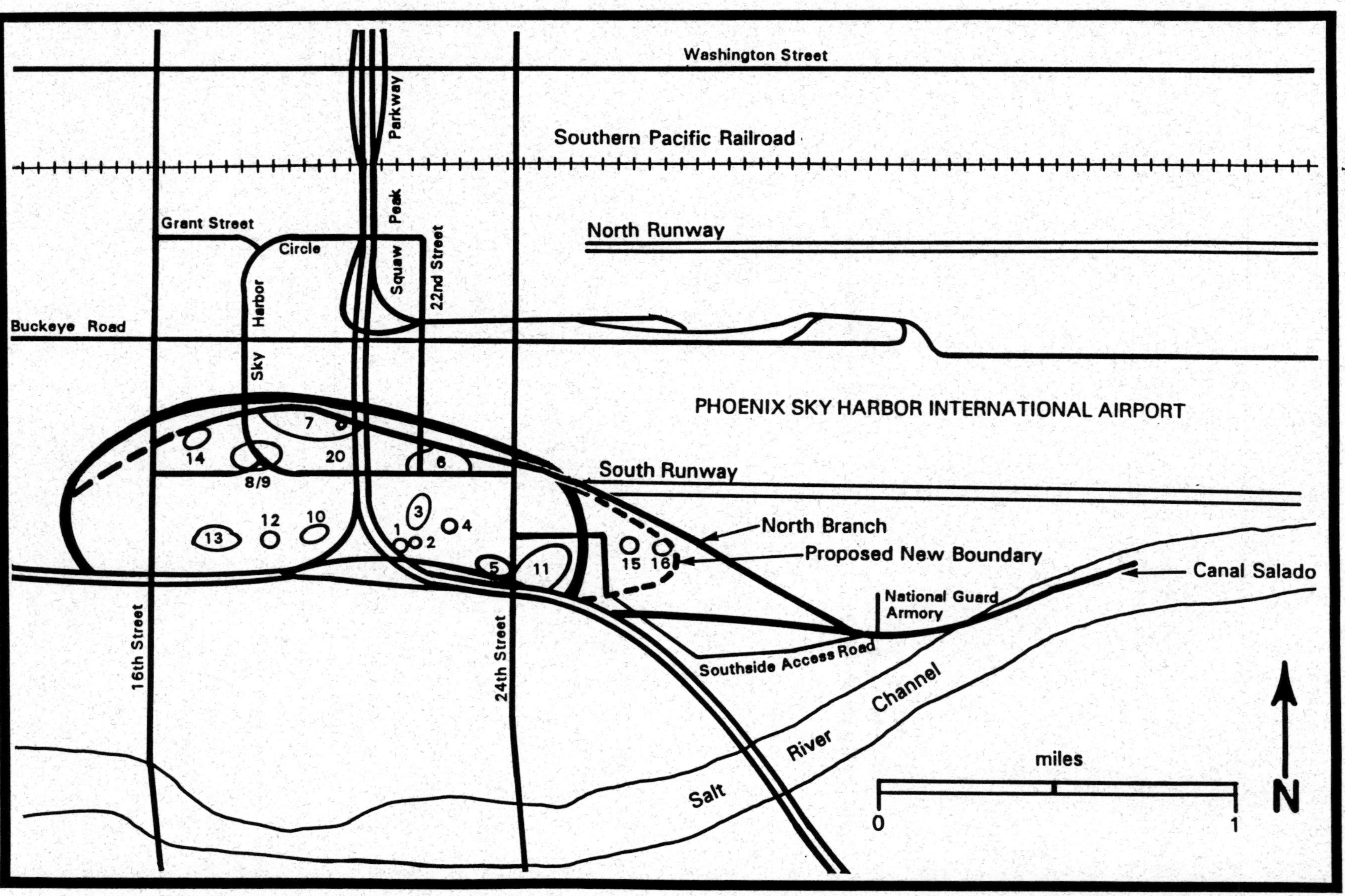

Figure 14.3. Plan view of Pueblo Salado, showing its boundaries, individual areas, the location of Canal Salado and the north branch of Canal Salado, and proposed adjusted boundaries with Area 15.

Civano Phase Settlement

Investigations at Area 8/9 determined that the Hohokam had not occupied this locus prior to the Civano phase (Volume 3:Chapter 2). The initial construction was a compound enclosure of massive adobe walls that measured 62 × 40 m, with a series of rooms built along the north and west walls and a plaza that had served as an outdoor activity area. Although preservation was poor, the extant evidence indicated that the initial compound probably contained the earliest courtyard group at Area 8/9.

Eventually, the inhabitants added at least five other courtyard groups to the initial compound, with a minimum of 36 rooms enclosed by a second compound wall. Double walls, spaced at intervals of between 5 m and 10 m, were built along the north and south sides of the compound, possibly for flood protection. The accretionary development of the compound created an enclosed area that eventually measured at least 85 × 80 m. The aggregation of these courtyard groups may represent the relocation of site residents and the descendants of the hamlets that were dispersed across the landscape during the Soho phase. The distribution of these residential groups within the compound constituted spatially segregated units, similar to the dispersed pattern observed among the Soho phase hamlets. Perhaps this aggregated settlement pattern developed in response to environmental or social pressures that caused restructuring of the earlier dispersed social units to provide increased protection from the threat of flooding or to compete better for agricultural lands and other resources. The result of population aggregation at Pueblo Salado was increased unity and a more defined sociopolitical structure, elements that would have been needed to construct the compound and maintain the canals that served this autonomous settlement.

Polvorón Phase Settlement

Occupation of the compound at Pueblo Salado extended beyond the end of the Civano phase (A.D. 1350), and architectural remains indicated a shift from courtyard groupings of contiguous adobe structures to individual semisubterranean structures, or pit houses. The general shift in house style following A.D. 1350 has been described in Chapter 8 as a post-Classic period phenomenon associated with the Polvorón phase as defined by Sires (1984). Hohokam archaeologists have not yet widely accepted the Polvorón phase, although researchers have steadily been adding to the body of data in support of a post-Classic period occupation. Project investigators assigned six pit structures at Pueblo Salado to the Polvorón phase.

This late pit house occupation represents a period of reduced population levels, a trend that appears at numerous late Classic period sites. Although investigators could not precisely date the shift in residential style and reduction in population levels at Pueblo Salado, the change appeared to have been gradual, which may be contrary to the pattern at other Classic period sites. Some structures may have been purposely abandoned, indicated by stacked vessels and other artifacts in positions that suggested residents had planned to return. Yet other structures appeared to have been reused, possibly until they were no longer desirable and were abandoned in favor of construction of a new house. Archaeologists not uncommonly find both adobe surface structures and pit houses in use during the Polvorón phase; in fact, this was frequently the case at Brady Wash (Greenwald and Ciolek-Torrello 1988b), and surface adobe structures were often remodeled (Chapter 8; Greenwald and Ciolek-Torrello 1988b, 1988c). With these examples in mind, and given the fact that field personnel recognized no hiatus in occupation between the Civano and Polvorón phases, investigators postulated that the compound had been continuously occupied, albeit with changes in settlement composition.

As indicated by the archaeomagnetic intervals and radiocarbon dates, Pueblo Salado was one of the latest occupied sites in the Hohokam sequence on the north side of the Salt River. Chronometric data supported the premise of an occupation that lasted until A.D. 1500–1550 and possibly as late as A.D. 1600. These dates were consistent with chronometric results from other projects, specifically, testing efforts by BRW (1989:86) in Area 7 and canal studies focusing on the north branch of Canal Salado by SWCA (Greenwald and Zyniecki 1993) in the extreme eastern portion of Pueblo Salado. The reduction in population at the Pueblo Salado compound may have followed the general abandonment or depopulation of Canal System 2, although at least a relict population remained at the compound well beyond A.D. 1450.

Agricultural Fields

Archaeologists have long recognized that the Hohokam relied on agriculture as a primary subsistence strategy. In fact, extensive canal systems have become one of the hallmarks of these early desert farmers. Much physical evidence (i.e., irrigation canals, fossil cultigens, and features associated with farming and fields) attests to the prehistoric practice of agriculture in southern Arizona. Rarely, however, has anyone found physical evidence—beyond botanical remains—of Hohokam agricultural fields, generally because of historic and modern disturbance associated with agriculture and urban development. The Pueblo Salado investigations discovered remnants of buried fields on the geologic floodplain. Although investigators could not establish the full extent of these fields, the research identified various physical, chemical, botanical, and archaeological characteristics, in essence compiling an attribute list of the morphology of fields associated with floodplains (Chapter 5). One of the problems in the study of agricultural fields has been the absence of previously undisturbed areas for comparative purposes. Historic and modern farming techniques, especially the application of fertilizers, have reduced our ability to develop a chemical profile for prehistoric fields. Despite this difficulty, researchers have gathered much data regarding buried fields, including information on soil properties; the effects of irrigation on soil; the occurrence of phytoliths, starch, and pollen; and correlations between soil particles and nutrient levels. These data and continued geomorphic consultation should prove valuable to future Hohokam and southwestern agricultural studies.

Canals

Various field investigations at Pueblo Salado since the testing efforts in the Squaw Peak Parkway (Euler and Sires 1984) have compiled considerable data regarding the north branch of Canal Salado. To date, the only information available on the main channel of Canal Salado comes from the early maps of Patrick (1903), Turney (1924, 1929), and Midvale (1968, n.d.). The mapped locations of Canal Salado correspond to the present location of Interstate 10/17, which forms the southern project boundary (Figure 14.1). Recent investigations east of the Phoenix Sky Harbor Center in a portion of the Phoenix Sky Harbor International Airport (Greenwald, Ballagh, and Zyniecki 1996; Greenwald and Zyniecki 1993) located the north branch of Canal Salado but not the main channel or the south branch; investigators could expect to find the main channel of Canal Salado in the extreme southern portion of the airport, extending west under Interstate 10/17, in the location indicated on the early canal maps.

The results of the Phoenix Sky Harbor Center and Airport projects demonstrate considerable variation in the canal exposures between the two project areas. These differences did not alter project interpretations regarding the function of the north branch of Canal Salado in the project area; in fact, the data from the two projects are complementary. The canal exposures in the project area typically exhibited coarse sediments and numerous rechanneling episodes. The coarse sediments were a result of the steep gradient of the canal in the project area, in contrast to the nearly level gradient within the airport. Rechanneling may have been caused by erosion from increased water velocity, as well as by frequent cleaning episodes. Channel migration was more frequent, increasing with distance toward the terminal (western) end of the project area, possibly due to less maintenance of the terminal portion of the canal. In the airport, the nearly level grade reduced channel erosion. The sediments within this portion of the canal led investigators to suggest that it had reached and maintained equilibrium and operated very efficiently in the extreme eastern portion of Pueblo Salado. Studies relating to canal discharge and arable lands (Chapter 6) indicate that the north branch had the potential to provide sufficient volumes of water to irrigate the arable soils on the alluvial island that Pueblo Salado occupied. By maintaining this small canal system, the residents of Pueblo Salado had the capability to attain agricultural self-sufficiency.

IMPLICATIONS RELATING TO POLVORÓN PHASE SETTLEMENT AND THE DECLINE OF HOHOKAM SOCIAL ORGANIZATION

Post-Classic period occupations dating to the Polvorón phase (A.D. 1350–1450) were evident at both Dutch Canal Ruin in Area 8 and at Pueblo Salado in Area 8/9. Both occupations were small, dispersed habitation settlements whose inhabitants subsisted on a diet similar to that of residents during the previous Classic period. Analysts compiled a list of traits characteristic of the Polvorón phase (Chapter 8; Volume 3:Chapter 13) using data from previously excavated sites, such as El Polvorón, Las Colinas, Casa Buena, and Locus E of Brady Wash and compared these characteristics with information obtained from the City of Phoenix Sky Harbor Center. These attributes were very consistent and, in fact, merit the designation of a separate phase, the Polvorón, rather than being regarded as a later subdivision of the Classic period Civano phase. Traits typical of this temporally discrete phase include the use of true pit house architecture; Roosevelt Red Ware frequencies between 85% and 88% of the total decorated assemblage; intrusive decorated ceramics that may include Hopi yellowware, San Carlos Red-on-brown, and Tanque Verde Red-on-brown; an increase in the frequency of obsidian over that of the Classic period compared to the rest of the flaked stone material types; a decrease in the amount of marine shell and a simplification of shell artifact design; and burial practices that include both inhumation and cremation. Many of the material culture traits, although clearly distinct from Classic period trends, indicate a modification rather than a fundamental change.

Changes in Group Size and Composition

Late Classic period settlements usually were aggregated, large in size and population, and based on a hierarchical organizational structure (Wilcox 1991) that probably helped to coordinate the water supply, build and maintain canals, and integrate ceremonial functions for irrigation communities and larger sociopolitical units. The Civano phase occupation at Pueblo Salado, although not part of the irrigation community of Canal System 2, was an aggregated settlement composed of three separate residential clusters comprising the six courtyard groups. This highly structured settlement pattern probably supported a community organization designed for building and maintaining Canal Salado and its north branch, as well as maintaining relations, including ceremonial and political alliances, with other intraregional systems.

The Polvorón phase occupation at Pueblo Salado exhibited a much different pattern from that of the Civano phase. Instead of the six courtyard groups, six pit structures were paired into three separate groups, each probably representing a nuclear or small extended family. Population and settlement size decreased dramatically, and, although occupation was confined to the compound area, the habitation areas included no courtyard group or residential group structure. The reason for the use of the compound was likely the threat of flood damage on the floodplain; the compound was constructed on a naturally elevated gravel bar and was protected in part by a double wall.

Area 8 at Dutch Canal Ruin represented a small hamlet of one true pit house, one pit structure, and associated features inhabited by a single nuclear or small extended family. The pit structures were situated less than 1 m apart; however, they may not have been either built or inhabited at the same time.

Both Polvorón phase occupations represent a significant decrease in population from the late Classic period, which limited the size of the settlement and the number of contemporaneous social units. Social units appeared to follow a paired pattern, noted earlier by Gregory (1988:41), without additional structures. Investigators found no evidence at Dutch Canal Ruin to indicate that residents had used the irrigation system; canal studies indicated construction and use only during the pre-Classic and Classic periods. However, the Pueblo Saladoans may have used canals well into the Polvorón phase. The practice of agriculture, attested by the macrobotanical and pollen records, may have included strategies other than irrigation, such as those practiced by nonriverine Hohokam. The Polvorón phase inhabitants of both project sites had a diverse diet, including domesticated crops and wild plant and animal resources, similar to the Classic period residents of Pueblo Salado, indicating that decisions and organization related to the subsistence base were continuous into the post-Classic period. Decreasing population levels probably would have been inefficient for managing irrigation systems as extensive as those of the Classic period; in addition, the smaller population would not have needed to sustain the earlier level of food production and may have relied more on native resources.

The use of both cremation and inhumation during the Polvorón phase may indicate a mixing of status within the confines of settlements. Interregional exchange continued and, in some cases, may have increased based on the concurrent decrease in population. Depopulated settlements lost their internal structure, but economic frameworks such as exchange routes still functioned, allowing residents to maintain economic, and possibly sociopolitical, ties as a residual benefit of the preceding Classic period system.

The Decline of Hohokam Social Organization

The Hohokam of the late Classic period in the Phoenix Basin were aggregated into settlements along irrigation canals, which were part of larger irrigation communities. During this time, irrigation communities appear to have formed local political systems, organized in a hierarchical manner and operating intra- and interregionally by forming sociopolitical alliances with other localized systems (Wilcox 1979:113–114). Consolidation of these communities through alliances increased security, the labor pool, and the productivity of the system. Salado interaction with this area, evident in the introduction of Salado polychromes, has been described as that of "a weakly integrated system of exchange among a large series of small-scale regional systems" (Wilcox and Sternberg 1983:244).

Changes documented in Polvorón phase settlements indicate that the Classic period regional structure was no longer in place. Ceramic evidence suggests that local affiliation shifted from irrigation communities to some other organizational structure; ceramic assemblages from the Polvorón phase settlements at Dutch Canal Ruin and Pueblo Salado were similar, crosscutting canal system affiliations. During this time the Hohokam were no longer building new canals and probably were only minimally using existing irrigation systems. Architecture changed not only in form but in method of construction. Perhaps the Classic period social structure was an integral part of the construction and maintenance of contiguous adobe structures; pit structures were the first house type to be constructed in the Southwest, when single nuclear and small extended families were the basic social units composing habitation settlements. Pit houses may indeed have been the most effective habitations for small social groups in the region, as they were less labor intensive to build than were adobe structures.

With the decline of Hohokam regional social organization, the coordination and management of major works, such as platform mounds and compounds, was unnecessary, and the local communities and polities that were based on irrigation systems no longer existed. A new form of interaction must have taken place. The increase in the frequency of Roosevelt Red Ware during the Polvorón phase suggests that the Salado influence grew in the Phoenix Basin, replacing former local alliances. A Salado system alliance probably helped to maintain the exchange network in the region, as evidenced by a continuation of the shell and obsidian supply and the presence of Roosevelt Red Ware and other intrusive wares in the ceramic assemblage. Although these exotic items were still available, their use and design became simpler during this time. At Dutch Canal Ruin and Pueblo Salado, Polvorón phase obsidian occurred most often as expedient used flakes, rather than as the formal tools, such as projectile points, usually seen during the Classic period. The shell artifact assemblages exhibited less artistic modification.

Settlements at Dutch Canal Ruin and Pueblo Salado during the Polvorón phase apparently were peripheral to the Salado system of which they were a part. McGuire (1991) has suggested that during the Classic period the Phoenix Basin settlements were peripheral to a Saladoan system, but other research has identified more evidence that this was the case during the post-Classic period. Strong ties to the Salado area are indicated by a high frequency of Roosevelt Red Ware compared to other decorated ceramics. The fact that the exchange system remained in place suggests that settlements were members of a larger system, but simplification of material culture, including architecture and shell and obsidian artifacts, and decrease and dispersal of the population indicate that the changes occurring at that time were related to a separation from the core system. Collapse of the hierarchical structure, of organized ritual, and of the sociopolitical organization responsible for activities related to platform mounds and irrigation systems would have been the precursor for the Polvorón phase way of life, which lacked large-scale community integration and was based on a simple, seemingly egalitarian organization of nuclear or extended family units.

CHANGING SUBSISTENCE STRATEGIES AND OCCUPATION PATTERNS

Investigations at Dutch Canal Ruin and Pueblo Salado afforded an opportunity to observe a variety of Hohokam occupational patterns and subsistence strategies through time and space, some of which were relatively new to prehistoric Hohokam studies. Settlement patterns were tied into agricultural activities and the exploitation of a diverse resource base as well as other logistical factors, such as the location of "parent" sites, the geology and topography of the landscape, and sociopolitical and economic alliances.

The Pre-Classic Period: The Dutch Canal Ruin Agricultural Zone

The pre-Classic period at Dutch Canal Ruin consisted of a settlement of field houses and farmsteads related primarily to agricultural activities, with residents using nearby canals that were part of the Canal System 2 network. Located on the northern portion of the geologic floodplain, the soils were deep and well drained, with high agricultural potential. Corn agriculture and canal construction and maintenance appeared to be the primary functions at the site. Occupation was temporary and seasonal, and the inhabitants of Dutch Canal Ruin likely had their permanent residence at neighboring hamlets or villages within Canal System 2, probably at higher elevations where flooding was less of a threat.

From the late Pioneer to the late Colonial periods, occupants of Dutch Canal Ruin constructed five canal alignments, despite the potential for flooding on the first terrace. Abandonment of the last canal occurred around the same time as a major flood episode, based on streamflow reconstructions for the Salt River (Nials, Gregory, and Graybill 1989). Agricultural specialization at this location indicates that although the risk of flood damage may have required extensive canal maintenance, the farming benefits must have been greater than the risk. This may have been due, in part, to the nondestructive annual floods of the Salt River that supplied nutrients both directly to the fields from the floodwaters and in the irrigation water (Brown 1869; Rea 1979, 1991).

The Classic and Post-Classic Periods: Permanent Settlement and Subsistence Diversity

Dutch Canal Ruin

During the Classic period, in Areas 9 and 10, the Dutch Canal Ruin site was again an agricultural zone. This time, however, the farmsteads appear to have been permanent settlements; settlement size and activities greatly increased from the pre-Classic periods, even though residents still faced the risk of flood damage. Classic period occupation probably was still tied to other neighboring settlements. The re-establishment of the canal network at Dutch Canal Ruin may have been an extension of the system in operation above the geologic floodplain. Regardless, the construction of canals through Dutch Canal Ruin would have involved interaction with other settlements along the system to effectively manage the water supply. The settlement represented an extension of the larger socioeconomic and political community utilizing Canal System 2.

Situated in the midst of three different environmental zones—a desert saltbush community, a creosotebush-bursage community, and a riparian habitat—the site location provided diverse economic species for supplementing the corn-based diet. Site inhabitants took advantage of their location by adding agave, cattail, little barley grass, and medium- and large-sized mammals to their diet. This strategy is in sharp contrast with that of the pre-Classic period occupation of Dutch Canal Ruin, which left evidence of few wild resources in the archaeological record.

The post-Classic period occupation at Dutch Canal Ruin, represented by the Polvorón phase at Area 8, was also a permanent settlement; however, it differed from the Classic period occupation in that it was not an extension of the larger Canal System 2 community. Investigators found no evidence of canal construction during this period, and they could not determine whether residents of the area had still used the existing canals. Although corn remained a staple in the post-Classic period diet, other foods, such as agave, gained importance. The landscape continued to be useful as an agricultural locus, with or without irrigation, and the inhabitants also utilized a variety of natural resources. Area 8 and Pueblo Salado were similar in their ceramic assemblages, suggesting close interaction between these two settlements. Perhaps residents of Area 8 were former residents of Pueblo Salado.

Pueblo Salado

Pueblo Salado, situated south of Dutch Canal Ruin, is one of the few documented sites on the floodplain of the Salt River. Although the site's proximity to the river enhanced the diversity of resources available for consumption, irrigation agriculture was still a major subsistence strategy. In fact, compared to Classic period sites in Canal System 2, the frequency of corn remains from Pueblo Salado was higher, indicating that corn agriculture had been pursued more intensively here than at contemporaneous sites that were not located on the floodplain. This may have been due to the increase in soil nutrients and decreased salinization of fields brought about by Salt River flooding (Ackerly 1982; Doolittle 1990). Riparian resources close by provided fish, cottonwood, willow, mesquite, and cattail; desertscrub land yielded economic plants such as grass, saltbush, hedgehog cactus, and other species. The inhabitants may have procured small- and medium-sized mammals from either environment, as well as from modified microenvironments created by canals, fields, field borders, and living areas.

The first settlement of Pueblo Salado was in the early Classic period in a rancheria-style pattern of small hamlets dispersed along Canal Salado and its north branch. The purpose of the settlement appears to have been the intensive production of corn during a time of expansion of settlements, increase in population, and agricultural intensification in the Phoenix Basin. The advantages of establishing a new canal system and new settlement in this area apparently outweighed the risks of severe flood damage.

During the Civano phase of the Classic period, a compound wall encompassed an aggregated settlement of at least five courtyard groups. Irrigation agriculture increased, with less exploitation of cacti and agave, compared with other Classic period sites (Volume 3:Chapter 8). The Polvorón phase of the post-Classic period was notable for a shift from adobe-walled compound architecture to the use of pit houses and by a large decrease in settlement and population size. Investigators identified little change in subsistence strategies or emphasis, except in a decline in cottonwood or willow wood and an increase in the use of mesquite wood. Wood from cottonwood or willow may have become less desirable or less accessible during the post-Classic period, or mesquite may have become a more important resource for both food and construction. Whereas the relatively straight woods of willow and cottonwood were necessary for adobe compound architecture, the more crooked branches of mesquite sufficed for pit house construction (Chapter 8). In addition, the available quantity of mesquite may have increased as local population, and thus pressure on this resource, decreased.

AUTONOMY AT PUEBLO SALADO

Although Pueblo Salado was similar to Dutch Canal Ruin as a settlement engaged in the intensive production of corn by irrigation agriculture, it was dissimilar in that its inhabitants built, maintained, and controlled their own canal system. Canal Salado may have had a canal head that was separate from Canal System 2, thus making it an independent system (Chapter 3); therefore, management of the Pueblo Salado water supply would not necessarily have entailed collaboration with neighboring communities. Canal Salado was a small system, but evidence indicated the capacity to irrigate the acreage available for agriculture at Pueblo Salado. Canal features, as well as accumulation and evidence of the removal of canal sediments, were testimony to some of the maintenance and engineering tasks undertaken prehistorically to keep the system functional.

Analysts found in the ceramic data another indication that the Pueblo Saladoans had been independent from the irrigation community of Canal System 2. During the Classic period at Pueblo

Salado, plainware temper derived from sources south of the Salt River was substantially more frequent than at Pueblo Grande (in Canal System 2), where the sources of plainware temper were mostly north of the Salt River (Abbott, ed. 1994). Because plainwares represent activities of the majority of the prehistoric population, researchers may infer that these distinctions between site temper sources represent basic differences in social, economic, or political relations between site inhabitants. The Polvorón phase ceramic assemblages from both Dutch Canal Ruin and Pueblo Salado were very similar. Data indicated that the frequency of temper derived from sources south of the Salt River was high for both sites. It is noteworthy that neither Polvorón phase settlement was an extension of Canal System 2 or part of the sociopolitical network north of the Salt River that was evident at Pueblo Grande during the Classic period. Perhaps this difference is related to changes in Hohokam settlement during the Polvorón phase, in which environmental conditions affected continued use of Canal System 2. This perspective is similar to that offered by Gregory (1991:187), in which he accounts for the abandonment of Canal System 2 by the destruction of canal heads through lateral erosion and downcutting of the Salt River channel in the area below the Papago Buttes. Such severe erosional changes in the Salt River would have resulted in bedrock exposures. The Hohokam, generally being unable to construct canals through bedrock, would have had no choice but to abandon large-scale irrigation north of the Salt River below the Papago Buttes. Canal System 1 was not limited by such physical constraints, and Pueblo Salado residents may have become aligned with groups on the south side of the river due to changes that were occurring to the north.

The location of Pueblo Salado in relation to physical features in the area also contributed to its ability to function as an autonomous settlement. The site was on an alluvial island, bordered on the south by the Salt River and on the north by a braided channel of the river, Turney's Gully. Turney's Gully would have effectively blocked efforts to construct any canals across its channel, logistically restricting the path of Canal Salado in that direction. Another feature was a naturally elevated gravel bar; the Pueblo Salado compound was built directly over it, suggesting that it provided protection against floods. In addition, the gravel bar area was not conducive to irrigation agriculture, limiting its potential for productive field use. The construction of a double-walled compound might represent another effort by site occupants to provide some flood protection, which would have required a substantial amount of intrasettlement organization. Perhaps the threat of flooding caused the shift from dispersed settlement during the Soho phase to aggregation during the Civano phase, with the compound offering the protection residents needed.

A final factor contributing to the autonomy of the site was that the landscape promoted self-sufficiency. Pueblo Salado's location on the floodplain enhanced field productivity, and its proximity to the Salt River increased the diversity of natural resources that could be exploited. Riverine resource zones provide abundant opportunities for hunting and gathering, and the extensive use of this environment was evident in the archaeological record.

FACTORS INFLUENCING HOHOKAM SETTLEMENT AND LAND USE IN THE GEOLOGIC FLOODPLAIN

Because the Phoenix Sky Harbor Center project area was located on the geologic floodplain of the Salt River, the prehistoric occupants would have had to develop adaptive strategies in response to the threat and effects of flooding. Most strategies appeared to have involved the utilization of the situation to the best advantage. Factors that influenced life on the floodplain were tied into the seasonal variability of the Salt River, the environmental zones that were available for human exploitation, and the sociopolitical implications of canal systems and irrigation communities.

Effects of Salt River Flooding on Land-Use Patterns

Data gathered during the City of Phoenix Sky Harbor Center Project indicated that flooding had been a serious consideration for Hohokam settlement and land use in this area. Few pre-Classic period settlements were on the geologic floodplain. In fact, settlement during the pre-Classic periods was confined principally to the lower bajada, above the level of all but the most severe floods. The Hohokam used the geologic floodplain for wild resource procurement (Cable and Doyel 1985c) and for agricultural fields, seasonally occupied structures, and irrigation canals during the late Pioneer and Colonial periods (Chapters 4, 6, and 7; Howard 1991b). During the Colonial period, the Hohokam began to change the orientation of the canal systems by directing them onto the lower bajada (Howard 1991b). The change from a pattern that paralleled the river to one that led away from the river provided two engineering benefits. First, canals located on the geologic floodplain were susceptible to in-filling through flooding along their entire lengths. Reorienting the direction of flow onto the lower bajada thus would have meant less canal cleaning. Second, irrigators could have better controlled discharge rates and erosion by following more optimal grades. Because canals located on the geologic floodplain often exhibit steep gradients that could result in basal and lateral channel erosion (Chapter 6; Greenwald 1988), either they would require higher maintenance, or discharge levels would need to be restricted, in comparison with canals that were built with gradients less susceptible to erosion, such as those on the lower bajada. The evidence from the canals at Dutch Canal Ruin indicated that by the end of the Colonial period (A.D. 950) the Hohokam had abandoned irrigation on that portion of the geologic floodplain.

Various lines of evidence associated with the compound and with sediment accumulation rates indicated the effects of flooding on land-use patterns at Pueblo Salado. The compound was located on a topographic rise, a natural gravel bar, reducing but by no means eliminating the threat of flooding. Thick alluvial sediments (in excess of 50 cm) formed at the site after the first (Soho phase) settlement, with accumulations continuing while the compound was under construction and throughout its occupation (including against the outside of the double compound wall, which may have been constructed in part to withstand high floods). Preservation of the double compound wall was restricted to the north, south, and west sides of the compound, with no remains on the east side. Perhaps, because the east side was the upstream side, floods had destroyed these walls, or the inhabitants may have used an earthen berm partly composed of trash to protect the portion of the compound that would have been most susceptible to flood damage.

The danger of flooding would have been greatest between November and April, when the annual flow was highest (Graybill and Nials 1989:23); therefore, inhabitants occupying the geologic floodplain during the growing season, on a temporary basis, would not have been as concerned with annual flooding events as those living in the area year-round or during the winter months. Potential damage to field areas and canals would have been similar, but permanent residents would have been faced with potential loss of life and damage to homes, personal property, and food and other stored goods.

If the Hohokam commonly practiced a single annual cropping strategy, as some researchers have argued (Nials, Gregory, and Graybill 1989:74), settlement location for temporary seasonal occupation would not have depended on the risk of flooding, given the relationship between periods of greatest discharge of the river and the growing season. Nials, Gregory, and Graybill (1989) suggested that the risks of maintaining canal systems through winter/spring high-magnitude flows would have been greater than the potential productivity of a dual growing season and that inferences from double-cropping in the historic period accounts do not take into consideration the existence of upstream dams that within the last century have largely controlled the river's flows. Year-round habitations would have been especially susceptible to damage from floods in excess of

50-year magnitudes. Prehistoric options to alleviate flood damage during permanent habitation would have meant occupying the floodplain only during periods of relatively stable or below-normal discharge, or taking measures to control flooding. Occupation of Pueblo Salado during the Civano and Polvorón phases may have been fortuitous as far as streamflow patterns were concerned, but the location and construction of the compound suggest that residents consciously chose the latter option in restricting their residential area to the elevated gravel bar and living within the confines of a double-walled compound. Considerable effort and planning is reflected in these strategies, as flooding still occurred and probably was a constant concern. The Pueblo Saladoans during the Soho phase probably settled the area based on the first option. During this time, the retrodicted streamflow pattern was stable (until A.D. 1197). The streamflow retrodiction suggested a high degree of variability in flows after that date; from A.D. 1197 to 1370 there were 14 above-mean flows, of which 6 exceeded a 40-year flood magnitude (Nials, Gregory, and Graybill 1989:Figure 5.2). Few data are available that researchers can use to reconstruct characteristics of Salt River streamflow after A.D. 1370. However, the Hohokam continued to live in the compound at Pueblo Salado and used canal irrigation well into the Polvorón phase of the post-Classic period. Population decreased during the post-Classic period occupation, and pit houses eventually replaced adobe architecture. Chronometric data suggested that Pueblo Salado may have been occupied as late as A.D. 1500 or 1550.

The Floodplain: A Renewable Resource

The geologic floodplain of the Salt River offered a number of resources to the prehistoric inhabitants that may have influenced settlement in that area. First among these was the river itself, supplying fish and other riparian plant and animal resources that were economic assets not as easily obtained from habitations on the bajada. Second, the location was well suited to irrigation systems, although the geomorphology of the floodplain may have restricted their size. Finally, periodic flooding added soil nutrients and probably washed excess salts from the soil, deterring the process of salt accumulation or salinization.

The inhabitants of both Dutch Canal Ruin and Pueblo Salado, particularly the latter, exploited riparian resources. The Pueblo Saladoans supplemented their corn-based diet with riverine resources at a higher frequency than did people from sites located farther from the river, emphasizing the adaptability of the Hohokam to local environmental zones. Fish, cattail, and wood from willow or cottonwood were some of the riparian products recovered most frequently during the project. Just as these resources were more abundant in the archaeological record at Pueblo Salado, plant and animal remains derived from mountain and bajada environmental zones were less frequent than at sites located at higher elevations and farther from the river. Settlement on the floodplain correlated with increased use of riparian resources, and the relative frequency of their use is associated at least with distance from the river. Other factors, such as affiliation with larger sociopolitical systems (e.g., canal systems), probably are related to the prevalence of particular habitat exploitation, although archaeologists presently have insufficient data to determine the influence of this variable on floodplain settlement.

Canal operations in the project area represented a span of over 850 years, from pre-A.D. 650 to possibly post-A.D. 1500, with a hiatus from approximately A.D. 900 to 1150. During this hiatus, the occupants used the project area as a resource exploitation zone, possibly attempting some dry farming at Dutch Canal Ruin using direct precipitation or runoff. The capacity of the irrigation canals was adequate for supplying fields in proximity to both Dutch Canal Ruin and Pueblo Salado (Chapter 6). In addition, retrodictions of velocities and discharges for canals indicated stable conditions for sections of the North Main Canal, the South Main Canal, and the north branch of

Canal Salado, specifically, Alignments 8001, 8002, 8003, and 3113. The ability to maintain irrigation canals at a stable state and obtain an adequate water supply for agricultural fields suggests that the irrigation systems on the floodplain were productive and were utilized as long as the labor investment was offset by the return. Site location on the deep, well-drained alluvium of the floodplain accounts for some of that success. The natural slope of the floodplain directly affected canal gradients; as the gradient became steeper, the chance that discharge rates could result in erosional damage to the canals was greater. Earlier canals in the project area reflected elevated erosional effect to a greater degree than did later ones, indicating that later canals had been engineered and regulated to reduce the potential effects of erosion.

Because the bed loads carried by floodwaters are often rich in nutrients, alluviation rejuvenates the soils of agricultural fields on the geologic floodplain. Flooding also serves to flush salts that accumulate through irrigation and evaporation. Soil conditions on the floodplain were good, and occasional flooding probably enabled farmers to cultivate fields on the floodplain for longer periods than those on the bajada (Ackerly 1988). Investigators need more botanical data from floodplain sites to determine if there is a correlation between location and prehistoric corn production. Macrobotanical studies from Pueblo Salado yielded more evidence of corn at this site than at sites investigated on the lower bajada in Canal System 2. Kwiatkowski and Miller suggest that site location on the geologic floodplain facilitated corn agriculture (Volume 3:Chapter 8), possibly by permitting extended cultivation of fields and because of the nutrient-rich alluvial soils. However, other economically important plants such as agave and cotton may not have been as well suited to the floodplain as corn, resulting in an emphasis on just one of the Hohokam staples. Although agave and cotton were present in the botanical record from both Dutch Canal Ruin and Pueblo Salado, their frequencies were extremely low.

Possible Social and Economic Factors

Settlement on the geologic floodplain may have been influenced by sociopolitical as well as environmental factors, with populations moving to this location based on requirements other than, or in addition to, a location offering high agricultural productivity. The ability to maintain and construct irrigation systems was dependent on an adequate labor pool and organizational structure, and the length and complexity of Canal System 2 may have required the presence of ditch tenders (zanjeros) to perform canal maintenance and regulate and allocate water in the area of Dutch Canal Ruin, with workers drawn from nearby settlements. If this was a factor in the settlement of Dutch Canal Ruin, it was complementary to the use of the land for agriculture. The pre-Classic period settlement of temporary, seasonally used field houses and farmsteads appeared to be primarily for the production of corn; occupants of these field house and farmstead loci would have resided at other locations, probably in permanently occupied hamlets and villages within Canal System 2.

Classic period settlement of the floodplain may have been the result of two factors: disrupted water distribution within Canal System 2 and encroachment or expansion into marginally settled areas. These explanations are equally difficult to demonstrate, given the intangible nature of the evidence; however, certain factors associated with water allocation may have been involved.

By the Sedentary period, Canal System 2 had grown into one of the largest and most complex irrigation systems in North America. During the Classic period, Canal System 2 reached its maximum areal extent, with a general trend toward northward expansion (Howard 1991b). This expansion may have been related to agricultural intensification, or it may have reflected a need to bring new lands under cultivation to replace overused field areas that exhibited diminished production levels caused by nutrient depletion, waterlogging, and increased levels of salinity. The retrodicted streamflow pattern for the Salt River (Nials, Gregory, and Graybill 1989) indicates that the Soho phase was a stable period, perhaps with decreased available moisture levels. If the Hohokam in Canal System 2 attempted to increase agricultural production at a time of stable to decreasing levels of effective moisture, the net outcome may have been periods of insufficient water for irrigation, much like that experienced by historic farmers during the late nineteenth century (Chapter 1). Despite a well-conceived and equitable water distribution plan, disputes over water allocations or shortages may have caused some individuals or settlements to receive less water than was necessary for subsistence. One response to such a water allocation shortage would be to relocate to another area with adequate amounts of water, whether to another settlement or to another canal or canal system. Pueblo Salado, on its alluvial island in the geologic floodplain, may have been settled by groups who were looking for new agricultural lands and a more reliable water source. By digging a canal that specifically supplied the spatially limited area of Pueblo Salado, occupants would have avoided dependence on the decisions of others regarding water allocation and would also have increased the reliability of their water supply. Although this scenario is dependent on environmental conditions, it offers one explanation for the establishment of a new canal system during the Classic period.

A second scenario for settlement on the floodplain during the Classic period considers encroachment into Canal System 2 or its margins by groups from other systems. This may have been the reason for the initial dispersed-settlement pattern at Pueblo Salado during the Soho phase that eventually evolved into the establishment of the compound. The ceramic temper analysis (Chapter 11) supported the hypothesis that Pueblo Saladoans had interacted with residents of Canal System 1. The fact that post-Classic period settlements remained on the floodplain after the "collapse" of the Classic period structure indicates that life on the geologic floodplain was viable and perhaps in some ways superior to that on the bajada.

CHRONOLOGY BUILDING

Archaeological projects typically have three goals relating to chronology. The first is to collect data that can be used to date events at the site or sites under investigation. The second objective is to determine the dates of those events, with reference both to the newly acquired data and to existing chronological frameworks. The third goal is to refine those frameworks, that is, to contribute to an interpretation of regional prehistory. The Phoenix Sky Harbor Center Project achieved all three of these goals. It should be noted that the chronological framework employed by the project was the "traditional" Hohokam cultural sequence, with the addition of Dean's (1991) calendrical dates of periods and phases.

Dutch Canal Ruin

The investigating team identified a number of temporal components at Dutch Canal Ruin; some of these related to occupation areas, others to canals. The Hohokam of the Phoenix Basin occupied Dutch Canal Ruin during the Pioneer, Colonial, Sedentary, Classic, and post-Classic periods; canals identified on this site were in use during both the pre-Classic and Classic periods (Volume 2:Figures 18.1–18.4). A hiatus in canal use occurred during the Sedentary period, with a resulting decline in subsistence activities. During the Phoenix Sky Harbor Center Project, researchers investigated all of these periods of use, with the exception of the site's Pioneer period component. Small numbers of decorated sherds recovered during testing indicated that two areas dated to the Classic period and suggested that the rest of the site dated to the pre-Classic period (Volume 1:Chapter 6). Evidence from excavation in most cases agreed with and strengthened the assignment of areas to particular periods, and in a number of cases allowed the dating of occupation areas to phases within those periods (Volume 2). This chronological evidence came primarily from ceramics and archaeomagnetic dates and, to a lesser extent, from radiocarbon dates and stratigraphic relationships. Excavation data did not, however, agree entirely with the conclusions drawn during the testing phase. Ceramics collected during testing had indicated a Classic period date for Area 8, but a combination of ceramic and architectural data obtained from excavation showed that the area actually dated to the post-Classic period. The investigations produced information on the sequence of canal construction and, to some extent, on the phases during which the canals had been built and used. Clearly, the canals had complex histories that belie any simple, synchronic reconstruction of canal systems in the Phoenix Basin (Volume 1:Chapter 5).

Most of the calibrated radiocarbon dates from Dutch Canal Ruin fell before A.D. 500 and were too early for the occupation areas and canals with which they were associated (Volume 2:Chapters 4 and 18). Many of the samples with early dates came from canal sediments, implying a particular bias in dates from this context. Most useful among the radiocarbon assays were three dates from Area 8 that clustered between A.D. 1280 and 1420. Ceramic data supported the assignment of this area to the post-Classic period Polvorón phase.

A small collection of archaeomagnetic dates contributed to the assignment of features and site areas at Dutch Canal Ruin to particular phases and periods. Most noteworthy were four dates from Area 3, including two almost-identical date ranges of A.D. 830–925 and 850–940 and two slightly later ranges of A.D. 900–975 and 940–1015. The two earlier dates were from extramural fire pits that may have shared absolute contemporaneity; all four dates were in agreement with the ceramic dating of Area 3 to an interval from the Gila Butte phase to the Sacaton phase (circa A.D. 775–1000).

Pueblo Salado

Unlike Dutch Canal Ruin, which dated primarily to the pre-Classic periods in the Hohokam chronological sequence, Pueblo Salado was inhabited during the Classic and post-Classic periods. The nearby Canal Salado presumably dated to the Classic period. Project investigators examined three periods of habitation at Pueblo Salado but not the canal. Data collected during the testing phase had indicated, first, that the site dated primarily to the Classic period; second, that Area 14 had been occupied earlier than Area 8/9; and third, that Area 8/9 dated to the late Classic period and, possibly, to the Polvorón phase (Volume 1:Chapter 7). Investigators generally confirmed these chronological inferences during the data recovery phase, with some elaboration, including the assignment of Area 14 to the Soho and early Civano phases, the dating of the compound in Area 8/9 to the Civano phase, and the confirmation of a Polvorón phase occupation of Area 8/9 (Volume 3:Chapter 12). Previous research had shown that a relatively high percentage of Roosevelt Red Ware is an indicator of a date in the late Civano phase or the Polvorón phase. Civano phase contexts in Area 8/9 yielded as high a percentage of Roosevelt Red Ware as any previously investigated Polvorón phase contexts, and Polvorón phase contexts in the area yielded even higher percentages of this ware. On the one hand, these data confirm the utility of percentages of Roosevelt Red Ware for identifying late Civano and Polvorón phase contexts; on the other hand, they suggest that these ceramic data are not sufficient for distinguishing between the two.

Both archaeomagnetic and radiocarbon dates from Pueblo Salado were consistent with the phase assignments indicated by ceramic and architectural data. In the case of Area 14, radiocarbon dates were earlier than archaeomagnetic dates, such that the former would agree with either a Sacaton or Soho phase assignment, whereas the latter were consistent with either the Soho or Civano phase. One structure in Area 14, a ramada, yielded three radiocarbon and two archaeomagnetic dates, and the radiocarbon assays were earlier than the archaeomagnetic ones. Thus, if both kinds of dates were given equal weight, the ramada would appear to have been in use around A.D. 1250, whereas the archaeomagnetic dates alone suggested use between A.D. 1250 and 1350. The radiocarbon dates from the compound, Area 8/9, also tended to be earlier than the associated archaeomagnetic dates; thus, the radiocarbon date ranges were predominantly pre–A.D. 1400, and the archaeomagnetic ranges were mostly post–A.D. 1400.

Area 8/9 produced 27 archaeomagnetic dates but only 5 radiocarbon dates. If one accepts all of the archaeomagnetic dates from Area 8/9, the Pueblo Saladoans would appear to have established the compound no later than A.D. 1400 and to have abandoned the area no earlier than A.D. 1550. A more conservative interpretation of the abandonment date makes use of the fact that 7 archaeomagnetic date ranges begin after about A.D. 1475. These data would indicate that residents occupied Area 8/9 into the final quarter of the fifteenth century, perhaps into the 1500s. These terminal dates presumably apply to the site's Polvorón phase occupation, although analysts have not demonstrated this association. Dean (1991), who treated the Civano phase as the end of the Hohokam cultural sequence, gave a maximum end date for this phase of A.D. 1550 and a preferred end date of 1500. Relative to either of these dates, Area 8/9 of Pueblo Salado would have been occupied at the very end of the Hohokam phenomenon as archaeologists recognize it today. Few chronometric date ranges from Hohokam contexts are as late as A.D. 1500 (Chapter 9). One example is a calibrated radiocarbon date of A.D. 1450 (one sigma A.D. 1431–1623 [Stuiver and Becker 1993]) from an horno found in Area 7 at Pueblo Salado (BRW 1989:86). This date, along with those from Area 8/9 of Pueblo Salado, suggests the possibility of closing the gap, in chronological terms, between the prehistoric and protohistoric periods of Native American settlement in the Phoenix Basin (see also Bostwick, Greenwald, and Walsh-Anduze 1996).

PROJECT RESULTS AND SUMMARY: AN EPILOGUE

Through the structured research plan carried out in the Phoenix Sky Harbor Center, the City of Phoenix made it possible to contribute substantially to a growing data base concerning prehistory in the Phoenix Basin. Although both Dutch Canal Ruin and Pueblo Salado were located on the geologic floodplain, they differed considerably in temporal association and function, opening up a wide spectrum of project research topics. The pre-Classic period occupation at Dutch Canal Ruin consisted of scattered field house and farmstead settlements associated with early canals and a subsistence strategy based on agriculture. During this early occupation of the project area, occupants probably lived at permanent village sites located farther north on the lower bajada, where the canals were some of the earliest in Canal System 2 (Turney 1929). During the late Pioneer and Colonial periods, use of the northern half of the Phoenix Sky Harbor Center was focused on agriculture, and investigators have referred to this area as the Dutch Canal Ruin agricultural zone. The project area inhabitants abandoned the canals during the Colonial period, possibly as early as A.D. 900, resulting in a hiatus in irrigation agriculture. Excavators observed no evidence of settlement or intensive land-use activities in the project area again until the Soho phase of the Classic period, when the Pueblo Saladoans established farmsteads and small hamlets. These settlements, scattered along the newly constructed Canal Salado and its north branch, constituted a new, albeit small, canal system that was restricted to an alluvial island between braided channels of the Salt River. During the Civano phase, residents of Pueblo Salado constructed a compound to which they attached clusters of rooms, and they further enclosed these habitation units by yet another wall. At this same time the Hohokam reoccupied Dutch Canal Ruin. They built new canals and settlements that exhibited greater complexity and duration of occupation than the previous field house and farmstead sites. Although settlement size may never have exceeded that of extended families, the occupations appeared to have been permanent and year-round.

In addition to examining pre-Classic and Classic period settlement and land-use patterns on the geologic floodplain, the Phoenix Sky Harbor Center Project afforded an opportunity to further study components of the post-Classic period Polvorón phase. Few sites have contained intact remains that date to the Polvorón phase, and this is one of the least understood periods of the Hohokam occupation of the Salt River Valley and the Phoenix Basin. Pueblo Salado contained the greatest concentration of Polvorón phase features; investigators also identified a distinct Polvorón phase component in Area 8 at Dutch Canal Ruin. The project has contributed to the identification of Polvorón phase remains and will assist with their identification during future projects elsewhere in the Phoenix Basin.

Chronometric studies indicated that Pueblo Salado had been occupied late in the Hohokam sequence, with results from multiple samples supporting the premise of an occupation that extended after A.D. 1400. In fact, many dates from this project suggested that Hohokam occupation of the area may have lasted until A.D. 1550. Recent efforts elsewhere on the geologic floodplain have produced late dates as well (Birnie 1994; BRW 1989; Greenwald et al. 1996; Greenwald and Zyniecki 1993), and it may be appropriate to scrutinize the results of previous projects relative to the chronometric interpretations. Perhaps other sites have produced data that support a late prehistoric/protohistoric continuum or at least demonstrate that the Hohokam culture can be traced in the Phoenix Basin into the sixteenth century. Recently the City of Phoenix Aviation Department sponsored excavations at Area 6 of Pueblo Salado. This investigation produced additional evidence in support of both the Polvorón phenomenon and a prehistoric/protohistoric continuum (Greenwald et al. 1996). The results of the Phoenix Sky Harbor International Airport investigations further demonstrate the data potential and importance of Pueblo Salado and floodplain archaeology in understanding the historical record of the Phoenix Basin.

One of the most significant recent contributions to Hohokam archaeology is the application of ceramic temper analysis and the implications of associated studies (Chapter 11; Abbott 1992; Abbott and Schaller 1991; Abbott and Walsh-Anduze 1991). Analysis of temper sources has enabled investigators to examine exchange strategies within the Salt River Valley and the canal systems at the site level. Despite the location of Pueblo Salado north of the river, the ceramic temper analysis for the site supported the concept that Pueblo Salado may have been autonomous and may indeed have established exchange networks with sites south of the Salt River in Canal System 1 as opposed to interaction spheres within Canal System 2. The application of ceramic temper analysis has tremendous potential for examining the social, political, and economic organization of the Hohokam in the Salt River Valley, further demonstrated through the results of the Papago Freeway investigations near Pueblo Grande (Abbott 1994). The application of such innovative techniques promises to expand our understanding of the peoples who preceded us in the Phoenix Basin, perhaps in ways that will contribute to our own welfare.

APPENDIXES

APPENDIX A

FLAKED STONE AND GROUND STONE RAW DATA FROM CORE AND PERIPHERY SITES

Dawn M. Greenwald

Table A.1. Flaked Stone Morphological Types and Attributes from Field House Sites

Morphological Type/Attribute[1]	Site							
	Block 28-North	Murphy's Addition	La Cuenca del Sedimento	New River	Salt-Gila Aqueduct	Waddell	TAP[2], Phase A	Dutch Canal Ruin
Debitage	73	532	210	1669	311	944	29	314
	*68.9	85.3	82.4	81.6	68.2	88.1	80.6	93.7
Used Flake	19	50	6	3	27	4	4	1
	17.9	8.0	2.4	0.1	5.9	0.4	11.1	0.3
Core	1	2	24	90	64	48		6
	0.9	0.3	9.4	4.4	14.0	4.5		1.8
Core Tool	2	2		260	20	16		
	1.9	0.3		12.7	4.4	1.5		
Core/Hammerstone				5	10	16		6
				0.2	2.2	1.5		1.8
Hammerstone	10	33	8	1	21			3
	9.4	5.3	3.1	tr	4.6			0.9
Uniface						3	2	1
						0.3	5.6	0.3
Biface				1	3	7	1	2
				tr	0.7	0.7	2.8	0.6
Projectile Point		1				1		
		0.2				0.1		
Other Tool	1	4+	7	16		33		2
	0.9	0.6	2.7	0.8		3.1		0.6
Total	106	624	255	2045	456	1072	36	335
Retouched Tool		*0.80	2.70	0.80	0.70	4.10	8.30	1.50
Unretouched/Retouched Tool		10.00	0.86	0.18	unk	0.36	1.33	0.20
Debitage/Core	24.33	133.00	8.75	4.70	3.31	11.80		26.17

[1]Attributes are presented below the morphological type totals and are derived from the morphological information.
[2]Tucson Aqueduct Project
*Column % tr = trace unk = unknown

Table A.2. Flaked Stone Morphological Types and Attributes from Farmstead Sites

Morphological Type/Attribute[1]	Site									
	New River	Hawk's Nest	Salt-Gila Aqueduct	Adobe Dam	Fastimes	La Cuenca del Sedimento	ANAMAX-Rosemont	TAP[2], Phase A	Waddell	Dutch Canal Ruin
Debitage	1064	1108	1296	1624	2311	1781	12,485	217	14,472	84
	*87.6	90.7	71.8	72.5	84.4	91.9	88.0	82.5	91.2	75.0
Used Flake	5	33	74	22	180	34	349	19	185	1
	0.4	2.7	4.1	1.0	6.6	1.8	2.5	7.2	1.2	0.9
Core	51	10	286	117	30	96	268	1	774	6
	4.2	0.8	15.8	5.2	1.1	5.0	1.9	0.4	4.9	5.4
Core Tool	82		23					1	19	4
	6.7		1.3					0.4	0.1	3.6
Core/Hammerstone	2	14	25	59			122		95	2
	0.2	1.2	1.4	2.6			0.9		0.6	1.8
Hammerstone[3]		7	77	417	54	4	25	13		10
		0.6	4.3	18.6	2.0	0.2	0.2	4.9		8.9
Uniface		26			124			5	10	
		2.1			4.5			1.9	0.1	
Biface	2	9	24		39			5	192	3
	0.2	0.7	1.3		1.4			1.9	1.2	2.7
Drill		1			(1)[4]			1		
		tr			tr			0.4		
Projectile Point	4	3			(16)				44	
	0.3	0.2			0.6				0.3	
Other Tool	5	11			(10)	23	938	1	78	2
	0.4	0.9			0.4	1.2	6.6	0.4	0.5	1.8
Total	1215	1222	1805	2239	2738	1938	14,187	263	15,869	112
Retouched Tool	*0.90	4.10	1.30		6.90		6.60	4.56	2.00	4.50
Unretouched/Retouched Tool	0.45	0.85	unk		0.95	1.48	0.37	1.58	0.75	0.33
Debitage/Core	7.88	46.17	3.88	9.22	77.03	18.55	32.01	14.47	16.30	7.00

[1]Attributes are presented below the morphological type totals and are derived from the morphological information. [2]Tucson Aqueduct Project
[3]In some assemblages, the term *hammerstone* includes core/hammerstones. [4]Numbers in parentheses ( ) represent a subset of another category and are not included in the total.
*Column % tr = trace unk = unknown

Table A.3. Flaked Stone Morphological Types and Attributes from Hamlet Sites

Morphological Type/Attribute[1]	Site							
	Salt-Gila Aqueduct	TAP[2], Phase A NA18,003, Locus I	TAP, Phase A NA18,030	TAP, Phase A Others	El Caserío	Block 24-East	La Lomita Pequeña	Pueblo Salado, Area 14
Debitage	1454	1862	1728	3032	2528	3703	1191	1552
	*71.8	80.2	85.3	74.8	90.1	77.6	70.1	88.8
Used Flake	82	257	89	523	30	362+	241	57
	4.0	11.1	4.4	12.9	1.1	7.6	14.2	3.3
Core	276	29	24	77	115	223	72	32
	13.6	1.2	1.2	1.9	4.1	4.7	4.2	1.8
Core Tool	38	8	2	38	6	2	31	11
	1.9	0.3	0.1	0.9	0.2	tr	1.8	0.6
Core/Hammerstone	36				55		83	32
	1.8				2.0		4.9	1.8
Hammerstone[3]	60	31	34	124	67	200	19	33
	3.0	1.3	1.7	3.1	2.4	4.2	1.1	1.9
Uniface		60	58	114				16
		2.6	2.9	2.8				0.9
Biface	80	52	60	101				5
	3.9	2.2	3.0	2.5				0.3
Drill		3	2	3				1
		0.1	0.1	0.1				0.1
Projectile Point		13	12	21	4	4	13	
		0.6	0.6	0.5	0.1	0.1	0.8	
Other Tool		7	17	21		277	50	8
		0.3	0.8	0.5		5.8	2.9	0.5
Total	2026	2322	2026	4054	2805	4771	1700	1747
Retouched Tool	*3.90	5.80	7.10	6.40	0.10	4.80	1.60	1.37
Unretouched/Retouched Tool	1.03	1.90	0.62	2.01	7.50	>1.29	8.93	2.38
Debitage/Core	4.15	27.38	28.80	12.69	14.36	>16.46	6.40	20.69

[1]Attributes are presented below the morphological type totals and are derived from the morphological information.
[2]Tucson Aqueduct Project [3]In some assemblages, the term *hammerstone* includes core/hammerstones.
*Column %
tr = trace

Table A.4. Flaked Stone Morphological Types and Attributes from Village Sites

Morphological Type/Attribute[1]	Site							
	Water World	Eastwing	Grand Canal Ruins	Cashion	New River	TAP[2], Phase A	ANAMAX-Rosemont	Pueblo Salado, Area 8/9, Civano Phase
Debitage	3996	6162	2585	6818	2457	2781	8314	10,214
	*87.1	94.0	84.8	78.7	86.6	78.8	87.7	93.2
Used Flake	217	9	134	579	6	302	107	312
	4.7	0.1	4.4	6.7	0.2	8.6	1.1	2.8
Core	67	276	99	207	113	71	184	145
	1.5	4.2	3.2	2.4	4.0	2.0	1.9	1.3
Core Tool			70		208	24		13
			2.3		7.3	0.7		0.1
Core/Hammerstone		89	57	498	9		63	86
		1.4	1.9	5.8	0.3		0.7	0.8
Hammerstone[3]	68	4	14	26	1	148	27	46
	1.5	0.1	0.5	0.3	tr	4.2	0.3	0.4
Uniface	162		8	231		86		50
	3.5		0.3	2.7		2.4		0.5
Biface	78	5	6	54	10	77		36
	1.7	0.1	0.2	0.6	0.4	2.2		0.3
Drill	(3)[4]		2		2	3		
	0.1		0.1		0.1	0.1		
Projectile Point	(27)	7	28	145	10	13		23
	0.6	0.1	0.9	1.7	0.4	0.4		0.2
Other Tool	1	4	44	101	21	21	790	35
	tr	0.1	1.4	1.2	0.7	0.6	8.3	0.3
Total	4589	6556	3047	8659	2837	3526	9485	10,960
Retouched Tool	*5.30	0.20	2.90	5.30	1.52	5.70	8.30	1.16
Unretouched/Retouched Tools	0.90	0.56	1.52	1.26	0.14	1.51	0.14	2.66
Debitage/Core	59.64	16.88	11.44	9.67	7.45	11.44	33.66	41.86

[1]Attributes are presented below morphological type totals and are derived from the morphological information. [2]Tucson Aqueduct Project
[3]In some assemblages, the term *hammerstone* includes core/hammerstones. [4]Numbers in parentheses ( ) represent a subset of another category and are not included in the total. *Column % tr = trace

Table A.5. Ground Stone Morphological Types from Field House Sites

Morphological Type	Site								Total
	Block 24-North	Murphy's Addition	La Cuenca del Sedimento	New River	Salt-Gila Aqueduct	Waddell	TAP[1], Phase A	Dutch Canal Ruin	
Mano	2	6		44	22	11	2	2	89
	*14.3	11.3		88.0	52.4	35.5	33.3	8.0	39.2
Handstone	5	12						2	19
	35.7	22.6						8.0	8.4
Metate	1	10	4	4	13	10	1	4	47
	7.1	18.9	66.7	8.0	31.0	32.3	16.7	16.0	20.7
Grinding Slab		2							2
		3.8							0.9
Abrader	4	17	2						23
	28.6	32.1	33.3						10.1
Anvil					2				2
					4.8				0.9
Axe/Maul		1		1				1	3
		1.9		2.0				4.0	1.3
Tabular Tool					5	7			12
					11.9	22.6			5.3
Polishing Stone								1	1
								4.0	0.4
Raw Material								1	1
								4.0	0.4
Mineral	2	1							3
	14.3	1.9							1.3
Other Ground Stone		4				3		2	9
		7.5				9.7		8.0	4.0
Indeterminate Ground Stone				1			3	12	16
				2.0			50.0	48.0	7.0
Total	14	53	6	50	42	31	6	25	227

[1]Tucson Aqueduct Project
*Column %

Table A.6. Ground Stone Morphological Types from Farmstead Sites

Morphological Type	Site										Total
	New River	Hawk's Nest	Salt-Gila Aqueduct	Adobe Dam	Fastimes	La Cuenca del Sedimento	ANAMAX-Rosemont	TAP[1], Phase A	Waddell	Dutch Canal Ruin	
Mano	31	49	48	20	52	25	386	18	324	13	966
	*64.6	55.1	38.4	44.4	39.1	71.4	44.7	31.0	62.1	31.7	49.3
Handstone							137	5			142
							15.9	8.6			7.2
Metate	12	22	48	15	10	3	150	10	109	7	386
	25.0	24.7	38.4	33.3	7.5	8.6	17.4	17.2	20.9	17.1	19.7
Grinding Slab										2	2
										4.9	0.1
Abrader			2		1	1	7	1			12
			1.6		0.8	2.9	0.8	1.7			0.6
Anvil			3								3
			2.4								0.2
Mortar	1	2				1	3				7
	2.1	2.2				2.9	0.3				0.4
Passive NFS										1	1
										2.4	tr
Pestle		2			4		23	2			31
		2.2			3.0		2.7	3.4			1.6
Axe/Maul			2	2		1	10		2	2	19
			1.6	4.4		2.9	1.2		0.4	4.9	1.0
Tabular Tool	1		22	6			19	2	48		98
	2.1		17.6	13.3			2.2	3.4	9.2		5.0
Polishing Stone		1			7	2	20	6		1	37
		1.1			5.3	5.7	2.3	10.3		2.4	1.9
Grinder NFS										12	12
										29.3	0.6
Palette	1				1		21	1			24
	2.1				0.8		2.4	1.7			1.2
Ornament					26	1	59	1			87
					19.5	2.9	6.8	1.7			4.4

Table A.6. Ground Stone Morphological Types from Farmstead Sites, continued

Morphological Type	Site										Total
	New River	Hawk's Nest	Salt-Gila Aqueduct	Adobe Dam	Fastimes	La Cuenca del Sedimento	ANAMAX-Rosemont	TAP[1], Phase A	Waddell	Dutch Canal Ruin	
Raw Material	1						3	1		1	6
	2.1						0.3	1.7		2.4	0.3
Perforated/Unperforated Disc							5				
							0.6				5
											0.3
Stone Bowl					6		8		1		15
					4.5		0.9		0.2		0.8
Stone Ring				1							1
				2.2							tr
Medicine Stone/Plummet				1							1
				2.2							tr
Other Ground Stone	1				6		11		38	2	58
	2.1				4.5		1.3		7.3	4.9	3.0
Indeterminate Ground Stone		13			20	1	2	11			47
		14.6			15.0	2.9	0.2	19.0			2.4
Total	48	89	125	45	133	35	864	58	522	41	1960

[1]Tucson Aqueduct Project
NFS = not further specified
*Column %
tr = trace

Table A.7. Ground Stone Morphological Types from Hamlet Sites

Morphological Type	Site								Total
	Salt-Gila Aqueduct	TAP[1], Phase A NA18,003, Locus I	TAP, Phase A NA18,030	TAP, Phase A Others	El Caserío	Block 24-East	La Lomita Pequeña	Pueblo Salado Area 14	
Mano	108	30	62	72	83	50	50	18	473
	*42.5	14.8	28.4	11.9	9.4	15.2	20.8	22.0	16.8
Handstone		8	6	40			18	5	77
		3.9	2.8	6.6			7.5	6.1	2.7
Metate	60	7	44	43	50	125	16	7	352
	23.6	3.4	20.2	7.1	5.6	37.9	6.7	8.5	12.5
Grinding Slab					10			1	11
					1.1			1.2	0.4
Abrader	7			3		84			94
	2.8			0.5		25.5			3.3
Anvil	13						6		19
	5.1						2.5		0.7
Mortar				1		2	1		4
				0.2		0.6	0.4		0.1
Passive NFS		6	2					3	11
		3.0	0.9					3.7	0.4
Pestle		1	5	24		3	3	1	37
		0.5	2.3	4.0		0.9	1.3	1.2	1.3
Axe/Maul	4			7			9		20
	1.6			1.2			3.8		0.7
Tabular Tool	62	46	13	129	16		38	1	305
	24.4	22.7	6.0	21.4	1.8		15.8	1.2	10.8
Polishing Stone		28	12	143					183
		13.8	5.5	23.7					6.5
Grinder NFS						22		4	26
						6.7		4.9	0.9
Palette			1				3	1	5
			0.5				1.3	1.2	0.2
Ornament		42	2	11	6		4	3	68
		20.7	0.9	1.8	0.7		1.7	3.7	2.4
Raw Material		11	16	8	637			20	692
		5.4	7.3	1.3	71.9			24.4	24.6

Table A.7. Ground Stone Morphological Types from Hamlet Sites, continued

Morphological Type	Site								Total
	Salt-Gila Aqueduct	TAP[1], Phase A NA18,003, Locus I	TAP, Phase A NA18,030	TAP, Phase A Others	El Caserío	Block 24-East	La Lomita Pequeña	Pueblo Salado Area 14	
Perforated/Unperforated Disc						1		1	2
						0.3		1.2	0.1
Stone Bowl						7	3		10
						2.1	1.3		0.4
Stone Ring						5	1		6
						1.5	0.4		0.2
Medicine Stone/Plummet							4		4
							1.7		0.1
Mineral				4	12		12		28
				0.7	1.4		5.0		1.0
Other Ground Stone		4	6	13	39		8	6	76
		2.0	2.8	2.2	4.4		3.3	7.3	2.7
Indeterminate Ground Stone		20	49	105	33	31	64	11	313
		9.9	22.5	17.4	3.7	9.4	26.7	13.4	11.1
Total	254	203	218	603	886	330	240	82	2816

[1]Tucson Aqueduct Project
NFS = not further specified
*Column %

Table A.8. Ground Stone Morphological Types from Village Sites

Morphological Type	Site										Total
	Water World	Eastwing	Grand Canal Ruins	Cashion	New River	TAP[1], Phase A	ANAMAX-Rosemont	Las Colinas Mound 8	Las Colinas 1982–1984 Excavations	Pueblo Salado Area 8/9 Civano Phase	
Mano	63 *38.2	117 44.3	59 4.3	179 16.7	128 57.1	67 10.9	145 54.7	214 31.3	557 31.8	28 11.6	1557 23.4
Handstone			2 0.2	98 9.2		13 2.1	11 4.2			10 4.1	134 2.0
Metate	36 21.8	84 31.8	26 1.9	424 39.6	72 32.1	32 5.2	53 20.0	94 13.8	453 25.8	8 3.3	1282 19.3
Grinding Slab								4 0.6	5 0.3	2 0.8	11 0.2
Abrader	12 7.3	10 3.8	7 0.5			2 0.3		2 0.3	3 0.2		36 0.5
Anvil					4 1.8						4 0.1
Mortar	1 0.6		1 0.1	1 0.1				1 0.1			4 0.1
Pestle	4 2.4		3 0.2	2 0.2	1 0.5	6 1.0	8 3.0		12 0.7		36 0.5
Axe/Maul	1 0.6	6 2.3	10 0.7	5 0.5	1 0.5	3 0.5	3 1.1	22 3.2	42 2.4	7 2.9	100 1.5
Tabular Tool		15 5.7			10 4.5	277 45.2	4 1.5	44 6.4	35 2.0	5 2.1	390 5.9
Polishing Stone	13 7.9				1 0.5	74 12.1	11 4.2		19 1.1	6 2.5	124 1.9
Grinder NFS									56 3.2	13 5.4	69 1.0
Palette	2 1.2			6 0.6			13 4.9		30 1.7		51 0.8
Ornament	5 3.0	2 0.8	1169 85.3	20 1.9		15 2.4	4 1.5	35 5.1	352 20.1	17 7.1	1619 24.3
Raw Material						35 5.7	4 1.5	126 18.4		89 36.9	254 3.8

Table A.8. Ground Stone Morphological Types from Village Sites, continued

Morphological Type	Site										Total
	Water World	Eastwing	Grand Canal Ruins	Cashion	New River	TAP[1], Phase A	ANAMAX-Rosemont	Las Colinas Mound 8	Las Colinas 1982–1984 Excavations	Pueblo Salado Area 8/9 Civano Phase	
Perforated/Unperforated Disc			21				2	77	29	1	130
			1.5				0.8	11.3	1.7	0.4	2.0
Stone Bowl	2	4		9	1		4		5	1	26
	1.2	1.5		0.8	0.5		1.5		0.3	0.4	0.4
Stone Ring		3		3	1			4	70		81
		1.1		0.3	0.5			0.6	4.0		1.2
Medicine Stone/Plummet		1		1				2	24		28
		0.4		0.1				0.3	1.4		0.4
Mineral		4	37					22			63
		1.5	2.7					3.2			0.9
Other Ground Stone	8		3		2	6	3	36	62	12	132
	4.8		0.2		0.9	1.0	1.1	5.3	3.5	5.0	2.0
Indeterminate Ground Stone	18	18	32	323	3	83				42	519
	10.9	6.8	2.3	30.2	1.3	13.5				17.4	7.8
Total	165	264	1370	1071	224	613	265	683	1754	241	6650

[1]Tucson Aqueduct Project
NFS = not further specified
*Column %

Table A.9. Ceramic, Flaked Stone, and Ground Stone Totals Based on Site Type Distribution

Site Type	Ceramics	Flaked Stone	Ground Stone
Field House			
Block 28-North	262	106	14
Murphy's Addition	1,809	624	53
La Cuenca del Sedimento	No data	255	6
New River	1,545	2,045	50
Salt-Gila Aqueduct	1,496	456	42
Waddell	1,397	1,072	31
TAP, Phase A	82	36	6
Dutch Canal Ruin	242	335	25
Farmstead			
New River	10,976	1,215	48
Hawk's Nest	1,895	1,222	89
Salt-Gila Aqueduct	5,792	1,805	125
Adobe Dam	5,902	2,239	45
Fastimes	12,441	2,738	133
La Cuenca del Sedimento	No data	1,938	35
ANAMAX-Rosemont	19,252	14,187	864
TAP, Phase A	7,761	263	58
Waddell	28,154	15,869	522
Dutch Canal Ruin	1,673	112	41
Hamlet			
Salt-Gila Aqueduct	22,506	2,026	254
TAP, Phase A			
(NA18,003, Locus I)	12,343	2,322	203
(NA18,030)	10,214	2,026	218
(Others)	44,111	4,054	603
El Caserío	18,000	2,805	886
Block 24-East	No data	4,771	330
La Lomita Pequeña	20,896	1,700	240
Pueblo Salado, Area 14	4,944	1,747	82
Village			
Water World	15,053	4,589	165
Eastwing	17,382	6,556	264
Grand Canal Ruins	37,400	3,047	1,370
Cashion	37,397	8,659	1,071
New River	23,009	2,837	224
TAP, Phase A	22,279	3,526	613
ANAMAX-Rosemont	12,020	9,485	265
Pueblo Salado, Area 8/9, Civano Phase	31,429	10,960	241
Las Colinas, Mound 8	134,382	7,035	683
Las Colinas, 1982–1984 Excavations	No data	No data	1,754

APPENDIX B

RAW FLOTATION DATA FROM THE POSSIBLE AGRICULTURAL FIELD AND CONTROL SAMPLES

Scott Kwiatkowski

Table B.1. Raw Flotation Data from the Possible Agricultural Field and Control Samples

Context/ Provenience	Bag No.	Volumes: Sample Size (lt. fraction >0.25 mm) [lt. fraction <0.25 mm]	Carbonized Plant Taxa	Uncarbonized Plant Taxa	Other
Adobe/Clay-Lined Pits Underlying the Possible Agricultural Field (control samples)					
Feature 8/9-52; N636.45 E809.84; 5.39-5.59 mbd	8/9-2369	4.0 L (3.0 ml) [1.5 ml]	1 Indeterminate seed fragment 1 Indeterminate spine fragment	1 cf. Cheno-am seed fragment 1 *Euphorbia* seed 1 Indeterminate inflorescence fragment (HF)	2 Charophyceae oogonia 1 Clam shell (HF) 21 Fecal pellets 1 Insect case 70 Insect exoskeleton fragments (18 <0.25 mm)[1,2] 643 Macrospores (243 <0.25 mm)[1,2] 2 Ostracods 16 Snail shells (1 HF) 6 Snail shell fragments (HF) 2 Unburned bone fragments (HF)
Feature 8/9-53; N634.18 E809.89; 5.49-5.59 mbd	8/9-2362	0.8 L (0.5 ml) [0.0 ml]	(Unproductive)	1 *Verbascum thapsus* seed	8 Insect exoskeleton fragments 132 Macrospores[1] 2 Unburned bone fragments (HF)
Feature 8/9-56; N629.72 E808.28; 5.64-5.81 mbd	8/9-2355	4.0 L (4.0 ml) [1.0 ml]	1 *Agave* fiber (TR; CaO) 1 Cheno-am seed fragment 4 Indeterminate seed fragments 9 Miscellaneous endosperm fragments 1 Miscellaneous fragment (CaO) 3 Miscellaneous round fibers (CaO) 1 *Suaeda* seed	2 Cheno-am seeds	1 Animal tooth fragment (HF) 2 *Chara* oogonia 2 Charophyceae oogonia 1 Charred termite pellet 30 Fecal pellets 248 Insect exoskeleton fragments (119 <0.25 mm)[1,2] 647 Macrospores (224 <0.25 mm)[1,2] 8 Ostracods 28 Snail shells 3 Snail shell fragments (HF) 4 Unburned bone fragments (HF)

Table B.1. Raw Flotation Data From the Possible Agricultural Field and Control Samples, continued

Context/ Provenience	Bag No.	Volumes: Sample Size (lt. fraction >0.25 mm) [lt. fraction <0.25 mm]	Carbonized Plant Taxa	Uncarbonized Plant Taxa	Other
Possible Agricultural Field					
N634 E809.8; 5.14-5.26 mbd	8/9-2359	4.0 L (5.0 ml) [1.0 ml]	2 Cheno-am seed fragments 1 cf. *Descurainia* seed fragment 1 Indeterminate seed fragment 2 Miscellaneous endosperm fragments 1 Miscellaneous fragment (CaO) 1 Miscellaneous round fiber (CaO)	1 Cheno-am cf. *Amaranthus palmeri* seed 3 Cheno-am seed fragments 5 cf. Cheno-am seed fragments 1 *Euphorbia* seed fragment 2 Indeterminate seed fragments 1 *Physalis* seed 3 *Schismus* grains	18 Fecal pellets Numerous[3] insect exoskeleton fragments 1288 Macrospores (315 <0.25 mm)[1,2] 2 Macrospore clusters 1 Red resin globule 1 Ostracod 1 Ostracod valve 26 Termite pellets 2 Unburned bone fragments (HF)
N633 E809.7; 5.14-5.24 mbd	8/9-2363	4.0 L (7.5 ml) [1.0 ml]	3 Cheno-am seed fragments 1 Indeterminate seed fragment 5 Miscellaneous endosperm fragments	6 cf. Cheno-am seed fragments 1 *Malva* seed 1 *Schismus* grain 1 *Tribulus terrestris* bur	4 Burned bone fragments (HF) Numerous[3] fecal pellets 1 Insect Numerous[3] insect exoskeleton fragments 1117 Macrospores (342 <0.25 mm)[1,2] 4 Unburned bone fragments (HF)
N636.7 E809.8; 5.13-5.25 mbd	8/9-2365	4.0 L (3.0 ml) [1.0 ml]	4 Indeterminate seed fragments 1 Miscellaneous fragment (CaO) 1 Miscellaneous round fiber (CaO) 1 cf. *Sporobolus*-type grain fragment	1 *Bromus*-type floret (HF) 3 Cheno-am seeds (1 cf. *Amaranthus palmeri*) 4 Cheno-am seed fragments 1 *Euphorbia* seed 2 *Euphorbia* seed fragments 1 *Schismus* grain	4 *Chara* oogonia 3 Charophyceae oogonia 1 Charred termite pellet 52 Fecal pellets[1] 138 Insect exoskeleton fragments (20 <0.25 mm)[1,2] 1220 Macrospores (420 <0.25 mm)[1,2] 1 Nacreous insect egg case 1 Nacreous insect egg case fragment (HF) 6 Ostracods 1 Ostracod valve 33 Snail shells 5 Snail shell fragments (HF) 1 Termite pellet 2 Unburned bone fragments (HF)

Table B.1. Raw Flotation Data From the Possible Agricultural Field and Control Samples, continued

Context/ Provenience	Bag No.	Volumes: Sample Size (lt. fraction >0.25 mm) [lt. fraction <0.25 mm]	Carbonized Plant Taxa	Uncarbonized Plant Taxa	Other
Possible Agricultural Field, continued					
N634.2 E810.5; 5.16-5.27 mbd	8/9-2368	4.0 L (3.5 ml) [1.5 ml]	1 Cheno-am seed fragment 1 cf. Cheno-am seed fragment 2 Indeterminate seed fragments 1 Miscellaneous endosperm fragment 1 cf. *Sphaeralcea* seed fragment	4 Cheno-am seeds 11 Cheno-am seed fragments 20 cf. Cheno-am seed fragments 10 *Euphorbia* seeds 10 *Euphorbia* seed fragments 3 cf. *Euphorbia* seed fragments 1 *Malva* seed 1 *Portulaca* seed 2 *Trianthema portulacastrum* seed fragments 2 cf. *Trianthema portulacastrum* seed fragments	1 Charred termite pellet 7 Fecal pellets Numerous[3] insect exoskeleton fragments 2119 Macrospores (912 <0.25 mm)[1,2] 1 Macrospore cluster 1 Obsidian lithic (HF) 3 Ostracods 46 Snail shells 8 Snail shell fragments (4 HF) 1 Termite pellet 2 Unburned bone fragments (HF)
N635.5-E810; 5.15-5.25 mbd	8/9-2372	4.0 L (5.0 ml) [1.0 ml]	1 cf. Cheno-am seed fragment 3 Indeterminate seed fragments 1 *Leptochloa*-type grain fragment 1 Miscellaneous endosperm fragment	2 Cheno-am seeds 5 Cheno-am seed fragments 24 *Euphorbia* seeds 16 *Euphorbia* seed fragments 7 cf. *Euphorbia* seed fragments 1 *Malva* seed 1 *Portulaca* seed 1 *Sphaeralcea* seed 1 *Trianthema portulacastrum* seed 2 *Trianthema portulacastrum* seed fragments 1 Unknown fruit	1 *Chara* oogonium 2 Charophyceae oogonia 1 Charred termite pellet 132 Fecal pellets[1] 270 Insect exoskeleton fragments (126 <0.25 mm)[1,2] 4944 Macrospores (2700 <0.25 mm)[1,2] 1 Macrospore cluster 1 Nacreous insect egg case 6 Ostracods (1 <0.25 mm)[2] 1 Ostracod valve (<0.25 mm)[2] 55 Snail shells 1 Unburned bone 3 Unburned bone fragments (1 HF)

Note: All samples contained less than 0.5 ml wood charcoal, none of which was identifiable.
HF = number of remains recovered from the heavy fraction. Material identified from the two fractions is quantified separately so that direct comparisons can be made with the samples from Dutch Canal Ruin and Pueblo Salado.
TR = trough-shaped cross-section; CaO = white styloid and/or raphide crystals present.
[1]Estimated number; [2]Number of remains recovered from the residue less than 0.25 mm; [3]*Numerous* is defined as more than 50 parts per L.

APPENDIX C

CERAMIC PETROGRAPHIC ANALYSIS

David M. Schaller

APPENDIX C

CERAMIC PETROGRAPHIC ANALYSIS

David M. Schaller

The temper in 30 ceramics has been identified with a polarized light microscope in order to determine the approximate geographical location of each sample's place of manufacture. Twelve of the sherds were from Dutch Canal Ruin, and 18 were from Pueblo Salado. Fourteen of the sherds are plainware, 7 are redware, 4 are buffware, and 5 are polychrome. Each sherd was cut into 25 × 50–mm chips and submitted to RA Petrographic of Los Angeles to be made into petrographic thin sections. The approximate abundance, maximum size, and shape of every mineral and rock fragment type present in each thin section has been identified and recorded on a separate data analysis form.

The identification of the temper in a pottery sherd is only the first step in determining the original production source of the ceramic vessel from which it was derived. It is essential to know the composition and geographical distribution of all mutually distinguishable sand varieties found within the potential source region of a ceramic. Fortunately, this has been determined for the lower Salt River Valley in central Arizona (Abbott, Schaller, and Birnie 1991). Within this region, several mutually distinguishable rock types occur. The bedrock crops out in isolated exposures surrounded by alluvium and separated by drainages. This geographic arrangement, along with the movement of sediments towards the drainages, has created zones containing sands of different compositions. Nine different sand composition zones have been defined for the region (Abbott, Schaller, and Birnie 1991). Five of these (South Mountain granodiorite, Estrella gneiss, Camelback granite, Squaw Peak schist, and Gila Butte schist) have been identified in the ceramics studied from Dutch Canal Ruin and Pueblo Salado. These sites are located on the floodplain of the Salt River, which consists of fine-grained silt. Coarse angular sand of the type used as temper in Hohokam ceramics is not available at these sites. The nearest sources of sand that could have been employed are the Phoenix Mountains (Squaw Peak schist), Papago Buttes (Camelback granite), and South Mountain (South Mountain granodiorite and Estrella gneiss).

The composition of the temper of each pottery sherd analyzed was recorded on a separate form. For each rock and mineral variety present, estimated abundance, maximum size, average sphericity, and average roundness is listed. The estimated abundance of each grain type is given as a relative volume percent of the entire sherd. To emphasize the approximate nature of the estimate, a range of values rather than a single number is listed. The maximum size and shape of a grain are dependent on both its original size and shape in the rock from which it weathered and the distance from its original source. The maximum size indicated on the analysis form is the maximum width of the largest grain of that type in the thin section. In the case of polycrystalline aggregates, the maximum size listed refers to the largest single crystal within the aggregate. Since each thin section represents a two-dimensional slice of a three-dimensional object, and given the fact that the thin section probably does not intercept the largest grain in the ceramic vessel, the stated maximum size represents a minimum upper size limit. The sphericity of a grain represents the degree to which a grain approaches a sphere in shape. A grain with low sphericity would be highly ellipsoidal, while a grain with a high degree of sphericity would be nearly spherical. The angularity of a grain refers to the degree of surface roughness. An angular grain has a very rough surface, while a rounded grain has a smooth surface.

Estimates for volume percentages, maximum size, sphericity, and roundness are given for the following categories: volcanic rock fragments, metamorphic rock fragments, sherd temper grains, quartz, albite-twinned feldspar, cross-twinned feldspar, muscovite, biotite, pyroxene, amphibole, opaque oxides, and cryptocrystalline quartz. These categories were selected because these are the most common rock and mineral varieties found in rocks cropping out in central Arizona, and it is a straightforward matter to assign an individual grain to one of these categories based on its appearance in thin section. Because of their fine grain size, volcanic and metamorphic rock fragments are readily recognized as such when found in sand-sized grains. Plutonic rocks, such as granite, are too coarse-grained to form rock fragments that can be recognized when broken up into sand-sized fragments. However, an abundance of large feldspar and quartz grains in a sherd would indicate that granite was present in the source region of the sand temper. Consolidated sedimentary rocks that could weather into recognizable sand-sized rock fragments are not found in central Arizona. Sherd temper is listed as a separate category, because sherd temper is fairly common in redware ceramics.

The categories for quartz, muscovite, and biotite are self-explanatory. Feldspar is divided into three separate categories on the basis of crystallographic twinning. Albite-twinned feldspars include plagioclase feldspars that display well-developed albite twinning. Cross-twinned feldspars are those microcline and anorthoclase feldspars that display quadrille twinning. Untwinned feldspars include orthoclase and other feldspars that do not display crystallographic twinning. Pyroxenes include augite and other pyroxene group minerals. Amphiboles include hornblende and other amphibole group minerals. Opaques include oxide and sulfide minerals that are opaque to transmitted light. Cryptocrystalline quartz includes chert, chalcedony, and other semicrystalline forms of silica. In the concluding section of the ceramic analysis form, the probable bedrock geology in the area from which the temper was derived is indicated on the basis of the composition of the mineralogy and petrology of the sand temper. The geographic area corresponding to this geology is also stated if it can be determined.

A summary of the temper composition of the sherds is presented in Table C.1. Based on the small sample studied, the most common temper variety found in the plainware from Dutch Canal Ruin and Pueblo Salado is South Mountain granodiorite. This is somewhat surprising, because this sand variety is found only to the south of the Salt River, which is on the opposite bank from the archaeological sites. The redware ceramics examined for this project are tempered with South Mountain granodiorite, frequently in conjunction with sherd temper. The buffware ceramics are tempered with Gila Butte schist fragments. This mica- and tourmaline-bearing schist crops out at Gila Butte, Pima Butte, and in the Santan Mountains, which are all located in the Gila River Valley, 30 km to the south of Phoenix. Gila and Tonto polychrome sherds are tempered with a mixture of basalt, rhyolite, and granite fragments. These sherds are made from a very pure clay almost totally lacking in silt. In contrast, all sherds known to have been locally made in the Phoenix area are made from very silty clays. Sand used to temper the polychrome sherds is exotic to the Phoenix area. Clearly, these sherds were imported into the lower Salt River Valley. At the present time, not enough is known about the composition and distribution of the rocks cropping out in central Arizona to be able to determine the exact geographic source of these sherds. It was not possible to distinguish Gila and Tonto polychrome sherds on the basis of temper composition; they were made either in the same geographic area or in different geographic areas characterized by very similar geology.

Table C.1. Ceramic Temper Composition

Major Temper Constituents	Sample Number
Plainware	
South Mountain Granodiorite	1-28 3-19 8-1 8-78 8/9-750 8/9-3879 14-83 14-190
Squaw Peak Schist	3-44 8/9-2203
Estrella Gneiss	8/9-1095
Estrella Gneiss + Squaw Peak Schist	8-82
Camelback Granite	3-131 (2-17?)
Redware	
South Mountain Granodiorite	8/9-1343 14-471
South Mountain Granodiorite + Sherd	8/9-1331 8/9-2223 (8/9-3880?) 14-268
Estrella Gneiss + Squaw Peak Schist + Sherd	14-454
Buffware	
Gila Butte Schist	2-5 3-72 3-78
Unknown Fine-Grained Sand	2-45
Gila Polychrome	
Granite + Basalt + Rhyolite	8/9-3025 1213
Tonto Polychrome	
Granite + Basalt + Rhyolite	8/9-1890 3270
Unidentified Polychrome	
Granite + Basalt + Rhyolite	8/9-1216

APPENDIX D

GLOSSARY OF SOILS ANALYSIS TERMS

APPENDIX D

GLOSSARY OF SOILS ANALYSIS TERMS

angular see *blocky*

argillan a horizon in which clay has accumulated, usually from an overlying horizon[1]

blocky soil structure in which aggregate separates out in angular chunks[2]

bioturbation disturbance of soil by plant growth or animal activity

cation ion with a positive charge

consistence the feel of the soil, ease with which a lump can be crushed by the fingers[3]

epipedon a soil horizon that has formed at the soil surface[4]

geomorphic relating to surface land features

hydroxyl OH ion

massive descriptive of structureless soils with no separate aggregates

millimho unit of conductance

mollic describing an epipedon that is relatively thick, dark colored, and humus rich, with moderate to strong structure[4]

ped individual natural soil aggregate (crumb, prism, block)[3]

pedogenic relating to the structure and development of soils

pedoturbation disturbance of the soil from any cause

pH indicator of degree of acidity or alkalinity; pH 7.0 is neutral, values below 7.0 are acidic, those above are alkaline[3]

prismatic soil structure in which aggregates separate into prism-like columnar chunks with clearly defined angular edges[2]

structure arrangement of primary soil particles into aggregates (peds) separated from adjoining aggregates and with distinct properties[3]

umbric similar to mollic, but with lower base saturation[4]

weak descriptive of soils with poorly formed, indistinct peds[2]

---

[1]from FitzPatrick 1972
[2]from Baver, Gardner, and Gardner 1972
[3]from Hartman 1977
[4]from Soil Survey Staff 1975

REFERENCES

REFERENCES

Abbott, David R.

1984 A Technical Assessment of Ceramic Variation in the Salt-Gila Aqueduct Area: Toward a Comprehensive Documentation of Hohokam Ceramics. In *Hohokam Archaeology along the Salt-Gila Aqueduct, Central Arizona Project,* edited by Lynn S. Teague and Patricia L. Crown, pp. 3–118. Arizona State Museum Archaeological Series No. 150, Vol. 8. The University of Arizona, Tucson.

1988 Form, Function, Technology, and Style in Hohokam Ceramics. In *The 1982–1984 Excavations at Las Colinas: Material Culture,* by David R. Abbott, Kim E. Beckwith, Patricia L. Crown, R. Thomas Euler, David A. Gregory, J. Ronald London, Marilyn B. Saul, Larry A. Schwalbe, Mary Bernard-Shaw, Christine R. Szuter, and Arthur W. Vokes, pp. 73–198. Arizona State Museum Archaeological Series No. 162, Vol. 4. The University of Arizona, Tucson.

1992 Ceramic Exchange and a Strategy for Reconstructing Organizational Developments among the Hohokam. Paper presented at the Third Southwest Symposium, Tucson.

1994a *Hohokam Social Structure and Irrigation Management: The Ceramic Evidence from the Central Phoenix Basin.* Unpublished Ph.D. dissertation, Arizona State University, Tempe.

1994b Appendix A: Rough-Sort Data. In *The Pueblo Grande Project: Ceramics and the Production and Exchange of Pottery in the Central Phoenix Basin*, edited by David R. Abbott, pp. 463-478. Soil Systems Publications in Archaeology No. 20, Vol. 3, Pt. 2. Phoenix.

Abbott, David R. (editor)

1994 *The Pueblo Grande Project: Ceramics and the Production and Exchange of Pottery in the Central Phoenix Basin.* Soil Systems Publications in Archaeology No. 20, Vol. 3. Phoenix.

Abbott, David R., and David A. Gregory

1988 Hohokam Ceramic Wares and Types. In *The 1982–1984 Excavations at Las Colinas: Material Culture,* by David R. Abbott, Kim E. Beckwith, Patricia L. Crown, R. Thomas Euler, David A. Gregory, J. Ronald London, Marilyn B. Saul, Larry A. Schwalbe, Mary Bernard-Shaw, Christine R. Szuter, and Arthur W. Vokes, pp. 5–28. Arizona State Museum Archaeological Series No. 162, Vol. 4. The University of Arizona, Tucson.

Abbott, David R., and David M. Schaller

1991 Electron Microprobe and Petrographic Analyses of Prehistoric Hohokam Pottery to Determine Ceramic Exchange within the Salt River Valley, Arizona. In *Materials Issues in Art and Archaeology II,* edited by Pamela B. Vandiver, James Druzik, and George Seagan Wheeler, pp. 441–453. Materials Research Society Symposium Proceedings No. 185. Pittsburgh.

Abbott, David R., David M. Schaller, and Robert I. Birnie

1991 Compositional Analysis of Hohokam Pottery from the Salt River Valley, Arizona. Paper presented at the 6 Annual Meeting of the Southwestern Anthropological Association, Tucson.

Abbott, David R., and Mary-Ellen Walsh-Anduze
1991 Reconciling Opposing Typologies: Multiple Manufacturing Traditions for Hohokam Redware Ceramics. Paper presented at the 56th Annual Meeting of the Society for American Archaeology, New Orleans.

Abbott, R. Tucker
1974 *American Seashells: The Marine Molluscs of the Atlantic and Pacific Coasts of North America.* 2nd ed. Van Nostrand Reinhold, New York.

Ackerly, Neal W.
1982 Irrigation, Water Allocation, and the Hohokam Collapse. *The Kiva* 47:91–106.

1988 False Causality in the Hohokam Collapse. *The Kiva* 53:305–319.

1991 Constructing Analog Models of Hohokam Irrigation Using Historic Documents. In *Prehistoric Irrigation in Arizona: Symposium 1988*, edited by Cory Dale Breternitz, pp. 41–60. Soil Systems Publications in Archaeology No. 17. Phoenix.

Ackerly, Neal W., and T. Kathleen Henderson
1989 *Prehistoric Agricultural Activities on the Lehi-Mesa Terrace: Perspectives on Hohokam Irrigation Cycles.* Northland Research, Flagstaff.

Ackerly, Neal W., Jerry B. Howard, and Randall H. McGuire
1987 *La Ciudad Canals: A Study of Hohokam Irrigation Systems at the Community Level.* Arizona State University Anthropological Field Studies No. 17. Tempe.

Ackerly, Neal W., JoAnn E. Kisselburg, and Richard J. Martynec
1989 Canal Junctions and Water Control Features. In *Prehistoric Agricultural Activities on the Lehi-Mesa Terrace: Perspectives on Hohokam Irrigation Cycles*, edited by Neal W. Ackerly and T. Kathleen Henderson, pp. 146–183. Northland Research, Flagstaff.

Adams, E. Charles
1978 The Function of Limited Activity Sites in the Settlement System of the Lower Piedra District, Colorado. In *Limited Activity and Occupation Sites: A Collection of Conference Papers*, compiled and edited by Albert E. Ward, pp. 99–107. Contributions to Anthropological Studies No. 1. Center for Anthropological Studies, Albuquerque.

Adams, E. D.
1974 Soil Survey of Eastern Maricopa County and Northern Pinal Counties Area, Arizona. U.S. Department of Agriculture, Soil Conservation Service. U.S. Government Printing Office, Washington, D.C.

Ahlstrom, Richard V. N., and David A. Phillips, Jr.
1983 *Archaeological Survey and Cultural Resources Evaluations for Various Properties in Phoenix and Vicinity, Maricopa County, Arizona.* New World Research, Inc., Western Division Report of Investigations No. 1. Tucson.

Allen, Wilma
1985 The Block 24-East Lithics. In *City of Phoenix, Archaeology of the Original Townsite: Block 24-East*, edited by John S. Cable, Kathleen S. Hoffman, David E. Doyel, and Frank Ritz, pp. 127–167. Soil Systems Publications in Archaeology No. 8. Phoenix.

Allen, Wilma, and John S. Cable

1983 The Lithic Assemblage from Block 28-North. In *City of Phoenix, Archaeology of the Original Townsite: Block 28-North,* edited by John S. Cable, Susan L. Henry, and David E. Doyel, pp. 125–139. Soil Systems Publications in Archaeology No. 2. Phoenix.

1984 The Murphy's Addition Lithic Assemblage. In *City of Phoenix, Archaeology of the Original Townsite: The Murphy's Addition,* edited by John S. Cable, Susan L. Henry, and David E. Doyel, pp. 141–167. Soil Systems Publications in Archaeology No. 3. Phoenix.

Amon, J., and S. Karhu

1991 Patterns of Enamel Hypoplasia in a Prehistoric Hohokam Population. *Ascent* 4(1):1–8.

Anderson, Keith M.

1986 Hohokam Cemeteries as Elements of Settlement Structure and Change. In *Anthropology of the Desert West: Essays in Honor of Jesse D. Jennings,* edited by C. J. Condle and D. O. Fowler, pp. 179–201. University of Utah Anthropological Papers No. 10. Salt Lake City.

Andresen, John M.

1985 Pottery and Architecture at Compound F, Casa Grande Ruins National Monument, Arizona. In *Proceedings of the 1983 Hohokam Symposium,* edited by Alfred E. Dittert, Jr., and Donald E. Dove, pp. 595–640. Arizona Archaeological Society Occasional Paper No. 2. Phoenix.

Antieau, John M.

1981 *The Palo Verde Archaeological Investigations. Hohokam Settlement at the Confluence: Excavations along the Palo Verde Pipeline.* Museum of Northern Arizona Research Paper No. 20. Flagstaff.

Antieau, John M., and David H. Greenwald

1981 Material Culture. In *The Palo Verde Archaeological Investigations. Hohokam Settlement at the Confluence: Excavations along the Palo Verde Pipeline,* by John M. Antieau, pp. 183–216. Museum of Northern Arizona Research Paper No. 20. Flagstaff.

Arizona Weekly Miner [Phoenix]

1868a Report on harvest and crop prices. 8 July. Phoenix.

1868b News item on canal conditions. 29 February. Phoenix.

Bamforth, Douglas B.

1986 Technological Efficiency and Tool Curation. *American Antiquity* 51:38–50.

Bandelier, Adolf F.

1884 Reports by A. F. Bandelier on His Investigations in New Mexico during the Years 1883–84. In *Fifth Annual Report of the Archaeological Institute of America,* pp. 55–98. Boston. J. Wilson and Son, Cambridge, Massachusetts.

Barney, James M.

1933 Phoenix: A History of Its Pioneer Days and People. *Arizona Historical Review* 5(4): 264–285.

Baver, L. D., Walter H. Gardner, and Wilford R. Gardner
1972 *Soil Physics*. 4th ed. John Wiley & Sons, New York.

Beck, Charlotte, and George T. Jones
1990 Toolstone Selection and Lithic Technology in Early Great Basin Prehistory. *Journal of Field Archaeology* 17:283–299.

Bequaert, Joseph C., and Walter B. Miller
1973 *The Mollusks of the Arid Southwest*. The University of Arizona Press, Tucson.

Bernard-Shaw, Mary
1983 The Stone Tool Assemblage of the Salt-Gila Aqueduct Project Sites. In *Hohokam Archaeology along the Salt-Gila Aqueduct, Central Arizona Project: Material Culture*, Parts 2–5, edited by Lynn S. Teague and Patricia L. Crown, pp. 373–443. Arizona State Museum Archaeological Series No. 150, Vol. 8. The University of Arizona, Tucson.

Binford, Lewis R.
1973 Interassemblage Variability—The Mousterian and the "Functional" Argument. In *The Explanation of Culture Change*, edited by Colin Renfrew, pp. 227–254. Duckworth, London.

1977 Forty-seven Trips. In *Stone Tools as Cultural Markers*, edited by R. V. S. Wright, pp. 24–36. Australian Institute of Aboriginal Studies, Canberra.

1979 Organization and Formation Processes: Looking at Curated Technologies. *Journal of Anthropological Research* 35:255–273.

Birkeland, Peter
1984 *Soils and Geomorphology*. Oxford University Press, New York.

Birnie, Robert I.
1991 *Archaeological Test Excavations at Pueblo Grande, AZ U:9:7(ASM): The Washington Street Expansion Project, Phoenix, Arizona*. Soil Systems Technical Report No. 91-3. Phoenix.

1994 Archaeological and Geological Investigations on the Lehi Terrace. In *The Pueblo Grande Project: Environment and Subsistence*, edited by Scott Kwiatkowski, pp. 67-125. Soil Systems Publications in Archaeology No. 20, Vol. 5. Phoenix.

Birnie, Robert I., and Mary-Ellen Walsh-Anduze
1991 *Archaeological Test Excavations at Pueblo Grande Cultural Park, AZ U:9:7 (ASM), Phoenix, Arizona: The Museum Expansion Project*. Soil Systems Technical Report No. 91-1. Phoenix.

Blanton, Richard E., Stephen A. Kowalewski, Gary Feinman, and Jill Appel
1981 *Ancient Mesoamerica*. Cambridge University Press, Cambridge.

Bohrer, Vorsila L.
1991 Recently Recognized Cultivated and Encouraged Plants among the Hohokam. *Kiva* 56:227–236.

Bostwick, Todd W.
1992 Platform Mound Ceremonialism in Southern Arizona: Possible Symbolic Meanings of Hohokam and Salado Platform Mounds. In *Proceedings of the Second Salado Conference, Globe, Arizona*, edited by Richard C. Lange and Stephen Germick, pp. 78–85. Occasional Paper, Arizona Archaeological Society, Phoenix.

1994 Trash Mounds. In *Archaeology of the Pueblo Grande Platform Mound and Surrounding Features: Features in the Central Precinct of the Pueblo Grande Community*, edited by Todd W. Bostwick and Christian E. Downum, pp. 1-46. Pueblo Grande Museum Anthropological Papers No. 1, Vol. 2. Phoenix.

Bostwick, Todd W., and Christian E. Downum
1994 *Archaeology of the Pueblo Grande Platform Mound and Surrounding Features: Features in the Central Precinct of the Pueblo Grande Community*. Pueblo Grande Museum Anthropological Papers No. 1., Vol. 2. Phoenix.

Bostwick, Todd W., David H. Greenwald, and Mary-Ellen Walsh-Anduze
1996 The Hohokam Post-Classic Period Occupation and a Piman Presence on the Salt River Floodplain. In Life on the Floodplain: Further Investigations at Pueblo Salado for Phoenix Sky Harbor International Airport. Data Recovery and Re-evaluation, edited by David H. Greenwald, Jean H. Ballagh, Douglas R. Mitchell, and Richard A. Anduze, pp. 461-490. Pueblo Grande Museum Anthropological Papers No. 4, Vol. 2. Phoenix. Draft.

Bostwick, Todd W., and Glen E. Rice
1987 *A Cultural Resources Inventory for the Southwest Loop Freeway Project*. Office of Cultural Resource Management Report No. 66. Arizona State University, Tempe.

Bostwick, Todd W., and M. Steven Shackley
1987 Settlement Strategies and Lithic Technology: An Examination of Variability in Hunter-Gatherer and Agriculturalist Chipped Stone Assemblages in the Sonoran Desert. Paper presented at the 1987 Hohokam Symposium, Arizona State University, Tempe.

Bradfield, Maitland
1971 *The Changing Patterns of Hopi Agriculture*. Royal Anthropological Institute of Great Britain and Ireland Occasional Papers No. 30. London.

Bradley, Bruce A.
1980 *Excavations at AZ BB:13:72, Santa Cruz Industrial Park*. Complete Archaeological Service Associates No. 1. Oracle, Arizona.

Brown, David E., Charles H. Lowe, and Janet F. Hausler
1977 Southwestern Riparian Communities: Their Biotic Importance and Management in Arizona. In *Importance, Preservation and Management of Riparian Habitat: A Symposium*, by the U.S. Department of Agriculture, Forest Service, pp. 201–211. U.S. Department of Agriculture, Forest Service General Technical Report RM-43. Fort Collins, Colorado.

Brown, J. R.
1869 *Adventures in Apache Country*. Harper and Brothers, New York.

Bruder, J. Simon, and Robert E. Gasser

1983 Lithics. In *Archaeological Investigations in the Adobe Dam Project Area*, by J. Simon Bruder, pp. 93–134. Museum of Northern Arizona Research Paper No. 27. Flagstaff.

Brunson, Judy L.

1989 *The Social Organization of the Los Muertos Hohokam: A Reanalysis of Cushing's Hemenway Expedition Data.* Unpublished Ph.D. dissertation, Department of Anthropology, Arizona State University, Tempe.

BRW, Inc.

1986 Sky Harbor Pedestrian Survey and Cultural Resources Program. Prepared for Sky Harbor Center, Community and Economic Development Department, City of Phoenix.

1989 Phase II Testing of Cultural Resources in the Combined MeraBank Phase2-3 Properties, Southwest Loop Road, and Adjoining Properties between 18th, 20th, Mohave, and Yuma Streets in the Sky Harbor Center, Phoenix, Arizona. Prepared for Sky Harbor Center, Community and Economic Development Department, City of Phoenix.

Bryan, Kirk

1954 *The Geology of Chaco Canyon, New Mexico, in Relation to the Life and Remains of the Prehistoric Peoples of Pueblo Bonito.* Smithsonian Miscellaneous Collections No. 122, Vol. 7. Smithsonian Institution, Washington, D.C.

Butzer, Karl W.

1982 *Archaeology as Human Ecology.* Cambridge University Press, Cambridge.

Bye, Robert A., Jr.

1981 Quelites: Ethnoecology of Edible Greens, Past, Present, and Future. *Journal of Ethnobiology* 1(1):109–123.

Cable, John S.

1985 An Archaeological Survey of the Southside Access Road at Sky Harbor Airport, Phoenix, Arizona. Prepared for the City of Phoenix Aviation Department (Report No. PGM-85-2). Pueblo Grande Museum, Phoenix.

1988 *Archaeological Testing along the Hohokam Expressway Corridor East of Pueblo Grande, Phoenix, Arizona: Results and Preliminary Research Design for Data Recovery.* Soil Systems Technical Report No. 88-21. Phoenix.

1991 The Role of Irrigation Agriculture in the Formation and Sociopolitical Development of Early Hohokam Villages in the Lowlands of the Phoenix Basin, Arizona. In *Prehistoric Irrigation in Arizona: Symposium 1988*, edited by Cory Dale Breternitz, pp. 107–137. Soil Systems Publications in Archaeology No. 17. Phoenix.

Cable, John S., and Wilma Allen

1982 Part 1: The Prehistoric Component. In *City of Phoenix, Archaeology of the Original Townsite: Blocks 1 and 2*, edited by John S. Cable, Susan L. Henry, and David E. Doyel, pp. 13–179. Soil Systems Publications in Archaeology No. 1. Professional Service Industries, Phoenix.

Cable, John S., and David E. Doyel

1984a Hohokam Research in the Central Phoenix Redevelopment Blocks. In *City of Phoenix, Archaeology of the Original Townsite: The Murphy's Addition,* edited by John S. Cable, Susan L. Henry, and David E. Doyel, pp. 11–33. Soil Systems Publications in Archaeology No. 3. Phoenix.

1984b The Implications of Field Houses for Modeling Hohokam Agricultural Systems. In *City of Phoenix, Archaeology of the Original Townsite: The Murphy's Addition,* edited by John S. Cable, Susan L. Henry, and David E. Doyel, pp. 259–280. Soil Systems Publications in Archaeology No. 3. Phoenix.

1985a Prehistoric Research Design. In *City of Phoenix, Archaeology of the Original Townsite: Block 24-East,* edited by John S. Cable, Kathleen S. Hoffman, David E. Doyel, and Frank Ritz, pp. 11–29. Soil Systems Publications in Archaeology No. 8. Phoenix.

1985b The Pueblo Patricio Sequence: Its Implications for the Study of Hohokam Origins, Pioneer Period Site Structure and the Processes of Sedentism. In *City of Phoenix, Archaeology of the Original Townsite: Block 24-East,* edited by John S. Cable, Kathleen S. Hoffman, David E. Doyel, and Frank Ritz, pp. 211–270. Soil Systems Publications in Archaeology No. 8. Phoenix.

1985c Hohokam Land-Use Patterns along the Terraces of the Lower Salt River Valley: The Central Phoenix Project. In *Proceedings of the 1983 Hohokam Symposium,* Part I, edited by Alfred E. Dittert, Jr., and Donald E. Dove, pp. 263–310. Arizona Archaeological Society Occasional Paper No. 2. Phoenix.

1986 The Archaeology of Swilling's Ditch: Phoenix's First Historic Canal. Ms. on file, Pueblo Grande Museum, Phoenix. Draft.

1987 Pioneer Period Village Structure and Settlement Pattern in the Phoenix Basin. In *The Hohokam Village: Site Structure and Organization,* edited by David E. Doyel, pp. 21–70. Southwestern and Rocky Mountain Division of the American Association for the Advancement of Science. Glenwood Springs, Colorado.

Cable, John S., and Ronald R. Gould

1988 The Casa Buena Ceramic Assemblage: A Study of Typological Systematics and Ceramic Change in Classic Period Assemblages. In *Excavations at Casa Buena: Changing Hohokam Land Use along the Squaw Peak Parkway,* edited by Jerry B. Howard, pp. 271–357. Soil Systems Publications in Archaeology No. 11. Phoenix.

Cable, John S., Kathleen S. Hoffman, David E. Doyel, and Frank Ritz

1985 *City of Phoenix, Archaeology of the Original Townsite: Block 24-East.* Soil Systems Publications in Archaeology No. 8. Phoenix.

Cable, John S., and Douglas R. Mitchell

1988 La Lomita Pequeña in Regional Perspective. In *Excavations at La Lomita Pequeña: A Santa Cruz/Sacaton Phase Hamlet in the Salt River Valley,* edited by Douglas R. Mitchell, pp. 95–177. Soil Systems Publications in Archaeology No. 10. Phoenix.

1989 Intrasite Structure, Chronology, and Community Organization of the Grand Canal Ruins. In *Archaeological Investigations at the Grand Canal Ruins: A Classic Period Site in Phoenix, Arizona,* edited by Douglas R. Mitchell, pp. 793–878. Soil Systems Publications in Archaeology No. 12. Phoenix.

1991 Settlement Growth and Integration in a Large Hohokam Canal System. In *Prehistoric Irrigation in Arizona: Symposium 1988,* edited by Cory Dale Breternitz, pp. 89–106. Soil Systems Publications in Archaeology No. 17. Phoenix.

Carothers, Steven W.
1977 Importance, Preservation and Management of Riparian Habitat: An Overview. In *Importance, Preservation and Management of Riparian Habitat: A Symposium,* by the U.S. Department of Agriculture, Forest Service, pp. 2–3. U.S. Department of Agriculture, Forest Service General Technical Report RM-43. Fort Collins, Colorado.

Castetter, Edward F., and Willis H. Bell
1942 *Pima and Papago Indian Agriculture.* Inter-American Studies No. 1. University of New Mexico Press, Albuquerque.

Castetter, Edward F., Willis H. Bell, and Alvin R. Grove
1938 *Ethnobiological Studies in the American Southwest: The Early Utilization and the Distribution of Agave in the American Southwest.* University of New Mexico Bulletin No. 6. Albuquerque.

Chisholm, M. D. I.
1962 *Rural Settlement and Land Use: An Essay in Location.* Hutchinson, London.

Chow, Ven Te
1959 *Open Channel Hydraulics.* McGraw-Hill, New York.

Ciolek-Torrello, Richard
1988 Chronology. In *Hohokam Settlement along the Slopes of the Picacho Mountains: Synthesis and Conclusions, Tucson Aqueduct Project,* edited by Richard Ciolek-Torrello and David R. Wilcox, pp. 42–120. Museum of Northern Arizona Research Paper No. 35, Vol. 6. Flagstaff.

Ciolek-Torrello, Richard, Martha M. Callahan, and David H. Greenwald (editors)
1988 *Hohokam Settlement along the Slopes of the Picacho Mountains: The Brady Wash Sites, Tucson Aqueduct Project.* Museum of Northern Arizona Research Paper No. 35, Vol. 2. Flagstaff.

City of Mesa
1965 *Our Town—Mesa, Arizona.* Tyler, Phoenix.

Colton, Harold S.
1952 Primitive Pottery Making: The Ceramic Methods of a Hopi Indian Potter. *Faenza* 38:135–138.

1956 *Pottery Types of the Southwest.* Museum of Northern Arizona Ceramic Series No. 3C. Flagstaff.

Colton, Mary Russell-Ferrell
1931 Technique of the Major Hopi Crafts. *Museum Notes* 2(3):1-7. Museum of Northern Arizona, Flagstaff.

Cook, Sherburne F.
1972 *Prehistoric Demography*. Addison-Wesley, Reading, Pennsylvania.

Cook, Sherburne F., and Robert F. Heizer
1965 *The Quantitative Approach to the Relation between Population and Settlement Size.* Contributions of the University of California Archaeological Research Facility No. 64. Berkeley.

Crosswhite, Frank S.
1981 Desert Plants, Habitat and Agriculture in Relation to the Major Pattern of Cultural Differentiation in the O'odham People of the Sonoran Desert. *Desert Plants* 3:47-76.

Crown, Patricia L.
1981 Analysis of the Las Colinas Ceramics. In *The 1968 Excavations at Mound 8, Las Colinas Ruin Group, Phoenix, Arizona,* edited by Laurens Hammack and Alan Sullivan, pp. 87-169. Arizona State Museum Archaeological Series No. 154. The University of Arizona, Tucson.

1983 Introduction: Field Houses and Farmsteads in South-Central Arizona. In *Hohokam Archaeology along the Salt-Gila Aqueduct, Central Arizona Project: Small Habitation Sites on Queen Creek,* edited by Lynn S. Teague and Patricia L. Crown, pp. 3-22. Arizona State Museum Archaeological Series No. 150, Vol. 5. The University of Arizona, Tucson.

1984a Classic Period Ceramic Manufacture: Exploring Variability in the Production and Use of Hohokam Vessels. In *Hohokam Archaeology along the Salt-Gila Aqueduct, Central Arizona Project: Material Culture,* edited by Lynn S. Teague and Patricia L. Crown, pp. 119-204. Arizona State Museum Archaeological Series No. 150, Vol. 8. The University of Arizona, Tucson.

1984b Adaptation through Diversity: An Examination of Population Pressure and Agricultural Techniques in the Salt-Gila Basin. In *Prehistoric Agricultural Strategies in the Southwest,* edited by Suzanne K. Fish and Paul R. Fish, pp.5-26. Anthropological Research Papers No. 33. Arizona State University, Tempe.

1985 Morphology and Function of Hohokam Small Structures. *The Kiva* 50:75-94.

1987 Classic Period Hohokam Settlement and Land Use in the Casa Grande Ruins Area. *Journal of Field Archaeology* 14:147-162.

1988 X-Ray Fluorescence Analysis of Materials Variability in Las Colinas Ceramics. In *The 1982-1984 Excavations at Las Colinas: Material Culture, Tucson Aqueduct Project*, by David R. Abbott, Kim E. Beckwith, Patricia L. Crown, R. Thomas Euler, David A. Gregory, J. Ronald London, Marilyn B. Saul, Larry A. Schwalbe, Mary Bernard-Shaw, Christine R. Szuter, and Arthur W. Vokes, pp. 29-72. Arizona State Museum Archaeological Series No. 162, Vol. 4. The University of Arizona, Tucson.

1991 The Hohokam: Current Views of Prehistory and the Regional System. In *Chaco and Hohokam: Prehistoric Regional Systems in the American Southwest,* edited by Patricia L. Crown and W. James Judge, pp. 135–157. School of American Research Press, Santa Fe.

Crown, Patricia L., and Ronald L. Bishop

1987 The Manufacture of the Salado Polychromes. *Pottery Southwest* 14(4):1–4.

1991 Manufacture of Gila Polychrome in the Greater American Southwest: An Instrumental Neutron Activation Analysis. In *Homol'ovi II: Archaeology of an Ancestral Hopi Village, Arizona,* edited by E. Charles Adams and Kelley Ann Hays, pp. 49–56. Anthropological Papers of the University of Arizona No. 55. The University of Arizona Press, Tucson.

Crown, Patricia L., Larry A. Schwalbe, and J. Ronald London

1988 X-ray Fluorescence Analysis of Materials Variability in Las Colinas Ceramics. In *The 1982–1984 Excavations at Las Colinas: Material Culture,* by David R. Abbott, Kim E. Beckwith, Patricia L. Crown, R. Thomas Euler, David A. Gregory, J. Ronald London, Marilyn B. Saul, Larry A. Schwalbe, Mary Bernard-Shaw, Christine R. Szuter, and Arthur W. Vokes, pp. 29-71. Arizona State Museum Archaeological Series No. 162, Vol. 4. The University of Arizona, Tucson.

Crown, Patricia L., and Earl W. Sires, Jr.

1984 The Hohokam Chronology and Salt-Gila Aqueduct Project Research. In *Hohokam Archaeology along the Salt-Gila Aqueduct, Central Arizona Project,* edited by Lynn S. Teague and Patricia L. Crown, pp. 73–85. Arizona State Museum Archaeological Series No. 150, Vol. 9. The University of Arizona, Tucson.

Cushing, Frank Hamilton

1890 Preliminary Notes on the Origin, Working Hypothesis, and Primary Researches of the Hemenway Southwestern Archaeological Expedition. In *Conves International des Americanistes, Compte-rendu de la septieme session,* pp. 151–194. Berlin.

1892 The Hemenway Southwestern Archaeological Expedition. Ms. on file, Peabody Museum, Harvard University, Cambridge.

Dean, Jeffrey S.

1991 Thoughts on Hohokam Chronology. In *Exploring the Hohokam: Prehistoric Desert Peoples of the American Southwest,* edited by George J. Gumerman, pp. 61–150. University of New Mexico Press, Albuquerque.

Deaver, William L.

1984 Pottery. In *Hohokam Habitation Sites in the Northern Santa Rita Mountains,* by Alan Ferg, Kenneth C. Rozen, William L. Deaver, Martyn D. Tagg, David A. Phillips, Jr., and David A. Gregory, pp. 237–419. Arizona State Museum Archaeological Series No. 147, Vol. 2. The University of Arizona, Tucson.

1989 Ceramics. In *Hohokam Archaeology along Phase B of the Tucson Aqueduct Central Arizona Project: Excavations at Water World (AZ AA:16:94), A Rillito Phase Ballcourt Village in the Avra Valley,* edited by Jon S. Czaplicki and John C. Ravesloot, pp. 153–211. Arizona State Museum Archaeological Series No. 178, Vol. 3. The University of Arizona, Tucson.

Debowski, Sharon S.
1974 Provenience and Description of Shell and Miscellaneous Artifacts from the Escalante Ruin Group. In *Excavations in the Escalante Ruin Group, Southern Arizona,* edited by David E. Doyel, pp. 276–298. Arizona State Museum Archaeological Series No. 37. The University of Arizona, Tucson.

DiPeso, Charles C.
1951 *The Babocomari Village Site on the Babocomari River, Southeastern Arizona.* Amerind Foundation Publications No. 5. Dragoon, Arizona.

1956 *The Upper Pima of San Cayetano del Tumacacori: An Archaeological Reconstruction of the Ootam of the Pimeria Alta.* The Amerind Foundation Publications No. 7. Dragoon, Arizona.

DiPeso, Charles C., John B. Rinaldo, and Gloria J. Fenner
1974 *Casas Grandes: A Fallen Trading Center of the Gran Chichimeca: Ceramics and Shell.* The Amerind Foundation Publications No. 9, Vol. 6. Dragoon, Arizona. Northland Press, Flagstaff.

Dodd, Walter A., Jr.
1979 The Wear and Use of Battered Tools at Armijo Rockshelter. In *Lithic Use-Wear Analysis,* edited by Bryan Hayden, pp. 231–242. Academic Press, New York.

Doelle, William H., and Henry D. Wallace
1990 The Transition to History in Pimería Alta. In *Perspectives on Southwestern Prehistory,* edited by Paul E. Minnis and Charles L. Redman, pp. 239–257. Westview Press, Boulder, Colorado.

Doolittle, William E.
1990 *Canal Irrigation in Prehistoric Mexico: The Sequence of Technological Change.* University of Texas Press, Austin.

Downum, Christian E., and Todd W. Bostwick (editors)
1994 *Archaeology of the Pueblo Grande Platform Mound and Surrounding Features: Introduction to the Archival Project and History of Archaeological Research.* Pueblo Grande Museum Anthropological Papers No. 1., Vol. 1. Phoenix.

Doyel, David E.
1974 *Excavations in the Escalante Ruin Group, Southern Arizona.* Arizona State Museum Archaeological Series No. 37. The University of Arizona, Tucson.

1980 Hohokam Social Organization and the Sedentary to Classic Transition. In *Current Issues in Hohokam Prehistory,* edited by David E. Doyel and Fred Plog, pp. 23–40. Anthropological Research Papers No. 23. Arizona State University, Tempe.

1981 *Late Hohokam Prehistory in Southern Arizona.* Contributions to Archaeology No. 2. Gila Press, Scottsdale.

1985 The New River–Palo Verde Community System. In *Hohokam Settlement and Economic Systems in the Central New River Drainage, Arizona,* edited by David E. Doyel and Mark D. Elson, pp. 681–699. Soil Systems Publications in Archaeology No. 4. Phoenix.

1991a Hohokam Cultural Evolution in the Phoenix Basin. In *Exploring the Hohokam: Prehistoric Desert Peoples of the American Southwest,* edited by George J. Gumerman, pp. 231–278. University of New Mexico Press, Albuquerque.

1991b Hohokam Exchange and Interaction. In *Chaco and Hohokam: Prehistoric Regional Systems in the American Southwest,* edited by Patricia L. Crown and W. James Judge, pp. 225–252. School of American Research Press, Santa Fe.

Doyel, David E., and Mark D. Elson (editors)

1985 Ceramic Analysis. In *Hohokam Settlement and Economic Systems in the Central New River Drainage, Arizona,* edited by David E. Doyel and Mark D. Elson, pp. 436–519. Soil Systems Publications in Archaeology No. 4, Vol. 2. Phoenix.

Effland, Richard W., Jr.

1988 An Examination of Hohokam Mortuary Practice from Casa Buena. In *Excavations at Casa Buena: Changing Hohokam Land Use along the Squaw Peak Parkway,* edited by Jerry B. Howard, pp. 693–794. Soil Systems Publications in Archaeology No. 11, Vol. 2. Phoenix.

Eidt, Robert C.

1984 *Advances in Abandoned Settlement Analysis: Application to Prehistoric Anthrosols in Colombia, South America.* The Center for Latin America, University of Wisconsin—Milwaukee.

Eighmy, Jeffrey L., and Kathleen A. Baker

1991 Archaeomagnetic Laboratory Results for Site AZ T:12:47 (ASM) with an Updated Analysis of Samples from NA19324. Ms. on file, SWCA, Inc., Environmental Consultants, Flagstaff.

Eighmy, Jeffrey L., J. Holly Hathaway, and Sharilee Counce

1987 *Independently Dated Virtual Geomagnetic Poles: The CSU Archaeomagnetic Data Base.* Technical Series No. 1. Archaeomagnetic Laboratory, Department of Anthropology, Colorado State University, Fort Collins.

Eighmy, Jeffrey L., J. Holly Hathaway, and Allen E. Kane

1985 Archaeomagnetic Secular Variation in the American Southwest between A.D. 700 and 900: Final Results from the Dolores Archaeological Program. Ms. on file, Archaeomagnetic Laboratory, Department of Anthropology, Colorado State University, Fort Collins.

Eighmy, Jeffrey L., and Randall H. McGuire

1988 *Archaeomagnetic Dates and the Hohokam Phase Sequence.* Technical Series No. 3. Colorado State University Archaeometric Laboratory, Fort Collins.

Ellis, Florence H.

1978 Small Structures Used by Historic Pueblo Peoples and Their Immediate Ancestors. In *Limited Activity and Occupation Sites: A Collection of Conference Papers,* compiled and edited by Albert E. Ward, pp. 59–68. Contributions to Anthropological Studies No. 1. Center for Anthropological Studies, Albuquerque.

El-Najjar, M. Y., M. V. DeSanti, and L. Ozebek
1978 Prevalence and Possible Etiology of Dental Enamel Hypoplasia. *American Journal of Physical Anthropology* 48:185–192.

Elstien, Elizabeth R.
1989 Chipped Stone Artifact Analysis, with contributions by Margerie Green, M. Steven Shackley, and James Burton. In *Settlement, Subsistence, and Specialization in the Northern Periphery: The Waddell Project,* edited by Margerie Green, pp. 643–735. Archaeological Consulting Services Cultural Resources Report No. 65, Vol. 2. Tempe.

Euler, R. Thomas, and David A. Gregory
1988 Pecked, Ground, and Polished Stone Artifacts. In *The 1982–1984 Excavations at Las Colinas: Material Culture,* by David R. Abbott, Kim E. Beckwith, Patricia L. Crown, R. Thomas Euler, David A. Gregory, J. Ronald London, Marilyn B. Saul, Larry A. Schwalbe, Mary Bernard-Shaw, Christine R. Szuter, and Arthur W. Vokes, pp. 299–317. Arizona State Museum Archaeological Series No. 162, Vol. 4. The University of Arizona, Tucson.

Euler, R. Thomas, and Earl W. Sires, Jr.
1984 Archaeological Testing at AZ T:12:47 ASM, A Hohokam Site in Phoenix, Arizona. Ms. on file, Cultural Resource Management Division, Arizona State Museum, The University of Arizona, Tucson.

Evermann, B. W., and Cloud Rutter
1895 The Fishes of the Colorado Basin. *Bulletin of the U.S. Fish Commission* 14(1894):473–486.

Faulk, Odie
1970 *Arizona: A Short History.* University of Oklahoma Press, Norman.

Ferg, Alan
1984a Nonutilitarian Ground Stone, Crystals, and Minerals. In *Hohokam Habitation Sites in the Northern Santa Rita Mountains,* by Alan Ferg, Kenneth C. Rozen, William L. Deaver, Martyn D. Tagg, David A. Phillips, Jr., and David A. Gregory, pp. 665–688. Arizona State Museum Archaeological Series No. 147, Vol. 2. The University of Arizona, Tucson.

1984b Discussion. In *Hohokam Habitation Sites in the Northern Santa Rita Mountains,* by Alan Ferg, Kenneth C. Rozen, William L. Deaver, Martyn D. Tagg, David A. Phillips, Jr., and David A. Gregory, pp. 725–822. Arizona State Museum Archaeological Series No. 147, Vol. 2. The University of Arizona, Tucson.

Fewkes, Jesse W.
1896 Pacific Coast Shell from Prehistoric Tusayan Pueblos. *American Anthropologist* (Old Series) 9:359–367.

Fink, T. Michael
1985 Tuberculosis and Anemia in a Pueblo II–III (ca. A.D. 900–1300) Anasazi Child from New Mexico. In *Health and Disease in the Prehistoric Southwest,* edited by C. F. Merbs and R. J. Miller, pp. 359–379. Arizona State University Anthropological Research Papers No. 34. Tempe.

1989a The Cremated Human Remains from Los Morteros. In *The 1979–1983 Testing at Los Morteros (AZ AA:12:57 ASM), a Large Hohokam Village Site in the Tucson Basin,* edited by Richard L. Lange and William L. Deaver, pp. 277–284. Arizona State Museum Archaeological Series No. 177. The University of Arizona, Tucson.

1989b The Human Skeletal Remains from the Grand Canal Ruins, AZ T:12:14 (ASU) and AZ T:12:16 (ASU). In *Archaeological Investigations at the Grand Canal Ruins: A Classic Period Site in Phoenix, Arizona,* edited by Douglas R. Mitchell, pp. 619–705. Soil Systems Publications in Archaeology No. 12, Vol. 2. Phoenix.

1990 Analysis of the Human Skeletal Remains from La Lomita (AZ U:9:67 [ASM]). In *The La Lomita Excavations: 10th Century Hohokam Occupation in South-Central Arizona,* edited by Douglas R. Mitchell, pp. 67–84. Soil Systems Publications in Archaeology No. 15. Phoenix.

1991 Prehistoric Irrigation Canals and Their Possible Impact on Hohokam Health. In *Prehistoric Irrigation in Arizona: Symposium 1988,* edited by Cory D. Breternitz, pp. 61–87. Soil Systems Publications in Archaeology No. 17. Phoenix.

1996 *Current Issues in Cremation Analysis: A Perspective from the American Southwest.* Unpublished Master's thesis, Arizona State University, Tempe.

Fink, T. Michael, and Charles F. Merbs
1991 Paleonutrition and Paleopathology of the Salt River Hohokam: A Search for Correlates. *Kiva* 56:293–318.

Fish, Suzanne K., and Paul R. Fish
1978 The Small House in a Social Network: Some Mesoamerican Examples. In *Limited Activity and Occupation Sites: A Collection of Conference Papers,* compiled and edited by Albert E. Ward, pp. 51–58. Contributions to Anthropological Studies No. 1. Center for Anthropological Studies, Albuquerque.

Fish, Suzanne K., Paul R. Fish, and John Madsen
1985 Analyzing Regional Agriculture: A Hohokam Example. Paper presented at the 50th Annual Meeting of the Society for American Archaeology, Denver.

Fish, Suzanne K., Paul R. Fish, Charles Miksicek, and John Madsen
1985 Prehistoric Agave Cultivation in Southern Arizona. *Desert Plants* 7:107–112.

Fish, Suzanne K., and Gary P. Nabhan
1991 Desert as Context: The Hohokam Environment. In *Exploring the Hohokam: Prehistoric Desert Peoples of the American Southwest,* edited by George J. Gumerman, pp. 29–60. University of New Mexico Press, Albuquerque.

FitzPatrick, Ewart Adsil
1972 *Pedology: A Systematic Approach to Soil Science.* Hafner, New York.

Fortier, S., and F. C. Scobey
1926 Permissible Canal Velocities. *American Society of Civil Engineers Transactions* 89:940–984.

Foster, Michael (editor)
1994 *The Pueblo Grande Project: Material Culture.* Soil Systems Publications in Archaeology No. 20, Vo. 4. Phoenix.

Fryman, Frank, James W. Woodward, and James W. Garrison
1977 An Initial Survey of Historic Resources within the Phoenix Metropolitan Area, Maricopa County, Arizona. Ms. on file, State Historic Preservation Office/Arizona State Parks, Phoenix.

Gardiner, Ronald, W. Bruce Masse, and Carl D. Halbirt
1987 Data Recovery at AZ U:9:71 (ASM). In *Archaeological Investigations of Portions of the Las Acequias–Los Muertos Irrigation System: Testing and Partial Data Recovery within the Tempe Section of the Outer Loop Freeway System, Maricopa County, Arizona,* edited by W. Bruce Masse, pp. 25–56. Arizona State Museum Archaeological Series No. 176. The University of Arizona, Tucson.

Gasser, Robert E.
1980 Exchange and the Hohokam Archaeobotanical Record. In *Current Issues in Hohokam Prehistory, Proceedings of a Symposium,* edited by David E. Doyel and Fred Plog, pp. 72–77. Anthropological Research Papers No. 23. Arizona State University, Tempe.

1981–1982 Hohokam Use of Desert Food Plants. *Desert Plants* 3:216–234.

Gasser, Robert E., and Scott M. Kwiatkowski
1991 Food for Thought: Recognizing Patterns in Hohokam Subsistence. In *Exploring the Hohokam: Prehistoric Desert People of the American Southwest,* edited by George J. Gumerman, pp. 417–459. University of New Mexico Press, Albuquerque.

Geib, Phil R., and Martha M. Callahan
1988 Clay Residue of Polishing Stones. *The Kiva* 53:357–362.

Giclas, Henry
1985 Stanley Sykes. *Journal of Arizona History* Summer:1–24.

Gifford, Edward W.
1947 Californian Shell Artifacts. *University of California Anthropological Records* 9(1):1–132.

Gladwin, Harold S., Emil W. Haury, E. B. Sayles, and Nora Gladwin
1937 *Excavations at Snaketown: Material Culture.* Medallion Papers No. 25. Gila Pueblo, Globe, Arizona.

1965 *Excavations at Snaketown: Material Culture.* Reprinted. The University of Arizona Press, Tucson. Originally published 1937, Medallion Papers No. 25, Gila Pueblo, Globe, Arizona.

Gladwin, Winifred, and Harold S. Gladwin
1935 *The Eastern Range of the Red-on-buff Culture.* Medallion Papers No. 16. Gila Pueblo, Globe, Arizona.

Goodman, A. H., G. J. Armelagos, and J. C. Rose
1984 The Chronological Distribution of Enamel Hypoplasias from Prehistoric Dickson Mounds Population. *American Journal of Physical Anthropology* 65(3):259–266.

Goodwin, James
1887 Map of Prehistoric Canals in the Salt River Valley. On file at the Peabody Museum of American Archaeology and Ethnology, Cambridge.

Goodyear, Albert C., III
1975 *Hecla II and III: An Interpretive Study of Archeological Remains from the Lakeshore Project, Papago Reservation, South Central Arizona.* Arizona State University Anthropological Research Paper No. 9. Tempe.

Grady, Mark A.
1976 *Aboriginal Agrarian Adaptation to the Sonoran Desert: A Regional Synthesis and Research Design.* Unpublished Ph.D. dissertation, Department of Anthropology, The University of Arizona, Tucson.

Graf, William L.
1983 Flood Related Channel Change in an Arid Region River. *Earth Surface Processes and Landforms* 8:125–139.

Graybill, Donald A.
1989 The Reconstruction of Prehistoric Salt River Streamflow. In *The 1982–1984 Excavations at Las Colinas: Environment and Subsistence,* by Donald A. Graybill, David A. Gregory, Fred L. Nials, Suzanne K. Fish, Robert E. Gasser, Charles H. Miksicek, and Christine R. Szuter, pp. 25–38. Arizona State Museum Archaeological Series No. 162, Vol. 5. The University of Arizona, Tucson.

Graybill, Donald A., and Fred L. Nials
1989 Aspects of Climate, Streamflow, and Geomorphology Affecting Irrigation Patterns in the Salt River Valley. In *The 1982–1984 Excavations at Las Colinas: Environment and Subsistence,* by Donald A. Graybill, David A. Gregory, Fred L. Nials, Suzanne K. Fish, Robert E. Gasser, Charles H. Miksicek, and Christine R. Szuter, pp. 5–23. Arizona State Museum Archaeological Series No. 162, Vol. 5. The University of Arizona, Tucson.

Grebinger, Paul
1976 Salado—Perspectives from the Middle Santa Cruz Valley. *The Kiva* 42:39–46.

Greenwald, David H.
1988 Water Conveyance Features. In *Archaeological Investigations at the Dutch Canal Ruin, Phoenix, Arizona: Archaeology and History along the Papago Freeway Corridor,* edited by David H. Greenwald and Richard Ciolek-Torrello, pp. 54–89. Museum of Northern Arizona Research Paper No. 38. Flagstaff.

Greenwald, David H., Jean H. Ballagh, Douglas R. Mitchell, and Richard A. Anduze (editors)
1996 *Life on the Floodplain: Further Investigations at Pueblo Salado for Phoenix Sky Harbor International Airport. Data Recovery and Re-evaluation.* Pueblo Grande Museum Anthropological Papers No. 4, Vol. 2. Phoenix.

Greenwald, David H., Jean H. Ballagh, and M. Zyniecki (editors)
1996 *Life on the Floodplain: Further Investigations at Pueblo Salado for Phoenix Sky Harbor International Airport. Preliminary Investigations and Research Design.* Pueblo Grande Museum Anthropological Papers No. 4, Vol. 1. Phoenix.

Greenwald, David H., and Richard Ciolek-Torrello
1988a *Archaeological Investigations at the Dutch Canal Ruin, Phoenix, Arizona: Archaeology and History along the Papago Freeway Corridor.* Museum of Northern Arizona Research Paper No. 38. Flagstaff.

1988b Locus E: Brady Wash Site NA18,003. In *Hohokam Settlement along the Slopes of the Picacho Mountains: The Brady Wash Sites, Tucson Aqueduct Project,* edited by Richard Ciolek-Torrello, Martha M. Callahan, and David H. Greenwald, pp. 254–365. Museum of Northern Arizona Research Paper No. 35, Vol. 2. Flagstaff.

1988c Locus C. In *Hohokam Settlement along the Slopes of the Picacho Mountains: The Brady Wash Sites, Tucson Aqueduct Project,* edited by Richard Ciolek-Torrello, Martha M. Callahan, and David H. Greenwald, pp. 63–179. Museum of Northern Arizona Research Paper No. 35, Vol. 2. Flagstaff.

Greenwald, David H., and M. Zyniecki
1993 *Archaeological Investigations in the Phoenix Sky Harbor International Airport: A Class I and II Survey for the Master Plan Improvements Project.* SWCA Archaeological Report No. 92-36. SWCA, Inc. Environmental Consultants, Flagstaff.

Greenwald, Dawn M.
1985 Preliminary Flaked Lithic Results from Brady Wash Locality, Tucson Aqueduct Project. Paper presented at the 58th Annual Pecos Conference, Salinas National Monument, New Mexico.

1988a Flaked Stone. In *Hohokam Settlement along the Slopes of the Picacho Mountains: Material Culture, Tucson Aqueduct Project,* edited by Martha M. Callahan, pp. 221–282. Museum of Northern Arizona Research Paper No. 35, Vol. 4. Flagstaff.

1988b Ground Stone. In *Hohokam Settlement along the Slopes of the Picacho Mountains: Material Culture, Tucson Aqueduct Project,* edited by Martha M. Callahan, pp. 127–220. Museum of Northern Arizona Research Paper No. 35, Vol. 4. Flagstaff.

1988c Lithic Studies. In *Investigations of the Baccharis Site and Extension Arizona Canal: Historic and Prehistoric Land Use Patterns in the Northern Salt River Valley,* by David H. Greenwald, pp. 103–138. Museum of Northern Arizona Research Paper No. 40. Flagstaff.

1990 *A Functional Evaluation of Hohokam Food Grinding Systems.* Unpublished Master's thesis, Department of Anthropology, Northern Arizona University, Flagstaff.

Gregory, David A.
1987 The Morphology of Platform Mounds and the Structure of Classic Period Hohokam Sites. In *The Hohokam Village: Site Structure and Organization,* edited by David E. Doyel, pp. 183–210. Southwestern and Rocky Mountain Division of the American Association for the Advancement of Science, Glenwood Springs, Colorado.

1988 The Changing Spatial Structure of the Mound 8 Precinct. In *The 1982–1984 Excavations at Las Colinas: The Mound 8 Precinct,* edited by David A. Gregory, David R. Abbott, Deni J. Seymour, and Nancy M. Bannister, pp. 25–50. Arizona State Museum Archaeological Series No. 162, Vol. 3. The University of Arizona, Tucson.

1991 Form and Variation in Hohokam Settlement Patterns. In *Chaco and Hohokam: Prehistoric Regional Systems in the American Southwest,* edited by Patricia L. Crown and W. James Judge, pp. 159–193. School of American Research Press, Santa Fe.

Gregory, David A., David R. Abbott, Deni J. Seymour, and Nancy M. Bannister (editors)
1988 *The 1982–1984 Excavations at Las Colinas: The Mound 8 Precinct,* Arizona State Museum Archaeological Series No. 162, Vol. 3. The University of Arizona, Tucson.

Gregory, David A., and Fred L. Nials
1985 Observations Concerning the Distribution of Classic Period Hohokam Platform Mounds. In *Proceedings of the 1983 Hohokam Symposium,* Part 1, edited by Alfred E. Dittert, Jr., and Donald E. Dove, pp. 373–388. Arizona Archaeological Society Occasional Paper No. 2. Phoenix.

Gross, G. Timothy
1989 Shell from the Grand Canal Ruins. In *Archaeological Investigations at the Grand Canal Ruins: A Classic Period Site in Phoenix, Arizona,* edited by Douglas R. Mitchell, pp. 469–496. Soil Systems Publications in Archaeology No. 12, Vol. 1. Phoenix.

Hack, John T.
1942 *The Changing Physical Environment of the Hopi Indians of Arizona.* Papers of the Peabody Museum of American Archaeology and Ethnology No. 35, Vol. 1. Harvard University Press, Cambridge.

Hackbarth, Mark R.
1992 Historic Occupation of the Lower Verde River Valley: Synthesis and Conclusions. In *Prehistoric and Historic Occupation of the Lower Verde River Valley: The State Route 87 Verde Bridge Project,* by Mark R. Hackbarth, pp. 489–501. Northland Research, Flagstaff.

Hackbarth, Mark R., and T. Kathleen Henderson
1992 The Study of the Verde Bridge Historic Resources. In *Prehistoric and Historic Occupation of the Lower Verde River Valley: The State Route 87 Verde Bridge Project,* by Mark R. Hackbarth, pp. 379–392. Northland Research, Flagstaff.

Hackbarth, Mark R., and Nicholas Lancaster
1992 Irrigation along the Lower Verde River: Investigations of the Velasco Ditch. In *Prehistoric and Historic Occupation of the Lower Verde River Valley: The State Route 87 Verde Bridge Project,* by Mark R. Hackbarth, pp. 423–449. Northland Research, Flagstaff.

Halbirt, Carl D.
1988 Ground Stone. In *Hohokam Archaeology along Phase B of the Tucson Aqueduct Central Arizona Project: Excavations at Fastimes (AZ AA:12:384), A Rillito Phase Site in the Avra Valley,* edited by Jon S. Czaplicki and John C. Ravesloot, pp. 201–221. Arizona State Museum Archaeological Series No. 178, Vol. 2. The University of Arizona, Tucson.

1989a Worked Stone Implements from La Cuenca del Sedimento. In *Prehistoric Agricultural Activities on the Lehi-Mesa Terrace: Excavations at La Cuenca del Sedimento*, edited by T. Kathleen Henderson, pp. 190–204. Northland Press, Flagstaff.

1989b Ground Stone. In *Hohokam Archaeology along Phase B of the Tucson Aqueduct Central Arizona Project: Small Sites and Specialized Reports*, edited by Jon S. Czaplicki and John C. Ravesloot, pp. 71–84. Arizona State Museum Archaeological Series No. 178, Vol. 4. The University of Arizona, Tucson.

1989c Ground Stone. In *Hohokam Archaeology along Phase B of the Tucson Aqueduct Central Arizona Project: Excavations at Water World (AZ AA:16:94), A Rillito Phase Ballcourt Village in the Avra Valley*, edited by Jon S. Czaplicki and John C. Ravesloot, pp. 213–235. Arizona State Museum Archaeological Series No. 178, Vol. 3. The University of Arizona, Tucson.

Halseth, Odd S.

1932 Prehistoric Irrigation in Central Arizona. *The Masterkey* 5(6):165–178.

Hammack, Laurens C., and Alan P. Sullivan (editors)

1981 *The 1968 Excavations at Mound 8, Las Colinas Ruins Group, Phoenix, Arizona*. Arizona State Museum Archaeological Series No. 154. The University of Arizona, Tucson.

Hammond, George P., and Agapito Rey

1928 *Obregón's History of 16th Century Explorations in Western America*. Wetzel, Los Angeles.

Hartman, George W.

1977 *Soil Survey of Maricopa County, Arizona, Central Part*. U.S. Department of Agriculture, Soil Conservation Service. U.S. Government Printing Office, Washington, D.C.

Hastings, James Rodney, and Raymond M. Turner

1965 *The Changing Mile: An Ecological Study of Vegetation Change with Time in the Lower Mile of an Arid and Semiarid Region*. The University of Arizona Press, Tucson.

Haury, Emil W.

1937 Shell. In *Excavations at Snaketown: Material Culture*, edited by Harold S. Gladwin, Emil W. Haury, E. B. Sayles, and Nora Gladwin, pp. 135–153. Medallion Papers No. 25. Gila Pueblo, Globe, Arizona.

1945 *The Excavation of Los Muertos and Neighboring Ruins in the Salt River Valley, Southern Arizona*. Papers of the Peabody Museum of Archaeology and Ethnology No. 24, Vol. 1. Harvard University Press, Cambridge.

1950 *The Stratigraphy and Archaeology of Ventana Cave*. The University of Arizona Press, Tucson.

1962 The Greater American Southwest. In *Courses toward Urban Life: Some Archaeological Considerations of Cultural Alternates*, edited by Robert J. Braidwood and Gordon R. Willey, pp. 106–131. Viking Fund Publications in Anthropology No. 32. New York.

1965 The Snaketown Canal. In *Excavations at Snaketown: Material Culture,* by Harold S. Gladwin, Emil W. Haury, E. B. Sayles, and Nora Gladwin, pp. 50–58. Reprinted. The University of Arizona Press, Tucson. Originally published 1937, Medallion Papers No. 25, Gila Pueblo, Globe, Arizona.

1976 *The Hohokam: Desert Farmers and Craftsmen.* The University of Arizona Press, Tucson.

Hayden, Irwin

1930 Notes on the Excavations, Compound F. Ms. on file, Arizona State Museum Library, Archives Folder No. A-180. Tucson.

Hayden, Julian D.

1957 *Excavations, 1940, at University Indian Ruin.* Southwestern Monuments Association Technical Series Vol. 5. Globe, Arizona.

Heacock, Laura A., and Martha M. Callahan

1988 Ceramics. In *Hohokam Settlement along the Slopes of the Picacho Mountains: Material Culture, Tucson Aqueduct Project,* edited by Martha M. Callahan, pp. 1–126. Museum of Northern Arizona Research Paper No. 35, Vol. 4. Flagstaff.

Hemmings, E. Thomas

1969 Salvage Excavations in a Buried Hohokam Site near Tucson, Arizona. *The Kiva* 34:199–206.

Henderson, T. Kathleen

1987 Building a Chronology for La Ciudad. In *Structure and Organization at La Ciudad,* edited by T. Kathleen Henderson, pp. 47–48. Arizona State University Anthropological Field Studies No. 18. Tempe.

Henderson, T. Kathleen, and Vera Morgan

1989 Archaeological Features within the Tempe Outer Loop Corridor. In *Prehistoric Agricultural Activities on the Lehi-Mesa Terrace: Excavations at La Cuenca del Sedimento*, edited by T. Kathleen Henderson, pp. 37-73. Northland Research, Tempe.

Heron, John, and Richard Ciolek-Torrello

1987 McClellan Wash Site, NA18,031. In *Hohokam Settlement along the Slopes of the Picacho Mountains: The Picacho Area Sites, Tucson Aqueduct Project,* edited by Richard Ciolek-Torrello, pp. 24–129. Museum of Northern Arizona Research Paper No. 35, Vol. 3. Flagstaff.

Herskovitz, Robert M.

1974 The Superstition Freeway Project. A Preliminary Report on the Salvage Excavation of a Dual Component Hohokam Site in Tempe, Arizona. Ms. on file, Arizona State Museum, Tucson.

Hill, James N.

1970 *Broken K Pueblo: Prehistoric Social Organization in the American Southwest.* Anthropological Papers of the University of Arizona No. 18. The University of Arizona Press, Tucson.

Hill, W. W.
1937 *Navajo Pottery Manufacture*. University of New Mexico Bulletin, Anthropological Series No. 2, Vol. 3. Albuquerque.

Hinkes, M. J.
1983 *Skeletal Evidence of Stress in Subadults: Trying to Come of Age at Grasshopper Pueblo*. Unpublished Ph.D. dissertation, Department of Anthropology, University of Arizona, Tucson.

Hjulstrom, F.
1938 Transportation of Detritus by Moving Water. In *Recent Marine Sediments*, edited by Parker D. Trask. American Association of Petroleum Geologists, Tulsa.

Hoffman, Theresa L.
1985 Pecked and Ground Stone Artifacts. In *Hohokam Settlement and Economic Systems in the Central New River Drainage, Arizona*, edited by David E. Doyel and Mark D. Elson, pp. 565–592. Soil Systems Publications in Archaeology No. 4, Vol. 2. Phoenix.

Hoffman, Theresa L., and David E. Doyel
1985 Analysis of Chipped Stone Artifacts. In *Hohokam Settlement and Economic Systems in the Central New River Drainage, Arizona*, edited by David E. Doyel and Mark D. Elson, pp. 593–650. Soil Systems Publications in Archaeology No. 4, Vol. 2. Phoenix.

Howard, Ann Valdo
1989 Pecked and Ground Stone Artifacts. In *Settlement, Subsistence, and Specialization in the Northern Periphery: The Waddell Project*, edited by Margerie Green, pp. 737–765. Cultural Resources Report No. 65, Vol. 2. Archaeological Consulting Services, Tempe.

Howard, Jerry B.
1987 The Lehi Canal System: Organization of a Classic Period Community. In *The Hohokam Village: Site Structure and Organization*, edited by David E. Doyel, pp. 211–222. Southwestern and Rocky Mountain Division of the American Association for the Advancement of Science, Glenwood Springs, Colorado.

1988a The Casa Buena Canals. In *Excavations at Casa Buena: Changing Hohokam Land Use along the Squaw Peak Parkway*, edited by Jerry B. Howard, pp. 245–269. Soil Systems Publications in Archaeology No. 11. Phoenix.

1988b Casa Buena Architectural Types and Descriptions. In *Excavations at Casa Buena: Changing Hohokam Land Use along the Squaw Peak Parkway*, edited by Jerry B. Howard, pp. 67–144. Soil Systems Publications in Archaeology No. 11. Phoenix.

1988c *Excavations at Casa Buena: Changing Hohokam Land Use along the Squaw Peak Parkway*. Soil Systems Publications in Archaeology No. 11. Phoenix.

1990 *Paleohydraulics: Techniques for Modeling the Operation and Growth of Prehistoric Canal Systems*. Unpublished Master's thesis, Department of Anthropology, Arizona State University, Tempe.

1991a System Reconstruction: The Evolution of an Irrigation System. In *The Operation and Evolution of an Irrigation System: The East Papago Canal Study*, by Jerry B. Howard and Gary Huckleberry, pp. 5.1–5.33. Soil Systems Publications in Archaeology No. 18. Phoenix.

1991b Charting the Past: Mapping the Prehistoric Canals and Sites of the Salt River Valley. In *The Operation and Evolution of an Irrigation System: The East Papago Canal Study*, by Jerry B. Howard and Gary Huckleberry, pp. 2.1–2.19. Soil Systems Publications in Archaeology No. 18. Phoenix.

1991c Regression Modeling of Canal Morphology and Paleohydraulics. In *The Operation and Evolution of an Irrigation System: The East Papago Canal Study*, by Jerry B. Howard and Gary Huckleberry, pp. 4.1–4.38. Soil Systems Publications in Archaeology No. 18. Phoenix.

Howard, Jerry B., and John S. Cable
1988 Intrasite Land-use Patterns and Community Organization. In *Excavations at Casa Buena: Changing Hohokam Land Use along the Squaw Peak Parkway*, edited by Jerry B. Howard, pp. 833–902. Soil Systems Publications in Archaeology No. 11. Phoenix.

Howard, Jerry B., and Gary Huckleberry
1991 *The Operation and Evolution of an Irrigation System: The East Papago Hohokam Canal Study*. Soil Systems Publications in Archaeology No. 18. Phoenix.

Howard, Jerry B., and David R. Wilcox
1988 The Place of Casa Buena and Locus 2 in the Evolution of Canal System 2. In *Excavations at Casa Buena: Changing Hohokam Land Use along the Squaw Peak Parkway*, edited by Jerry B. Howard, pp. 903–939. Soil Systems Publications in Archaeology No. 11, Vol. 2. Phoenix.

Hubbard, John P.
1977 Importance of Riparian Ecosystems: Biotic Considerations. In *Importance, Preservation and Management of Riparian Habitat: A Symposium*, by U.S. Department of Argriculture Forest Service, pp. 5–9. U.S. Department of Argriculture, Forest Service General Technical Report RM-43. Fort Collins, Colorado.

Huckell, Bruce B.
1986 *A Ground Stone Implement Quarry on the Lower Colorado River, Northwestern Arizona*. Arizona Bureau of Land Management Cultural Resources Series Monograph No. 3. Arizona State Office of the Bureau of Land Management, Phoenix.

1988 Flaked Stone. In *Hohokam Archaeology along Phase B of the Tucson Aqueduct Central Arizona Project: Excavations at Fastimes (AZ AA:12:384), a Rillito Phase Site in the Avra Valley*, edited by Jon S. Czaplicki and John C. Ravesloot, pp. 175–200. Arizona State Museum Archaeological Series No. 178, Vol. 2. The University of Arizona, Tucson.

1989a Flaked Stone. In *Hohokam Archaeology along Phase B of the Tucson Aqueduct Central Arizona Project: Small Sites and Specialized Reports*, edited by Jon S. Czaplicki and John C. Ravesloot, pp. 63–70. Arizona State Museum Archaeological Series No. 178, Vol. 4. The University of Arizona, Tucson.

1989b Flaked Stone. In *Hohokam Archaeology along Phase B of the Tucson Aqueduct Central Arizona Project: Excavations at Water World (AZ AA:16:94), a Rillito Phase Ballcourt Village in the Avra Valley*, edited by Jon S. Czaplicki and John C. Ravesloot, pp. 193–211. Arizona State Museum Archaeological Series No. 178, Vol. 3. The University of Arizona, Tucson.

Huckleberry, Gary

1988 Relict Irrigation Canals in the East Papago Freeway Corridor. In *Arizona Department of Transportation Archaeological Testing Program: Part 2, East Papago Freeway*, edited by Daniel G. Landis, pp. 109–167. Soil Systems Publications in Archaeology No. 13. Phoenix.

1989 Prehistoric Canal Hydraulics and Sedimentology at El Caserío. In *El Caserío: Colonial Period Settlement along the East Papago Freeway*, edited by Douglas R. Mitchell, pp. 41–75. Soil Systems Publications in Archaeology No. 14. Phoenix.

1991 A Geoarchaeological Study of Canal System 2. In *The Operation and Evolution of an Irrigation System: The East Papago Canal Study*, by Jerry B. Howard and Gary Huckleberry, pp. 3.1–3.75. Soil Systems Publications in Archaeology No. 18. Phoenix.

1992 Soil Evidence of Hohokam Irrigation in the Salt River Valley, Arizona. *Kiva* 57:237–249.

1993 Geomorphology, Stratigraphy, and Soils at Phoenix Sky Harbor International Airport. In *Archaeological Investigations in the Phoenix Sky Harbor International Airport: A Class I and II Survey for the Master Plan Improvements Project*, by David H. Greenwald and M. Zyniecki, pp. 79–87. SWCA Archaeological Report No. 92-36. Flagstaff.

Huckleberry, Gary, and Scott Kwiatkowski

1989 Physical and Natural Environment. In *Archaeological Investigations at the Grand Canal Ruins: A Classic Period Site in Phoenix, Arizona*, edited by Douglas R. Mitchell, pp. 21–35. Soil Systems Publications in Archaeology No. 12. Phoenix.

Hutchinson, D. L., and C. S. Larsen

1988 Determination of Stress Episode Duration from Linear Enamel Hypoplasia: A Case Study from St. Catherines Island, Georgia. *Human Biology* 60:93–110.

Israelsen, O., and V. Hansen

1962 *Irrigation Principles and Practices*. 3rd ed. John Wiley and Sons, New York.

Janick, Jules, Robert W. Schery, Frank W. Woods, and Vernon W. Ruttan

1974 *Plant Science: An Introduction to World Crops*. W. H. Freeman, San Francisco.

Jewett, Roberta A.

1989 Distance, Interaction, and Complexity: The Spatial Organization of Pan-Regional Settlement Clusters in the American Southwest. In *The Sociopolitical Structure of Prehistoric Southwest Societies*, edited by Steadman Upham, Kent G. Lightfoot, and Roberta A. Jewett, pp. 363–388. Westview Press, Boulder, Colorado.

Judd, Neil M.

1931 Arizona's Prehistoric Canals from the Air. In *Explorations and Fieldwork of the Smithsonian Institution in 1930*, pp. 157–166. Smithsonian Institution, Washington, D.C.

Keen, A. Myra
1971 *Sea Shells of Tropical West America: Marine Mollusks from Baja California to Peru.* 2nd ed. Stanford University Press, Palo Alto.

Kent, Edward
1910 *Decision and Decree No. 4564.* Court of the Third Judicial District, Maricopa County, Territory of Arizona.

Kidder, Alfred V., J. D. Jennings, and E. M. Shook
1946 *Excavations at Kaminaljuyú, Guatemala.* Carnegie Institute Report No. 561. Washington, D.C.

Kisselburg, JoAnn E.
1985 Hohokam Plainware at La Ciudad: Solving Site-Specific Problems in Chronology and Community Patterns. Paper presented at the 50th Annual Meeting of the Society for American Archaeology, Denver.

Kluckhohn, Clyde
1939 Discussion. In Preliminary Report on the 1937 Excavations, Bc 50-51, Chaco Canyon, New Mexico. *The University of New Mexico Bulletin* 3(2):151-163.

Landis, Daniel G.
1989a The Lithic Assemblages of El Caserío. In *El Caserío: Colonial Period Settlement along the East Papago Freeway,* edited by Douglas R. Mitchell, pp. 103-141. Soil Systems Publications in Archaeology No. 14. Phoenix.

1989b Chipped Stone Assemblages of the Grand Canal Ruins. In *Archaeological Investigations at the Grand Canal Ruins: A Classic Period Site in Phoenix, Arizona,* edited by Douglas R. Mitchell, pp. 385-441. Soil Systems Publications in Archaeology No. 12, Vol. 1. Phoenix.

Landye, J. Jerry
1981 *Current Status of Endangered, Threatened, and/or Rare Mollusks of New Mexico and Arizona.* Report submitted to the U.S. Department of the Interior, U.S. Fish and Wildlife Service, Office of Rare and Endangered Species, Albuquerque.

Lane, Anne Marie
1989 The Grand Canal Ceramic Assemblage: Ceramic Type Descriptions. In *Archaeological Investigations at the Grand Canal Ruins: A Classic Period Site in Phoenix, Arizona,* edited by Douglas R. Mitchell, pp. 249-293. Soil Systems Publications in Archaeology No. 12, Vol. 1. Phoenix.

Lekson, Steven H.
1984 Dating Casas Grandes. *The Kiva* 50:55-60.

Linares, Olga F.
1976 "Garden Hunting" in the American Tropics. *Human Ecology* 4:331-349.

Lombard, James
1988 Paleohydraulics and Sedimentation in Hohokam Canals at the Dutch Canal Ruin. In *Archaeological Investigations at the Dutch Canal Ruin, Phoenix, Arizona: Archaeology and History along the Papago Freeway Corridor*, edited by David Greenwald and Richard Ciolek-Torrello, pp. 90–105. Museum of Northern Arizona Research Papers No. 38. Flagstaff.

Lowe, Charles H.
1959 *Arizona's Natural Environment*. The University of Arizona Press, Tucson.

Luckingham, Bradford
1989 *Phoenix: The History of a Southwestern Metropolis*. The University of Arizona Press, Tucson.

McAllister, Martin E.
1980 Hohokam Social Organization: A Reconstruction. *The Arizona Archaeologist* No. 14. Phoenix.

McClintock, James H.
1921 *Mormon Settlement in Arizona*. The University of Arizona Press, Tucson.

McGuire, Randall H.
1987 *Death, Society and Ideology in a Hohokam Community: Colonial and Sedentary Period Burials from La Ciudad*. Office of Cultural Resource Management Report No. 68. Arizona State University, Tempe.

1991 On the Outside Looking in: The Concept of Periphery in Hohokam Archaeology. In *Exploring the Hohokam: Prehistoric Desert Peoples of the American Southwest*, edited by George J. Gumerman, pp. 347–382. University of New Mexico Press, Albuquerque.

Magne, Martin P. R.
1981 Controlled Lithic Reduction Experiments towards an Understanding of Lithic Utilization in Large Scale Settlement-Subsistence Systems of the Interior Plateau. Paper presented at the 46th Annual Meeting of the Society for American Archaeology, San Diego.

Marcus, Joyce
1973 Territorial Organizations of the Lowland Classic Maya. *Science* 180:911–916.

Maricopa County Immigration Union
1887 *Salt River Valley, the Garden of the Southwest*. Rand McNally, Chicago.

Mariella, Patricia
1982 Preliminary Draft, Description and Significance Items, Fort McDowell District, Nomination to the National Register of Historic Places. Ms. on file, Fort McDowell Mohave-Apache Indian Community, Fountain Hills, Arizona.

Martin, Debra L., Alan H. Goodman, George H. Armelagos, and Ann L. Magennis
1991 *Black Mesa Anasazi Health: Reconstructing Life from Patterns of Death and Disease*. Southern Illinois University at Carbondale Center for Archaeological Investigations Occasional Paper No. 14.

Masse, W. Bruce
1976 *The Hohokam Expressway Project: A Study of Prehistoric Irrigation in the Salt River Valley, Arizona.* Arizona State Museum Contributions to Highway Salvage Archaeology in Arizona No. 43. The University of Arizona, Tucson.

1980 *Excavations at Gu Achi: A Reappraisal of Hohokam Settlement and Subsistence in the Arizona Papagueria.* Western Archaeological Center Publications in Anthropology No. 12. Tucson.

1987 *Archaeological Investigations of Portions of the Las Acequias–Los Muertos Irrigation System: Testing and Partial Data Recovery within the Tempe Section of the Outer Loop Freeway System, Maricopa County, Arizona.* Arizona State Museum Archaeological Series No. 176. The University of Arizona, Tucson.

1988 Appendix C: Archaeological Sediments of the Hohokam Expressway Canals. In *The 1982–1984 Excavations at Las Colinas: The Site and its Features,* by David R. Abbott, Kim E. Beckwith, Patricia L. Crown, R. Thomas Euler, David A. Gregory, J. Ronald London, Marilyn B. Saul, Larry A. Schwalbe, Mary Bernard-Shaw, Christine R. Szuter, and Arthur W. Vokes, pp. 333–353. Arizona State Museum Archaeological Series No. 162, Vol. 2. The University of Arizona, Tucson.

1991 The Quest for Subsistence Sufficiency and Civilization in the Sonoran Desert. In *Chaco and Hohokam: Prehistoric Regional Systems in the American Southwest,* edited by Patricia L. Crown and W. James Judge, pp. 195–223. School of American Research Press, Santa Fe.

Matthews, Washington, J. L. Wortman, and John S. Billings
1893 The Human Bones of the Hemenway Collection in the United States Army Medical Museum at Washington. *Memoirs of the National Academy of Sciences* 6(7):141–286.

Mawn, Geoffrey
1977 Promoters, Speculators, and the Selection of the Phoenix Townsite. *Arizona and the West* 19(3):207–224.

Merbs, Charles F.
1987 Uncremated Burials at La Ciudad. In *La Ciudad: Specialized Studies in the Economy, Environment and Culture of La Ciudad,* edited by JoAnn E. Kisselburg, Glen E. Rice, and Brenda L. Shears, pp. 201–214. Office of Cultural Resource Management Anthropological Field Studies No. 20. Arizona State University, Tempe.

Merbs, Charles F., and Judy Brunson
1987 Burial Orientation and Concepts of Afterlife in the American Southwest and Canadian Arctic. Paper presented at the 5 Annual Meeting of the Society for American Archaeology, Toronto.

Midvale, Frank
1934 The Frank Midvale Collection. Ms. on file, Site Record Collections, Department of Anthropology, Arizona State University, Tempe.

1965 Prehistoric Irrigation of the Casa Grande Ruins Area. *The Kiva* 30:82–86.

1966 Prehistoric Irrigation of the Salt River Valley. Map on file, Arizona Collection, Hayden Library, Arizona State University, Tempe.

1968 Prehistoric Irrigation in the Salt River Valley, Arizona. *The Kiva* 34:28–32.

1970 Prehistoric Canal-Irrigation in the Buckeye Valley and Gila Bend Areas in Western Maricopa County, Arizona. Paper presented at the Pecos Conference Water Control Systems Symposium, Santa Fe.

1974 Prehistoric Ruins and Irrigation in the Eastern Buckeye Valley. *The Arizona Archaeologist* 8:37–39.

n.d. The Frank Midvale Collection. Ms. on file, Site Record Collections, Department of Anthropology, Arizona State University, Tempe.

Miller, Robert Rush
1955 Fish Remains from Archaeological Sites in the Lower Colorado River Basin, Arizona. *Papers of the Michigan Academy of Science, Arts, and Letters* 40:125–136.

Mindeleff, Cosmos
1900 Localization of Tusayan Clans. In *Nineteenth Annual Report of the Bureau of American Ethnology*, pp. 637–653. Smithsonian Institution, Washington, D.C.

Mindeleff, Victor
1891 A Study of Pueblo Architecture: Tusayan and Cibola. In *Eighth Annual Report of the Bureau of American Ethnology*, pp. 12–228. Smithsonian Institution, Washington, D.C.

Mitchell, Douglas R.
1988 Analysis of the La Lomita Pequeña Lithic Assemblage. In *Excavations at La Lomita Pequeña: A Santa Cruz/Sacaton Phase Hamlet in the Salt River Valley*, edited by Douglas R. Mitchell, pp. 179–230. Soil Systems Publications in Archaeology No. 10. Phoenix.

1989a *Archaeological Investigations at the Grand Canal Ruins: A Classic Period Site in Phoenix, Arizona.* Soil Systems Publications in Archaeology No. 12. Phoenix.

1989b Settlement Patterns and Social Organization for the Phoenix Area Classic Period. In *Archaeological Investigations at the Grand Canal Ruins: A Classic Period Site in Phoenix, Arizona*, edited by Douglas R. Mitchell, pp. 859–878. Soil Systems Publications in Archaeology No. 12, Vol. 2. Phoenix.

1989c The Grand Canal Ground Stone Assemblage. In *Archaeological Investigations at the Grand Canal Ruins: A Classic Period Site in Phoenix, Arizona*, edited by Douglas R. Mitchell, pp. 443–467. Soil Systems Publications in Archaeology No. 12, Vol. 1. Phoenix.

Mitchell, Douglas R., T. Michael Fink, and Wilma Allen
1989 Disposal of the Dead: Explorations of Mortuary Variability and Social Organization at the Grand Canal Ruins. In *Archaeological Investigations at the Grand Canal Ruins: A Classic Period Site in Phoenix, Arizona*, edited by Douglas R. Mitchell, pp. 705–774. Soil Systems Publications in Archaeology No. 12, Vol. 2. Phoenix.

Mixson, Ben, and R. E. White
1991 Skywatchers of the Salt River Hohokam. *Astronomy Quarterly* 84(4):245–260.

Moore, Bruce M.
1978 Are Pueblo Field Houses a Function of Urbanization? In *Limited Activity and Occupation Sites: A Collection of Conference Papers*, compiled and edited by Albert E. Ward, pp. 9–16. Contributions to Anthropological Studies No. 1. Center for Anthropological Studies, Albuquerque.

Morris, Donald H.
1969 Red Mountain: An Early Pioneer Period Hohokam Site in the Salt River Valley of Central Arizona. *American Antiquity* 34:40–54.

Munsell Soil Color Chart
1988 *Munsell Color*. MacBeth Division of Kollmorgen Corporation, Baltimore.

Myers, Cindy L., and A. E. Rogge
1986 Research Design for Historical Archaeological Studies, Regulatory Storage Division (Plan 6), Central Arizona Project. Ms. on file, Arizona Projects Office/U.S. Bureau of Reclamation, Phoenix.

Nabhan, Gary P., Amadeo M. Rea, Karen L. Reichhardt, Eric Mellink, and Charles F. Hutchinson
1982 Papago Influences on Habitat and Biotic Diversity: Quitovac Oasis Ethnoecology. *Journal of Ethnobiology* 2(2):124–143.

Naroll, Raoul
1962 Floor Area and Settlement Population. *American Antiquity* 27:587–596.

Neitzel, Jill
1987 The Sociopolitical Implications of Canal Irrigation: A Reconsideration of the Hohokam. In *Coasts, Plains and Deserts: Essays in Honor of Reynold J. Ruppe*, edited by Sylvia W. Gaines, pp. 205–211. Arizona State University Anthropological Research Papers No. 38. Tempe.

Nelson, Richard S.
1981 *The Role of the Puchtecha System in Hohokam Exchange*. Unpublished Ph.D. dissertation, Department of Anthropology, New York University, New York.

1991 *Hohokam Marine Shell Exchange and Artifacts*. Arizona State Museum Archaeological Series No. 179. The University of Arizona, Tucson.

Nials, Fred L., and David A. Gregory
1989 Irrigation Systems in the Lower Salt River Valley. In *The 1982–1984 Excavations at Las Colinas: Environment and Subsistence*, by Donald A. Graybill, David A. Gregory, Fred L. Nials, Suzanne K. Fish, Robert E. Gasser, Charles H. Miksicek, and Christine R. Szuter, pp. 39–58. Arizona State Museum Archaeological Series No. 162, Vol. 5. The University of Arizona, Tucson.

Nials, Fred L., David A. Gregory, and Donald A. Graybill
1989 Salt River Streamflow and Hohokam Irrigation Systems. In *The 1982–1984 Excavations at Las Colinas: Environment and Subsistence,* by Donald A. Graybill, David A. Gregory, Fred L. Nials, Suzanne K. Fish, Robert E. Gasser, Charles H. Miksicek, and Christine R. Szuter, pp. 59–76. Arizona State Museum Archaeological Series No. 162, Vol. 5. The University of Arizona, Tucson.

Nicholas, Linda
1981 *Irrigation and Sociopolitical Development in the Salt River Valley, Arizona: An Examination of Three Prehistoric Canal Systems.* Unpublished Master's thesis, Department of Anthropology, Arizona State University, Tempe.

Nicholas, Linda, and Jill Neitzel
1984 Canal Irrigation and Sociopolitical Organization in the Lower Salt River Valley: A Diachronic Analysis. In *Prehistoric Agricultural Strategies in the Southwest,* edited by Suzanne K. Fish and Paul R. Fish, pp. 161–178. Arizona State University Anthropological Research Papers No. 33. Tempe.

Officer, James E.
1978 Shell. In *The Hodges Ruin: A Hohokam Community in the Tucson Basin,* by Isabel Kelly, pp. 110–120. Anthropological Papers of the University of Arizona No. 30. The University of Arizona Press, Tucson.

O'Rourke, Mary Kay
1983 Pollen from Adobe Brick. *Journal of Ethnobiology* 3(1):39–48.

Patrick, Herbert H.
1903 *The Ancient Canal Systems and Pueblos of the Salt River Valley, Arizona.* Phoenix Free Museum Bulletin No. 1. Phoenix.

Pearsall, Deborah M.
1989 *Paleoethnobotany, A Handbook of Procedures.* Academic Press, San Diego.

Peterson, Jane D., and David R. Abbott
1991 Indices of Salado Polychrome Ceramic Production: the Evidence from Pueblo Grande, Arizona. Paper presented at the 56th Annual Meeting of the Society for American Archaeology, New Orleans.

Phagan, Carl J.
1980 Lithic Technology: Flake Analysis. In *Prehistory of the Ayacucho Basin, Peru,* by Richard S. MacNeish, Robert K. Vierra, Antoinette Nelkin-Terner, and Carl J. Phagan, pp. 233–281, Vol. 3. University of Michigan Press, Ann Arbor.

Phelan, John T., and Wayne D. Criddle
1955 Surface Irrigation Methods. In *The Yearbook of Agriculture 1955: Water,* U.S. Department of Agriculture, pp. 258–266. U.S. Government Printing Office, Washington, D.C.

Pielou, E. C.
1975 *Ecological Diversity.* John Wiley and Sons, New York.

Piperno, Dolores R.
1988 *Phytolith Analysis: An Archaeological and Geological Perspective.* Academic Press, San Diego.

Pokotylo, David Leslie
1978 *Lithic Technology and Settlement Patterns in Upper Hat Creek Valley, British Columbia.* Unpublished Ph.D. dissertation, Department of Anthropology, University of British Columbia, Vancouver.

Price, James E.
1977 Anticipated Impacts of the Little Black River Watershed Project on the Finite Cultural Resource Base. In *Conservation Archaeology: A Guide for Cultural Resource Management Studies*, edited by M. B. Schiffer and G. J. Gumerman, pp. 302–308. Academic Press, New York.

Ravesloot, John C., Jeffrey S. Dean, and Michael S. Foster
1986 A New Perspective on the Casas Grandes Tree-Ring Dates. Paper presented at the Fourth Mogollon Conference, University of Arizona, Tucson.

Rea, Amadeo M.
1979 The Ecology of Pima Fields. *Environment Southwest* 484:8–13.

1983 *Once a River: Bird Life and Habitat Changes on the Middle Gila.* The University of Arizona Press, Tucson.

1991 Gila River Pima Dietary Reconstruction. *Arid Lands Newsletter* 31:3–10.

Reed, Bill
1977 *The Last Bugle Call: A History of Fort McDowell, Arizona Territory 1865–1890.* McClain, Parsons, West Virginia.

Rehder, Harald A.
1981 *The Audubon Society Field Guide to North American Seashells.* Alfred A. Knopf, New York.

Reinhard, K. J.
1985 Parasitism at Antelope House, a Puebloan Village in Canyon de Chelly, Arizona. In *Health and Disease in the Prehistoric Southwest*, edited by C. F. Merbs and R. J. Miller, pp. 220–233. Arizona State University Anthropological Research Papers No. 34. Tempe.

Reinhard, K. J., R. H. Hevly, and G. A. Anderson
1987 Helminth Remains from Prehistoric Indian Coprolites on the Colorado Plateau. *Journal of Parasitology* 73:630–639.

Rice, Glen E.
1987 The Architecture of Pre-Classic Hohokam Houses. In *A Spatial Analysis of the Hohokam Community of La Ciudad*, edited by Glen E. Rice, pp. 103–118. Arizona State University Anthropological Field Studies No. 16. Tempe.

Rice, Glen E., Owen Lindauer, and John C. Ravesloot
1990 Gila Polychrome: An Investigation of the Hallmark of the Salado Tradition. In *A Design for Salado Research*, edited by Glen E. Rice, pp. 85–102. Arizona State University Anthropological Field Series No. 22. Tempe.

Roberts, Frank H. H.
1935 *A Folsom Complex: Preliminary Report on Investigations at the Lindenmeier Site in Northern Colorado*. Smithsonian Miscellaneous Collections, Vol. 94, No. 4. Washington, D.C.

Rodgers, James B.
1987 *Studies along the Lower Agua Fria River: The Eastwing Site and the Marinette Canal*. Museum of Northern Arizona Research Paper No. 37. Flagstaff.

Rodgers, James B., and David H. Greenwald
1988a Archival Research and Field Investigations of the Extension Arizona Canal and Other Historic Waterworks. In *Investigations of the Baccharis Site and Extension Arizona Canal: Historic and Prehistoric Land Use Patterns in the Northern Salt River Valley*, by David H. Greenwald, pp. 9–34. Museum of Northern Arizona Research Paper No. 40. Flagstaff.

1988b Historic Resources. In *Archaeological Investigations at the Dutch Canal Ruin, Phoenix, Arizona: Archaeology and History along the Papago Freeway Corridor*, edited by David H. Greenwald and Richard Ciolek-Torrello, pp. 24–53. Museum of Northern Arizona Research Paper No. 38. Flagstaff.

Rodiek, Jon
1981 Wetland Trees of Arizona for Possible Oasis Use in Arid Regions. *Desert Plants* 3:88–91.

Rose, Jerome C., Keith W. Condon, and Alan H. Goodman
1985 Diet and Dentition: Developmental Disturbances. In *The Analysis of Prehistoric Diets*, edited by Robert I. Gilbert, Jr., and James H. Mielle, pp. 281–305. Academic Press, Orlando.

Rozen, Kenneth C.
1981 Patterned Associations among Lithic Technology, Site Content, and Time: Results of the TEP St. Johns Project Lithic Analysis. In *Prehistory of the St. Johns Area, East-Central Arizona: The TEP St. Johns Project*, by Deborah A. Westfall, pp. 157–232. Arizona State Museum Archaeological Series No. 153. The University of Arizona, Tucson.

1984 Flaked Stone. In *Hohokam Habitation Sites in the Northern Santa Rita Mountains*, by Alan Ferg, Kenneth C. Rozen, William L. Deaver, Martyn D. Tagg, David A. Phillips, Jr., and David A. Gregory, pp. 421–604. Arizona State Museum Archaeological Series No. 147, Vol. 2, Pt. 1. The University of Arizona, Tucson.

Ruppé, Patricia A., Linda Scott Cummings, and David H. Greenwald
1988 Botanical Remains. In *Archaeological Investigations at the Dutch Canal Ruin, Phoenix, Arizona: Archaeology and History along the Papago Freeway Corridor*, edited by David H. Greenwald and Richard Ciolek-Torrello, pp. 168–196. Museum of Northern Arizona Research Paper No. 38. Flagstaff.

Rusling, James E.
1874 *The Great West and the Pacific Coast.* Sheldon, New York.

Russell, Frank
1975 *The Pima Indians.* 26th Annual Report of the Bureau of American Ethnology, 1904–1905. Reprinted. University of Arizona Press, Tucson. Originally published 1908, Smithsonian Institution, Washington, D.C.

Salt River Project
1979 *The Taming of the Salt.* Salt River Project, Phoenix.

n.d.a *SRP Canals.* Salt River Project, Phoenix.

n.d.b *A Valley Reborn: The Story of the Salt River Project.* Salt River Project, Phoenix.

Salt River Valley Water Users' Association
1921 Salt River Project Map. On file, Salt River Project, Phoenix.

Savage, S. H.
1991 A Spatial Analysis of Pueblo Grande Burial Groups 13–15. Ms. in possession of author, Arizona State University, Tempe.

Sayles, E. B.
1965 Stone Implements and Bowls. In *Excavations at Snaketown: Material Culture,* by Harold S. Gladwin, Emil W. Haury, E. B. Sayles, and Nora Gladwin, pp. 101–120. Reprinted. The University of Arizona Press, Tucson. Originally published 1937, Medallion Papers No. 25, Gila Pueblo, Globe, Arizona.

Schaller, David M.
1991 Petrographic Analysis of Ceramics from Dutch Canal Ruin and Pueblo Salado, Phoenix Sky Harbor Project. Ms. on file, SWCA, Flagstaff.

1993 Geographic Sources of Phoenix Basin Hohokam Plainware Based on Petrographic Analysis. In *The Pueblo Grande Project: Ceramics and the Production and Exchange of Pottery in the Central Phoenix Basin,* edited by David R. Abbott, pp. 17-90. Soil Systems Publications in Archaeology No. 20, Vol. 3, Pt. 1. Phoenix.

Schiffer, Michael B.
1987 *Formation Processes of the Archaeological Record.* University of New Mexico Press, Albuquerque.

Schiffer, Michael B., and George J. Gumerman
1977 Forecasting Impacts. In *Conservation Archaeology: A Guide for Cultural Resource Management Studies,* edited by Michael B. Schiffer and George J. Gumerman, pp. 291–301. Academic Press, New York.

Schiffer, Michael B., and John H. House
1977 Assessing Impacts: Examples from the Cache Project. In *Conservation Archaeology: A Guide for Cultural Resource Management Studies,* edited by Michael B. Schiffer and George J. Gumerman, pp. 309–320. Academic Press, New York.

Schroeder, Albert H.

1940 *A Stratigraphic Survey of Pre-Spanish Trash Mounds of the Salt River Valley, Arizona.* Unpublished Master's thesis, Department of Anthropology, The University of Arizona, Tucson.

1980 Discussion. In *Current Issues in Hohokam Prehistory: Proceedings of a Symposium*, edited by David Doyel and Fred Plog, pp. 176–179. Arizona State University Anthropological Research Papers No. 23. Tempe.

Schroeder, Kelly James

1990 *Socio-Spatial Centrality of Pre-Classic Hohokam Cremation Cemeteries and Their Relationship to House Cluster Integrity.* Unpublished Master's thesis, Arizona State University, Tempe.

1991 Second Preliminary Report on the Pioneer and Military Memorial Park Archaeological Project in Phoenix, Arizona. Ms. on file, Pueblo Grande Museum, Phoenix. Draft.

Shackley, M. Steven

1988 Sources of Archaeological Obsidian in the Southwest: An Archaeological, Petrological, and Geochemical Study. *American Antiquity* 53:752–772.

Sheets, Payson D.

1978 From Craftsman to Cog: Quantitative Views of Mesoamerican Lithic Technology. In *Papers on the Economy and Architecture of the Ancient Maya*, edited by Raymond Sidrys, pp. 40–71. Institute of Archaeology Monograph No. 8. University of California at Los Angeles.

Shephard, Anna O.

1956 *Ceramics for the Archaeologist.* Carnegie Institution of Washington Publication No. 609. Washington, D.C.

1980 *Ceramics for the Archaeologist.* Reprinted. Carnegie Institution of Washington Publication No. 609. Washington, D.C.

Silsbee, Joan M.

1958 Determining the Source of California *Olivella* Shells. In *Reports of the University of California Archaeological Survey*, No. 41, pp. 10–11. Papers in California Archaeology No. 64. Department of Anthropology, University of California, Berkeley.

Sires, Earl W., Jr.

1984 Excavations at El Polvorón. In *Hohokam Archaeology along the Salt-Gila Aqueduct, Central Arizona Project*, edited by Lynn S. Teague and Patricia L. Crown, pp. 221–326. Arizona State Museum Archaeological Series No. 150, Vol. 9. The University of Arizona, Tucson.

Slaughter, Mark C., Lee Fratt, Kirk Anderson, and Richard V. N. Ahlstrom

1992 *Making and Using Stone Artifacts: A Context for Evaluating Lithic Sites in Arizona.* SWCA Archaeological Report No. 92-5. Flagstaff and Tucson.

Soil Conservation Service
1934 Aerial Survey Maps. On file, Cartographic Division, Salt River Project, Phoenix.

Soil Survey Staff
1975 *Soil Taxonomy.* U.S. Department of Agriculture, Soil Conservation Service, Handbook No. 436. U.S. Government Printing Office, Washington, D.C.

Sposito, Garrison
1989 *The Chemistry of Soils.* Oxford University Press, New York.

Sprague, Roderick
1968 A Suggested Terminology and Classification for Burial Description. *American Antiquity* 33:479–485.

Stein, Pat H.
1979 *Archeological Investigations along the Salt-Gila Aqueduct.* Technical Report No. DI-BR-APO-CCRS 79-9. Prepared for U.S. Bureau of Reclamation, Phoenix, Arizona. Submitted by Museum of Northern Arizona, Flagstaff.

Stodder, A. W.
1987 The Physical Anthropology and Mortuary Practice of the Dolores Anasazi: An Early Pueblo Population in Local and Regional Context. In *Dolores Archaeology Program Supporting Studies: Settlement and Environment,* edited by K. L. Peterson and J. D. Orcutt, pp. 335–470. U.S. Department of the Interior, Bureau of Reclamation, Denver.

Stuiver, Minze, and Bernard Becker
1993 High-Precision Calibration of the Radiocarbon Time Scale, AD 1950–6000 BC. *Radiocarbon* 35:35–65.

Stuiver, Minze, and Paula J. Reimer
1993 Extended ^{14}C Data Base and Revised CALIB 3.0 ^{14}C Age Calibration Program. *Radiocarbon* 35:215-230.

Sullivan, Alan P., III
1988 Prehistoric Southwestern Ceramic Manufacture: The Limitations of Current Evidence. *American Antiquity* 53:23–35.

Sullivan, Alan P., III, and Kenneth C. Rozen
1985 Debitage Analysis and Archaeological Interpretation. *American Antiquity* 50:755–779.

1989 Measurement, Method, and Meaning in Lithic Analysis: Problems with Amick and Mauldin's Middle-Range Approach. *American Antiquity* 54:169–175.

Swidler, Nina B.
1989 Variability in the Chipped Stone Assemblage from La Cuenca del Sedimento. In *Prehistoric Agricultural Activities on the Lehi-Mesa Terrace: Excavations at La Cuenca del Sedimento,* edited by T. Kathleen Henderson, pp. 154–189. Northland Research, Flagstaff.

Szuter, Christine R.
1991 Hunting by Hohokam Farmers. *Kiva* 56:277–291.

Tagg, Martyn D.
1984 Utilitarian Ground Stone. In *Hohokam Habitation Sites in the Northern Santa Rita Mountains*, by Alan Ferg, Kenneth C. Rozen, William L. Deaver, Martyn D. Tagg, David A Phillips, Jr., and David A. Gregory, pp. 605–664. Arizona State Museum Archaeological Series No. 147, Vol. 2. The University of Arizona, Tucson.

Teague, George A.
1981 The Nonflaked Stone Artifacts from Las Colinas. In *The 1968 Excavations at Mound 8, Las Colinas Ruins Group, Phoenix, Arizona*, edited and assembled by Laurens C. Hammack and Alan P. Sullivan, pp. 201–247. Arizona State Museum Archaeological Series No. 154. The University of Arizona, Tucson.

Teague, Lynn S.
1984 The Organization of Hohokam Economy. In *Hohokam Archaeology along the Salt-Gila Aqueduct, Central Arizona Project: Synthesis and Conclusions*, edited by Lynn S. Teague and Patricia L. Crown, pp. 187–249. Arizona State Museum Archaeological Series No. 150, Vol. 9. The University of Arizona, Tucson.

1985 The Organization of Hohokam Exchange. In *Proceedings of the 1983 Hohokam Symposium*, Part 2, edited by Alfred E. Dittert, Jr., and Donald E. Dove, pp. 397–418. Arizona Archaeological Society Occasional Paper No. 2. Phoenix.

1988 The History of Occupation at Las Colinas. In *The 1982–1984 Excavations at Las Colinas: The Site and Its Features*, by David A. Gregory, William L. Deaver, Suzanne K. Fish, Ronald Gardiner, Robert W. Layhe, Fred L. Nials, and Lynn S. Teague, pp. 121–152. Arizona State Museum Archaeological Series No. 162, Vol. 2. The University of Arizona, Tucson.

1989 The Postclassic and the Fate of the Hohokam. In *The 1982–1984 Excavations at Las Colinas: Syntheses and Conclusions*, edited by Lynn S. Teague and William L. Deaver, pp. 145–167. Arizona State Museum Archaeological Series No. 163, Vol. 6. The University of Arizona, Tucson.

Teague, Lynn S., and Patricia L. Crown
1980 Archaeological Data Collection Studies along the Salt-Gila Aqueduct, Central Arizona Project. Research Design and Report of Test Excavations and Related Studies. Ms on file, Arizona State Museum, The University of Arizona, Tucson.

Thomas, David Hurst
1976 *Figuring Anthropology: First Principles of Probability and Statistics*. Holt, Rinehart, and Winston, New York.

Thompson, Louis M.
1957 *Soils and Soil Fertility*. McGraw Hill, New York.

Thorne, D. W., and H. B. Peterson
1954 *Irrigated Soils: Their Fertility and Management*. The Blackiston Company, New York.

Trimble, Marshall
1986 *Roadside History of Arizona*. Mountain Press, Missoula.

Tuohy, Donald R.
1968 Stone Sinkers from Western Nevada. *American Antiquity* 33:211–215.

Turner, Raymond M.
1959 Evolution of the Vegetation of the Southwestern Desert Region. In *Arid Lands Colloquia, 1958–59*, pp. 46–53. The University of Arizona Press, Tucson.

1974 Map Showing Vegetation in the Phoenix Area, Arizona. Folio of the Phoenix area, map I-845-I. U.S. Geological Survey, Denver.

Turney, Omar
1924 *Land of the Stone Hoe*. Arizona Republican Print Shop, Phoenix.

1929 *Prehistoric Irrigation in Arizona*. Arizona Historical Review No. 2, Vol. 5. Phoenix.

Upham, Steadman, and Glen E. Rice
1980 Up the Canal without a Pattern: Modeling Hohokam Interaction and Exchange. In *Current Issues in Hohokam Prehistory*, edited by David E. Doyel and Fred Plog, pp. 78–105. Arizona State University Anthropological Research Papers No. 23. Tempe.

Urban, Sharon F.
1981 The Las Colinas Shell Assemblage. In *The 1968 Excavations at Mound 8, Las Colinas Ruins Group, Phoenix, Arizona*, edited and assembled by Laurens C. Hammack and Alan P. Sullivan, pp. 303–335. Arizona State Museum Archaeological Series No. 154. The University of Arizona, Tucson.

U.S. Geological Survey
1912 Phoenix, Arizona, 15 minute series topographic map. On file, Northern Arizona University Special Collections Library, Flagstaff.

U.S. Geological Survey, Reclamation Service
1902–1903 Salt River Project, Salt River Valley, Arizona, Topographic Map. On file, Salt River Project, Phoenix.

Vokes, Arthur W.
1984 The Shell Assemblages of the Salt-Gila Aqueduct Sites. In *Hohokam Archaeology along the Salt-Gila Aqueduct, Central Arizona Project: Material Culture*, edited by Lynn S. Teague and Patricia L. Crown, pp. 463–574. Arizona State Museum Archaeology Series No. 150, Vol. 8. The University of Arizona, Tucson.

1986 The Tanque Verde Wash Shell Assemblage. In *Archaeological Investigations at the Tanque Verde Wash Site: A Middle Rincon Settlement in the Eastern Tucson Basin*, by Mark D. Elson, pp. 313–324. Institute for American Research Anthropological Papers No. 7. Tucson.

1988 Shell Artifacts. In *The 1982–1984 Excavations at Las Colinas: Material Culture,* by David R. Abbott, Kim E. Beckwith, Patricia L. Crown, R. Thomas Euler, David A. Gregory, J. Ronald London, Marilyn B. Saul, Larry A. Schwalbe, Mary Bernard-Shaw, Christine R. Szuter, and Arthur W. Vokes, pp. 319–384. Arizona State Museum Archaeological Series No. 163, Vol. 4. The University of Arizona, Tucson.

Vokes, Arthur W., and Charles H. Miksicek

1987 Snails, Clams, and Canals: An Analysis of Nonmarine Molluscan Remains. In *Archaeological Investigations of Portions of the Las Acequias-Los Muertos Irrigation System: Testing and Partial Data Recovery within the Tempe Section of the Outer Loop Freeway System, Maricopa County, Arizona,* edited by W. Bruce Masse, pp. 177–187. Arizona State Museum Archaeological Series No. 176. The University of Arizona, Tucson.

Wagoner, Jay J.

1970 *Arizona Territory 1863–1912: A Political History.* The University of Arizona Press, Tucson.

1983 *Arizona's Heritage: Revised Edition.* Peregrine Smith Books, Salt Lake City.

Walsh-Anduze, Mary-Ellen

1991a Functional Variability in Hohokam Burial Ceramics. Paper presented at the Arizona-Nevada Academy of Science Annual Meeting, Flagstaff.

1991b Ceramic Analysis. In *Archaeological Test Excavations at Pueblo Grande, AZ U:9:7(ASM): The Washington Street Expansion, Phoenix, Arizona,* by Robert I. Birnie, pp. 45–54. Soil Systems Technical Report No. 91-3. Phoenix.

1993 Ceramic Analysis. In *Archaeological Investigations in the Phoenix Sky Harbor International Airport: A Class I and II Survey for the Master Plan Improvements Project,* prepared by David H. Greenwald and M. Zyniecki, pp. 55–62. SWCA Archaeological Report No. 92-36. Flagstaff.

1994 The Pueblo Grande Whole Vessel Study: An Examination of Production and Short-Distance Exchange. In *The Pueblo Grande Project: Ceramics and the Production and Exchange of Pottery in the Central Phoenix Basin,* edited by David R. Abbott, pp. 149-259. Soil Systems Publications in Archaeology No. 20, Vol. 3. Phoenix.

Ward, Albert E.

1978 Sinagua Farmers before the "Black Sands" Fell: An Example from Wupatki National Monument. In *Limited Activity and Occupation Sites: A Collection of Conference Papers,* compiled and edited by Albert E. Ward, pp. 135–146. Contributions to Anthropological Studies No. 1. Center for Anthropological Studies, Albuquerque.

Weaver, Donald E., Jr.

1972 A Cultural-Ecological Model for the Classic Hohokam Period in the Lower Salt River Valley, Arizona. *The Kiva* 38:57–94.

1977 *Investigations Concerning the Hohokam Classic Period in the Lower Salt River Valley, Arizona.* The Arizona Archaeologist No. 9. Arizona Archaeological Society, Phoenix.

Wilcox, David R.
1978 The Theoretical Significance of Field Houses. In *Limited Activity and Occupation Sites: A Collection of Conference Papers,* compiled and edited by Albert E. Ward, pp. 25–32. Contributions to Anthropological Studies No. 1. Center for Anthropological Studies, Albuquerque.

1979 The Hohokam Regional System. In *An Archaeological Test of Sites in the Gila Butte–Santan Region, South-Central Arizona,* edited by Glen Rice, pp. 77-116. Anthropological Research Papers No. 18. Arizona State University, Tempe.

1984 Site Structure and Maximum Extent of the Pueblo Grande Site. Ms. on file, Pueblo Grande Museum, Phoenix.

1987 *The Frank Midvale Investigations of the Site of La Ciudad.* Office of Cultural Resources Management Anthropological Field Studies No. 19. Department of Anthropology, Arizona State University, Tempe.

1991 Hohokam Social Complexity. In *Chaco and Hohokam: Prehistoric Regional Systems in the American Southwest,* edited by Patricia L. Crown and W. James Judge, pp. 253–291. School of American Research Press, Santa Fe.

Wilcox, David R., and Lynette O. Shenk
1977 *The Architecture of the Casa Grande and Its Interpretation.* Arizona State Museum Archaeological Series No. 115. The University of Arizona, Tucson.

Wilcox, David R., and Charles Sternberg
1983 *Hohokam Ballcourts and Their Interpretation.* Arizona State Museum Archaeological Series No. 160. The University of Arizona, Tucson.

Willey, Gordon R., and Philip Phillips
1958 *Method and Theory in American Archaeology.* University of Chicago Press.

Willey, P., and Bob Mann
1986 The Skeleton of an Elderly Woman from the Crow Creek Site and Its Implications for Paleodemography. *Plains Anthropologist* 31(112):141–152.

Wolfman, D.
1991 Retrospect and Prospect. In *Archaeomagnetic Dating,* edited by Jeffrey L. Eighmy and Robert S. Sternberg, pp. 313–364. The University of Arizona Press, Tucson.

Woodbury, Richard B.
1960 The Hohokam Canals at Pueblo Grande, Arizona. *American Antiquity* 26:267–270.

1961a *Prehistoric Agriculture at Point of Pines, Arizona.* Memoirs of the Society for American Archaeology No. 17. Salt Lake City.

1961b A Reappraisal of Hohokam Irrigation. *American Anthropologist* 63:550–560.

Woodbury, Richard B., and James A. Neely
1972 Water Control Systems of the Tehuacán Valley. In *The Prehistory of the Tehuacán Valley. Vol. 4, Chronology and Irrigation*, edited by Frederick Johnson, pp. 81–153. University of Texas Press, Austin and London.

Woodbury, Richard B., and John Q. Ressler
1962 Effects of Environmental and Cultural Limitations upon Hohokam Agriculture, Southern Arizona. In *Civilization in Desert Lands*, edited by Richard B. Woodbury, pp. 41–55. University of Utah Anthropological Papers No. 62. Salt Lake City.

Zar, Jerrold H.
1984 *Biostatistical Analysis*. 2nd ed. Prentice-Hall, New Jersey.

Zarbin, Earl
1979 *The Swilling Legacy*. Salt River Project, Phoenix.

Zier, Christian J.
1983 The Ceren Site: A Classic Period Maya Residence and Agricultural Field in the Zapotitlán Valley. In *Archaeology and Volcanism in Central America*, edited by Payson D. Sheets, pp. 119–143. University of Texas Press, Austin.

Zyniecki, M.
1993 Discussion and Summary. In *Pueblo Viejo: Archaeological Investigations at a Classic Period Cemetery in El Reposo Park, Phoenix, Arizona*, edited by M. Zyniecki, pp. 147–158. SWCA Archaeological Report No. 92-75. Flagstaff.

Zyniecki, M., Thomas N. Motsinger, and David H. Greenwald
1990 A Cultural Resources Survey of the Proposed Granite Reef Recharge Project, Salt River Pima and Maricopa Indian Community, Maricopa County, Arizona. Ms. on file, Salt River Project, Phoenix.